MTP International Review of Science

Volume 1

Theoretical Chemistry

Edited by **W. Byers Brown**
University of Manchester

Butterworths · London
University Park Press · Baltimore

THE BUTTERWORTH GROUP

ENGLAND
Butterworth & Co (Publishers) Ltd
London: 88 Kingsway, WC2B 6AB

AUSTRALIA
Butterworths Pty Ltd
Sydney: 586 Pacific Highway 2067
Melbourne: 343 Little Collins Street, 3000
Brisbane: 240 Queen Street, 4000

NEW ZEALAND
Butterworths of New Zealand Ltd
Wellington: 26–28 Waring Taylor Street, 1

SOUTH AFRICA
Butterworth & Co (South Africa) (Pty) Ltd
Durban: 152–154 Gale Street

ISBN 0 408 70262 1

UNIVERSITY PARK PRESS

U.S.A. and CANADA
University Park Press Inc
Chamber of Commerce Building
Baltimore, Maryland, 21202

Library of Congress Cataloging in Publication Data

Byers Brown, William, 1929–
    Theoretical chemistry.

    (Physical chemistry, series one, v. 1) (MTP
international review of science)
    1. Chemistry, Physical and theoretical. I. Title.
QD453.2.P58 Vol. 1 541'.3'08s [541'.2] 72–4333
ISBN 0–8391–1015–4

First Published 1972 and © 1972
MTP MEDICAL AND TECHNICAL PUBLISHING CO. LTD.
Seacourt Tower
West Way
Oxford, OX2 OJW
and
BUTTERWORTH & CO. (PUBLISHERS) LTD.

Filmset by Photoprint Plates Ltd., Rayleigh, Essex
Printed in England by Redwood Press Ltd., Trowbridge, Wilts
and bound by R. J. Acford Ltd., Chichester, Sussex

Consultant Editor's Note

The MTP International Review of Science is designed to provide a comprehensive, critical and continuing survey of progress in research. The difficult problem of keeping up with advances on a reasonably broad front makes the idea of the Review especially appealing, and I was grateful to be given the opportunity of helping to plan it.

This particular 13-volume section is concerned with Physical Chemistry, Chemical Crystallography and Analytical Chemistry. The subdivision of Physical Chemistry adopted is not completely conventional, but it has been designed to reflect current research trends and it is hoped that it will appeal to the reader. Each volume has been edited by a distinguished chemist and has been written by a team of authoritative scientists. Each author has assessed and interpreted research progress in a specialised topic in terms of his own experience. I believe that their efforts have produced very useful and timely accounts of progress in these branches of chemistry, and that the volumes will make a valuable contribution towards the solution of our problem of keeping abreast of progress in research.

It is my pleasure to thank all those who have collaborated in making this venture possible – the volume editors, the chapter authors and the publishers.

Cambridge   A. D. Buckingham

MTP International Review of Science

Theoretical Chemistry

MTP International Review of Science

Publisher's Note

The MTP International Review of Science is an important new venture in scientific publishing, which we present in association with MTP Medical and Technical Publishing Co. Ltd. and University Park Press, Baltimore. The basic concept of the Review is to provide regular authoritative reviews of entire disciplines. We are starting with chemistry because the problems of literature survey are probably more acute in this subject than in any other. As a matter of policy, the authorship of the MTP Review of Chemistry is international and distinguished; the subject coverage is extensive, systematic and critical; and most important of all, new issues of the Review will be published every two years.

In the MTP Review of Chemistry (Series One), Inorganic, Physical and Organic Chemistry are comprehensively reviewed in 33 text volumes and 3 index volumes, details of which are shown opposite. In general, the reviews cover the period 1967 to 1971. In 1974, it is planned to issue the MTP Review of Chemistry (Series Two), consisting of a similar set of volumes covering the period 1971 to 1973. Series Three is planned for 1976, and so on.

The MTP Review of Chemistry has been conceived within a carefully organised editorial framework. The over-all plan was drawn up, and the volume editors were appointed, by three consultant editors. In turn, each volume editor planned the coverage of his field and appointed authors to write on subjects which were within the area of their own research experience. No geographical restriction was imposed. Hence, the 300 or so contributions to the MTP Review of Chemistry come from many countries of the world and provide an authoritative account of progress in chemistry.

To facilitate rapid production, individual volumes do not have an index. Instead, each chapter has been prefaced with a detailed list of contents, and an index to the 13 volumes of the MTP Review of Physical Chemistry (Series One) will appear, as a separate volume, after publication of the final volume. Similar arrangements will apply to the MTP Review of Organic Chemistry (Series One) and to subsequent series.

Butterworth & Co. (Publishers) Ltd.

Physical Chemistry
Series One

Consultant Editor
A. D. Buckingham
Department of Chemistry
University of Cambridge

Volume titles and Editors

1 THEORETICAL CHEMISTRY
Professor W. Byers Brown, *University of Manchester*

2 MOLECULAR STRUCTURE AND PROPERTIES
Professor G. Allen, *University of Manchester*

3 SPECTROSCOPY
Dr. D. A. Ramsay, F.R.S.C.,
National Research Council of Canada

4 MAGNETIC RESONANCE
Professor C. A. McDowell, F.R.S.C.,
University of British Columbia

5 MASS SPECTROMETRY
Professor A. Maccoll, *University College, University of London*

6 ELECTROCHEMISTRY
Professor J. O'M Bockris, *University of Pennsylvania*

7 SURFACE CHEMISTRY AND COLLOIDS
Professor M. Kerker, *Clarkson College of Technology, New York*

8 MACROMOLECULAR SCIENCE
Professor C. E. H. Bawn, F.R.S.,
University of Liverpool

9 CHEMICAL KINETICS
Professor J. C. Polanyi, F.R.S.,
University of Toronto

10 THERMOCHEMISTRY AND THERMODYNAMICS
Dr. H. A. Skinner, *University of Manchester*

11 CHEMICAL CRYSTALLOGRAPHY
Professor J. Monteath Robertson, F.R.S.,
University of Glasgow

12 ANALYTICAL CHEMISTRY – PART 1
Professor T. S. West, *Imperial College, University of London*

13 ANALYTICAL CHEMISTRY – PART 2
Professor T. S. West, *Imperial College, University of London*

INDEX VOLUME

**Inorganic Chemistry
Series One**
Consultant Editor
H. J. Emeléus, F.R.S.
*Department of Chemistry
University of Cambridge*

Volume titles and Editors

1 **MAIN GROUP ELEMENTS—
HYDROGEN AND GROUPS I–IV**
Professor M. F. Lappert, *University of
Sussex*

2 **MAIN GROUP ELEMENTS—
GROUPS V AND VI**
Professor C. C. Addison, F.R.S. and
Dr. D. B. Sowerby, *University of
Nottingham*

3 **MAIN GROUP ELEMENTS—
GROUP VII AND NOBLE GASES**
Professor Viktor Gutmann, *Technical
University of Vienna*

4 **ORGANOMETALLIC DERIVATIVES
OF THE MAIN GROUP
ELEMENTS**
Dr. B. J. Aylett, *Westfield College,
University of London*

5 **TRANSITION METALS—PART 1**
Professor D. W. A. Sharp, *University of
Glasgow*

6 **TRANSITION METALS—PART 2**
Dr. M. J. Mays, *University of
Cambridge*

7 **LANTHANIDES AND ACTINIDES**
Professor K. W. Bagnall, *University of
Manchester*

8 **RADIOCHEMISTRY**
Dr. A. G. Maddock, *University of
Cambridge*

9 **REACTION MECHANISMS IN
INORGANIC CHEMISTRY**
Professor M. L. Tobe, *University College,
University of London*

10 **SOLID STATE CHEMISTRY**
Dr. L. E. J. Roberts, *Atomic Energy
Research Establishment, Harwell*

INDEX VOLUME

**Organic Chemistry
Series One**
Consultant Editor
D. H. Hey, F.R.S.
*Department of Chemistry
King's College, University of London*

Volume titles and Editors

1 **STRUCTURE DETERMINATION
IN ORGANIC CHEMISTRY**
Professor W. D. Ollis, F.R.S.,
University of Sheffield

2 **ALIPHATIC COMPOUNDS**
Professor N. B. Chapman,
Hull University

3 **AROMATIC COMPOUNDS**
Professor H. Zollinger, *Swiss Federal
Institute of Technology*

4 **HETEROCYCLIC COMPOUNDS**
Dr. K. Schofield, *University of Exeter*

5 **ALICYCLIC COMPOUNDS**
Professor W. Parker, *University of
Stirling*

6 **AMINO ACIDS, PEPTIDES AND
RELATED COMPOUNDS**
Professor D. H. Hey, F.R.S. and
Dr. D. I. John,
King's College, University of London

7 **CARBOHYDRATES**
Professor G. O. Aspinall, *University of
Trent, Ontario*

8 **STEROIDS**
Dr. W. D. Johns, *G. D. Searle & Co.,
Chicago*

9 **ALKALOIDS**
Professor K. F. Wiesner, F.R.S.,
University of New Brunswick

10 **FREE RADICAL REACTIONS**
Professor W. A. Waters, F.R.S.,
University of Oxford

INDEX VOLUME

Physical Chemistry Series One

Consultant Editor
A. D. Buckingham

Preface

The avowed aim of this series is to review each field of science comprehensively every two years. In practice I have found it impossible to plan an interesting review of the whole of theoretical chemistry which can be fitted into a single volume. I have therefore selected certain subjects which seem to me particularly appropriate at the present time; I hope the areas omitted can be reviewed in later volumes in the series. My second confession is that rather than encouraging my authors to adopt a strictly impartial and impersonal point of view, I have suggested that they express their opinions, including criticisms. Not all of them have availed themselves of this liberty, but where they have I share the responsibility.

Chapters 1, 2, 3 and 5 deal broadly with the basic problem in chemistry of calculating the energy and other properties of atoms and molecules in their ground states. This is a major field of continuing interest in which the computer is essential and often dominant. Chapters 6 and 8 are devoted to statistical mechanics, which is of such importance in chemistry that it would be criminal to omit it from this volume. For the remaining articles I have sought reviews of active areas of theory closely related to experiment. The long article by Certain and Bruch describes both the recent calculations of intermolecular forces and their deduction from experiment. The article by Levine is on a field in which, as he says, experimentalists often refuse to recognise the division of labour accepted by theoreticians. I had also planned to include a review entitled 'Radiationless Processes in Molecules and Crystals' by Jortner, who has been a major contributor to this expanding field, but unfortunately he had to withdraw owing to illness. In spite of this and other omissions, I hope that the present volume gives a useful coverage of theoretical chemistry at the present time.

Manchester   W. Byers Brown

Contents

1
Electron Correlation in Atoms

J. I. MUSHER

Belfer Graduate School of Science, Yeshiva University, New York

1.1 INTRODUCTION AND SURVEY

1.1.1 The correlation problem

The Schrödinger equation for an N-electron non-relativistic atom is a $3N$-dimensional eigenvalue equation.

$$H\Psi = E\Psi \tag{1.1}$$

An applied mathematician would consider the solution of this equation to any relatively high degree of accuracy a most unlikely possibility. Fortunately, Nature has been kind, and the solutions to this equation, at least for ground and certain excited states of atoms, are essentially governed by the Coulomb attraction of the nucleus for the surrounding electrons. Thus the Periodic

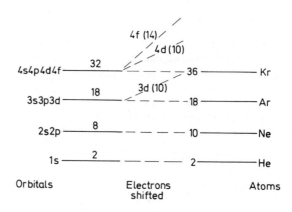

Figure 1.1   Numbers of electrons in hydrogen-like orbitals and exact descriptions

Table of the elements can be reproduced if electrons are filled into pure hydrogen-like orbitals, using the Pauli principle, with the proviso that higher angular momentum orbitals be appropriately shifted into higher rows to provide the various transition series as shown in Figure 1.1. This is the justification for assuming Ψ to be well-approximated in an independent-particle description such as

$$\Psi_0 = \mathscr{A} \, a(1)b(2)c(3) \ldots \tag{1.2}$$

where $a(1)$, $b(2)$, etc. are spin orbitals, generally products of spatial orbitals, each doubly occupied, with spin functions; or for open-shell states as linear combinations of the degenerate functions constructed from the same orbitals, such as $\Psi_0 = (0^+, 0^-) + (1^+, -1^-) + (-1^+, 1^-)$, for the ^1S state of carbon in the usual notation for the separate determinants.

The significance of this essentially empirical observation cannot be over-estimated as it is on the basis of it alone that the solution of equation (1.1) becomes a conceivable possibility. Conceptually, this follows by recognising that if an independent-particle description gives an accurate initial approximation, a cluster expansion in which corrections include two-particle functions, three-particle functions, etc. in term-by-term order, ought to be rapidly convergent. Mathematically, this follows from a perturbation theoretic argument wherein Ψ_0 is recognised to be an eigenfunction of a separable Hamiltonian, H_0, of the form

$$H_0 = \sum_{i=1}^{N} h_0(i) \tag{1.3}$$

so that if the perturbation is defined by

$$\lambda H_1 \equiv H(\lambda) - H_0 = \lambda \sum_{i<j} g(ij) \tag{1.4}$$

then the perturbation corrections in the series

$$\Psi = \Psi_0 + \lambda \Psi_1 + \lambda^2 \Psi_2 + \ldots \tag{1.5}$$

are such that Ψ_1 can be constructed from pair functions and unexcited orbitals. Ψ_2 from three-particle functions, pair functions and unexcited orbitals, etc. Thus the perturbation theoretical description provides the intuitively-expected cluster expansion albeit not a unique expansion and surely not an 'exact cluster' expansion in which lower-order clusters do not appear in higher-order terms and the theory of small perturbations assures that the contributions of the successive clusters decrease in size appropriately. The actual magnitudes of these contributions are governed by the choice of the orbitals in Ψ_0, and, in the perturbative expansion, by the choice of the H_0 of which it is an eigenfunction. The criteria on which these choices should be made are the pragmatic conditions which provide most rapid convergence with least computational effort. Unfortunately there has been far less investigation of this problem than its importance warrants.

As there is no analogy to Figure 1.1 for molecules, molecular theorists, who apply atomic techniques to polyatomic molecular problems, are making an extrapolation whose validity is by no means assured. For the record I list the arguments in favour of such an extrapolation of the atomic independent-particle model:

(i) inner-shell electrons should behave similarly in molecules as they do in atoms;

(ii) valence electrons seem susceptible of qualitative description in terms of electron-pair bonds and (so-called) delocalised π-electron systems;

(iii) the atomic experience with paired electrons is assumed to apply directly to delocalised π-electrons and does not lead to any violent disagreement with experiment;

(iv) valence electrons in ground and excited states of diatomic molecules seem to be well represented (both qualitatively and quantitatively) by diatomic orbitals;

(v) calculations on polyatomic molecules using the Hartree–Fock (HF) procedure give estimates of molecular energies of approximately the same accuracy as that given by their atomic analogues.

These arguments, however, are by no means conclusive. In particular an independent-particle description in which the ('canonical') orbitals have a spatial extent of more than two atoms is assumed to be valid only on an *ad hoc* basis. The fact that such a description is often obtained from a variational description is irrelevant, as the latter only implies that the solution obtained is the best approximation of the *given* orbital structure.

As long as we are aware of the potential limitations involved, we can define the problem of *electron correlation* as the means of improving a simple independent-particle description by the inclusion of pair and higher-electron-cluster terms in the wave function. This is a rather general definition which explicitly *avoids* the association of electron correlation with a particular independent-particle description, such as the Hartree–Fock solution, which is taken as the reference point for electron correlation in most of the literature. This has the following advantages:

(i) It avoids relating detailed calculations to an arbitrary reference point which is not always accessible, and which is usually irrelevant: even though electron correlation is almost invariably *discussed* in terms of the HF Ψ_0, it is almost never *calculated* from the HF Ψ_0;

(ii) It obviates any assumption about the nature and multiplicity of the non-linear HF equations, about which little is known and which are, in any case, virtually never solved without restrictions of one kind or another[1];

(iii) It obviates the requirement that all molecules be well described by the same types of independent-particle descriptions which are known to be valid for atoms, even though we only consider such descriptions in the present article, with some brief exception in Section 1.4, and

(iv) It avoids the further arbitrariness which arises when electron correlation is measured from the 'better', but more complex orbital descriptions, such as the EHF[2], the different GI[3] and the BRNO[4, 4a] schemes which have been developed over the past years in order to obtain 'HF correlation energy' within an orbital description; none of these can be considered as a simple independent-particle scheme, and furthermore, the BRNO (also called SOGI[5] and SEHF[6]) which is the best of these is clearly not practicable, or even advantageous, for large systems.

Electron correlation is a problem which concerns all atoms and molecules from helium and the hydrogen molecule onwards. Therefore a great deal of the physics and mathematics of the problem can be discussed with reference to the simplest systems, those containing two electrons, and this has been done in a most comprehensive way by Bethe and Salpeter[7] for the helium atom (although there is almost no literature discussing electron correlation in the hydrogen molecule). The present article therefore considers exclusively the complications which arise when there are more than two electrons, which complications do not involve new principles in any way (since the Hamiltonian contains no three-particle interactions) but only additional computations. Since the Bethe–Salpeter studies of the numerous small terms in the Hamiltonian are almost invariably given in the independent-particle description, these can be immediately applied to the N-electron problem using the description given here and double perturbation theory.

The solutions to the Schrödinger equation (1.1) for an N-electron system are obtained from the single working hypothesis that some independent-particle description, as yet unspecified, provides a good initial approximation. The approach is of the type that would be taken by an engineer, i.e. first, the problem is assumed to be soluble in terms of an independent particle Ψ_0, secondly, the Ψ_0 and its Hamiltonian H_0 are chosen on the basis of both convenience and expected rapid convergence, thirdly, the cluster corrections are calculated by whatever means are most suitable, without prejudice, on grounds of either physical significance or upper- or lower-bound properties and finally, the problem is considered as solved if the series, however defined, converges sufficiently rapidly and/or the result corresponds with sufficient accuracy to the experimental value. There is nothing profound in this approach and the question is only one of technique, not of physics. The actual simplicity of the problem has the explicit advantage of permitting the application of typical engineering procedures such as trying different sets of orbitals, different H_0s, different summations of higher-order terms, etc, and allowing the pragmatic criterion of the most accurate result with least work to govern the actual methods used in the solution.

The plan of this article is as follows: after some remarks on the history of the problem and the utility of computations which conclude this section, the perturbation theoretical description of electron correlation is considered in detail in Section 1.2 and the choice of H_0 is discussed in Section 1.3. The various techniques that can be used in computation are discussed relatively briefly in Section 1.4 although the details are left for the more specialised literature. In the Appendices several examples showing the convergence of different calculations are presented.

The reader's attention is called to the numerous reviews of the correlation problem which, more often than not, describe the personal views and computational procedures of the authors in a far more comprehensive manner than the space limitations of the present article permit me to do here. I would particularly cite the articles of Pitzer[8], Löwdin[9], Sinanoğlu[10, 11], Kelly[12, 13], Nesbet[14, 15], Wahl and G. Das[16], McWeeny and Steiner[17], Freed[18] and the book of Sinanoğlu and Brueckner[19]. The important contributions of Jucys, T. P. Das, Harris, Tolmachev and Safronova among others are presented in the proceedings of the Frascati summer school *cum* symposium of 1967 published as Volume 14 of *Advances in Chemical Physics*, in the proceedings of the Vilnius symposium of 1969 published by the Lithuanian Academy of Sciences and in the proceedings of the Gif-sur-Yvette symposium of 1970 published as the Volume 31-C4 of the *Journal de Physique*.

Apologies are made for any lack of completeness of the present article, whether intentional, or unintentional.

1.1.2 Some historical remarks

I believe that in an article such as this a few historical remarks can be of interest*. The calculation of accurate atomic wave functions has been of significant concern to physicists and chemists since the early days of quantum mechanics. Yet for 30 years electron correlation was studied almost solely on the He atom — the most notable exception was the work of Fock, Veselov and Petrashen[21] — beginning with the pioneer work of Hylleraas[22] and culminating in the calculations of Pekeris[23], Schwartz[24], Scherr[25] and Midtdal[26] which give the energy eigenvalues of the lowest states to the phenomenal accuracy of eight significant figures. The many-electron correlation problem in atoms was perhaps first formally recognised in 1959 with the appearance of the review article of Löwdin[9] and a related article by Pitzer[8], despite much historical precedent both in the theory of metals† and in the many-body physics developed in the 1950s[19]. Immediately thereafter, Sinanoğlu[28, 29], a student of Pitzer, showed that in a perturbation theoretical description, the first-order corrections are given explicitly in terms of independently-determined pair functions. This key contribution appeared at approximately the same time that K. Watson noted that the diagrammatic methods developed for the electron gas and for infinite nuclear matter provided an explicit procedure for evaluating the lowest-order two-electron and other higher-order corrections in atoms and suggested to Kelly[30] that he attempt some calculations on the problem. Thus two approaches to electron correlation which were to have much success were initiated almost simultaneously. A third approach, that of direct configuration interaction (C.I.), which is essentially a brute-force variational method, had of course been in existence much longer, and the fairly obvious fact that perturbational C.I. will mix into the Ψ_0 determinants with at most two-excited orbitals was explicitly discussed by Møller and Plesset[31] and Brillouin[32] in 1934 and later by Nesbet[33] in 1955. Nevertheless the immediate implication that lowest-order correlation could be described as the sum of independently determined pair-correlations seems to have been uniformly missed‡, and it was only following the initial argument of Sinanoğlu that the C.I. procedure gave way to the cluster expansion§ developed by Nesbet[14] and Sinanoğlu's perturbational description was itself modified into a cluster expansion, his many electron theory[34]. The equally obvious implication that the contributions of the doubly excited determinants in perturbative C.I. or many-body theory could be summed explicitly to give the pair-corrections as solutions to two-particle inhomogeneous partial differential equations (p.d.e.s) seems also to have been missed, despite the fact that Dalgarno[35] and Schwartz[36] had recently used such techniques on one-electron perturbation problems. The simplest of these three approaches remains, however, the perturbational description of Sinanoğlu — with the simplified derivation of Schulman and myself[37] utilised here — in which the inhomogeneous p.d.e.s of perturbation theory are reduced directly to sets of two-electron, three-

---

*Discussions of this have been recently given by Condon[20] and Nesbet[182].

†For some early references on this see Reference 27.

‡Although see some of the papers of Goudsmit and co-workers in the early 1930s.

§An independent derivation of a cluster expansion is that of Szasz[34a].

electron, etc., equations as appropriate to the order involved, and which then can be solved by whatever techniques desired, including the expansion in excited orbitals.

1.1.3 On the utility of computations

The physicist tends to think of a problem as solved once he has a correct prescription for performing a computation. As a number of such prescriptions are now available for the many-electron problem, all of which are well-understood and differ only in the order in which higher-order terms are included, the question arises as to what physical problems require the use of computation, i.e. wherein are accurate wave functions and energies of scientific value? Unfortunately, the utility of accurate atomic and molecular wave functions is rather small, and, in fact, an examination of the recent computational literature shows an essential lack of concern with the physical world, which should imply the relative sterility of the field. There is, of course, the implicit argument that 'we are now only developing the methods and when these methods are accurate enough for large systems, then all of chemistry will open up for us'. It is not possible to refute such an argument, but for the moment we are indeed far from that utopia, if utopia it is, where computers will completely replace chemical and biochemical laboratories. It seems that only in astrophysics, where the location of spectral lines for systems inaccessible to laboratory spectroscopy is of primary concern, that accurate computations for systems of more than two electrons can find immediate application to problems of real physical interest. Of course, when nuclear physics becomes sufficiently refined to develop a real use for accurate nuclear moments, or conversely, when N-electron calculations become sufficiently refined to be able to distinguish the existence or absence of three-electron forces, then such a statement would have to be revised. In principle, there should be explicit problems of significant concern to spectroscopists which might utilise accurate computational results in one way or another, but I know of no such problems at the present time. It is also possible that atomic computations can be of use in studying the mathematical structure of the multi-dimensional eigenvalue equations of which these are perhaps the most accessible examples and also in assessing the validity of the different independent-particle models, such as the HF or the random-phase approximation, regularly used by the theoretical physicist.

1.2 PERTURBATION THEORETICAL APPROACH

1.2.1 Perturbation theory

In order to be able to write a perturbation expansion for atomic wave functions, we define a Hamiltonian as a function of λ

$$H(\lambda) \equiv H_0 + \lambda H_1 \qquad (1.6)$$

as a sum of two terms, one independent of λ and one linear in λ, which coincides with the true Hamiltonian of the physical system for $\lambda = 1$; the separation into H_0 and H_1 is otherwise arbitrary. With this definition, each exact eigenfunction and eigenvalue of the model Schrödinger equation

$$[H(\lambda) - E(\lambda)]\Psi(\lambda) = 0 \qquad (1.7)$$

which coincides with the Schrödinger equation (1.1) of the physical problem for $\lambda = 1$, can be expanded in a series in λ as in equation (1.5). Similarly, the expectation value of any operator, and any second-order property can be expanded in such series.

The fact that equation (1.7) is a well-defined equation for all values of λ with $\Psi(\lambda)$ and $E(\lambda)$ being continuous functions of λ enables the solution of the desired problem, for which $\lambda = 1$, to be carried out perturbatively, even though there is not necessarily a variable linear parameter such as Z^{-1} or ε within H_1 itself. This point is of importance since there has been some prejudice against using perturbation theory in the absence of such a parameter. The argument is simply that if the expansions converge for $\lambda \leqslant \rho$ and $\rho > 1$, then they converge for $\lambda = 1$ to the unique solutions. Of course absolute convergence is not really necessary, and semi-convergence can be satisfactory as it accords with the engineering character of the present approach.

It is important to note that the choice of H_0, the unperturbed Hamiltonian, and its eigenfunction, Ψ_0, can only affect (i) the radius of convergence of the expansions, (ii) the rate of convergence of the expansions; and (iii) the ease with which the individual terms can be calculated.

It is thus desirable to choose an H_0 such that the lowest order energies, E_n, are easily calculated and such that

$$R_n(E) \equiv |E_n/E_{n-1}| \ll \lambda^{-1} = 1$$

for low values of n.

1.2.2   The separable Hamiltonian H_0

The choice of Hamiltonians so that the perturbation expansion is tractable is essentially limited to separable H_0s taking the form of equation (1.3), the sum of identical one-electron operators, or to Hamiltonians of the matrix form

$$H_0 = \sum_n \varepsilon_n |n\rangle \langle n| \qquad (1.8)$$

where $|n\rangle$ are a complete set of N-electron functions, usually Slater determinants. In actual computations these two are intimately related[38], the former giving p.d.e.s with the latter giving perturbative matrix diagonalisation. For example, if $\varepsilon_n = \langle n|H|n\rangle \equiv H_{nn}$ then the perturbation expansion is equivalent to the expansion of the secular equation[39, 40] as

$$E = H_{00} + \sum_n{}' \frac{|H_{0n}|^2}{H_{00} - H_{nn}} + \dots \qquad (1.9)$$

Because of this equivalence, the remaining discussion is limited to H_0s of the form of equation (1.3) but it should be noted that the matrix technique is a useful one when small basis-set calculations are to be performed on large molecules as it permits the specification of contributing terms and the neglect of others (if this can actually be justified) as utilised so successfully in the PCILO method[41].

With an H_0 of the form of equation (1.3), the perturbation, λH_1, can be written as the sum of identical two-electron terms as in equation (1.4), and this leads to two essential simplifications of the perturbation expansion. The first is that Ψ_0 which satisfies

$$(H_0 - E_0)\Psi_0 = 0 \tag{1.10}$$

can be written as a single determinant of the form of equation (1.2) — or a linear combination of degenerate determinants — which is constructed from spin orbitals a,b, etc. that are eigenfunctions of h_0 satisfying the relationship

$$[h_0(i) - \varepsilon_a]a(i) = 0 \tag{1.11}$$

of eigenvalues ε_a, ε_b, etc. such that $E_0 = \Sigma\varepsilon_a$. The second simplification is that the nth-order corrections, Ψ_n, can be obtained from solutions of *uncoupled* inhomogeneous partial differential equations involving at the most $n+1$ electrons, i.e. from at most $(n+1)$ electron clusters rather than from N-electron or all-electron functions.

1.2.3 The solutions

In order to obtain an explicit expression for Ψ_1, the solution to*

$$(H_0 - E_0)\Psi_1 = (E_1 - H_1)\Psi_0 \tag{1.12}$$

the term $-H_1\Psi_0$ is written as

$$-H_1\Psi_0 = -\sum_{i<j} g(ij)\mathscr{A}\,a(1)b(2)c(3)\ldots$$

$$= -\mathscr{A}\sum_{i<j} g(ij)a(1)b(2)c(3)\ldots \tag{1.13}$$

By the usual procedure for considering inhomogeneous p.d.e.s it is seen that there is a contribution to Ψ_1 arising from each term on the right hand side of equation (1.13) introduced by the two operators $\mathscr{A}$ and Σ. Thus Ψ_1 can be written as

$$\Psi_1 = \mathscr{A}\sum_{a<b} u_{ab}(12)c(3)\ldots \tag{1.14}$$

with the spin-pair function, u_{ab}, the solution to the p.d.e.

$$[h_0(1) + h_0(2) - \varepsilon_a - \varepsilon_b]u_{ab}(12) = -g(12)a(1)b(2) + \text{b.c. terms} \tag{1.15}$$

where a and b are meant to be indices which extend over the occupied orbitals of Ψ_0 and where the boundary condition (b.c.) terms arise from

---

*The relationship of this derivation to the original one of Sinanoğlu[29] for the HF Ψ_0, also used by Byron and Joachain[42] is discussed in Appendix A of Reference 37. The Ψ_1 values are, of necessity, the same but the individual pair-functions are different and the number of different pair-functions that have to be calculated in this procedure is smaller than that described in the original treatment.

the decomposition of $E_1\Psi_0$ on the right hand side of equation (1.12). Since the $g(12)$ is usually spin independent, u_{ab} will have electrons 1 and 2 in the same spin states respectively as they are in the spin orbitals $a(1)$ and $b(2)$. The solution to this equation can be written in the usual sum-over-states manner as

$$u_{ab}(12) = \sum_{k,k'}{}' \frac{|k(1)k'(2)\rangle \langle kk'|g|ab\rangle}{\varepsilon_a+\varepsilon_b-\varepsilon_k-\varepsilon_{k'}} \tag{1.16}$$

where the prime on the Σ refers to k and k' not simultaneously being a and b. Notice that the sum does not exclude k or k' from being any of the other occupied orbitals, as is usual in p.t. but as different from many-body theory. This assures the independence of these pair functions, even though it means the inclusion of some terms in u_{ab} whose contribution to the total energy will be cancelled out by some terms of u_{ad}, etc.

The second- and third-order energies can be obtained by substituting Ψ_1 into the p.t. equations for E_2 and E_3 so that E_2 is given by

$$E_2 = \sum_{a<b} \varepsilon_{ab}$$

$$= \sum_{a<b} \{\langle P^{(2)}ab|g|u_{ab}\rangle + \sum_{c\neq a,b} \langle P^{(3)}a(1)b(2)c(3)|g(13)+g(23)|u_{ab}(12)c(3)\rangle\} \tag{1.17}$$

if $u_{ab}(12)$ is orthogonalised to $a(1)b(2)$ and $b(1)a(2)$ as in equation (1.16) where

$$P^{(n)} = \sum_{q=0}^{n!} (-1)^q P_q$$

is the projector onto the antisymmetric representation of the symmetric group of n particles. Notice that while the sum is over pairs of orbitals there is actually an explicit dependence on the other orbitals through the second summation and that this dependence cannot be removed in any way whatsoever, although in the Hartree–Fock case, for which there are no single excitations, two-thirds of these terms vanish identically.

1.2.4   Discussion

This derivation of the second-order energy has the virtue of being straightforward in that it leaves nothing for the uninitiated reader to accept on faith. It has the added virtue of providing a transparent demonstration of the fact that pair energies are not independent but rather explicitly involve the remaining occupied orbitals, which is not properly appreciated.

It is immediately recognised that the term in $\mathscr{A}u_{ab}(12)c(3)...$ in Ψ_1 gives no contribution for the part of u_{ab} constructed from the other occupied orbitals $c, d, ...$. The summation of (16) can thus be taken as

$$\hat{u}_{ab} = \sum_{k,k'}{}'' \frac{|kk'\rangle \langle kk'|g|ab\rangle}{\varepsilon_a+\varepsilon_b-\varepsilon_k-\varepsilon_{k'}} \tag{1.18}$$

The double prime indicates exclusion of $k, k' = c, d, ...$ as well as k and k'

not simultaneously equal to a and b, although, of course, $\hat{u}_{ab}$ includes *implicitly* the non-vanishing 'exclusion effects'. The energy remains in the form of equation (1.17) except that one-third of the terms in the second sum vanish, as they have been cancelled out by terms in the first summation. Only in the Hartree-Fock case of no single excitations can E_2 be written in the simple form

$$E_2 = \sum_{a<b} \langle ab - ba \,|\, g \,|\, \hat{u}_{ab} \rangle = \sum_{a<b} \sum_{k,k'}{''} \frac{\langle ab - ba \,|\, g \,|\, kk' \rangle \langle kk' \,|\, g \,|\, ab \rangle}{\varepsilon_a + \varepsilon_b - \varepsilon_k - \varepsilon_{k'}} \qquad (1.19)$$

there remain otherwise the 'single-particle' corrections corresponding to the diagram:

and its exchange analogue.

What is deceptive about this argument is that not only do the single-excitation terms usually remain, but also that the pair functions u_{ab} are not, actually, independent, but depend explicitly on the exclusion of the occupied orbitals from the summation. There is thus little reason to expect any quantitative 'transferability' of pair energies from one system to another unless the occupied orbitals are either the same in both systems or have near-vanishing matrix elements in equation (1.16).

The p.d.e. derivation possesses a few further advantages:

(i) It is compact. For example, part of the quadruply excited term of Ψ_2 can be written as

$$\sum_{a<b\,\neq\,c<d} \mathscr{A}\, u_{ab}(12) u_{cd}(34) e(5) f(6)\ldots \qquad (1.20)$$

which provides the terms deduced by Sinanoğlu from Watson's calculation on Be to be important; also the 'infinite-order' pair function can be written as the solution to

$$[h_0(1) + h_0(2) - \varepsilon_a - \varepsilon_b] u_{ab}^{(\infty)}(12) = -g(12)[a(1)b(2) + u_{ab}^{(\infty)}(12)] + \text{b.c. terms} \qquad (1.21)$$

which on substitution into equation (1.17) gives an 'infinite-order' pair energy.

(ii) It avoids the representation in terms of explicit diagrams which are all too often given a physical significance that is unwarranted, particularly in view of the dependence of the various contributions on the H_0 employed. Of course any term in the p.d.e. can be immediately developed in terms of diagrams if this proves desirable in computation and this might be useful to incorporate the matrix techniques for higher-order corrections. Notice that although the third- and higher-order energies can be derived in a straightforward way using the p.d.e. approach, the actual computational expressions become rather cumbersome when three-and-more-particle terms are to be included; and

(iii) Symmetry takes care of itself naturally. There is thus no need to symmetrise the u_{ab} or collect terms which are eigenfunctions of two-particle L^2, L_z, S^2 or S_z as is often assumed to be the case[43-45]. If H_0 commutes with total L^2, L_z, S^2 and S_z such that Ψ_0 is taken as an eigenfunction of these operators, then H_1 can never mix in states of undesired symmetry and seemingly non-symmetric terms are always cancelled out at the end*.

1.2.5 Electromagnetic properties

Perturbative expansions for electromagnetic properties are obtained by using double perturbation theory[47, 47a]. Thus, for example, if an external electric field, $\mathcal{E}$, is applied along the z-axis, the dipole moment can be written in a λ-series as

$$\mu_z = -\left.\frac{\partial E}{\partial \mathcal{E}}\right|_{\mathcal{E}=0} = \langle 0|Z|0\rangle + \lambda\langle 0|Z|1\rangle + \lambda^2[\langle 1|Z|1\rangle + 2\langle 0|Z|2\rangle]$$
$$+ \dots \tag{1.22}$$

where Z symbolises Σz_i. Similarly the zz-component of the polarisability tensor can be written as

$$\alpha_{zz} = -\frac{1}{\mathcal{E}}\left.\frac{\partial E}{\partial \mathcal{E}}\right|_{\mathcal{E}=0} = 2\langle 00|Z|01\rangle$$
$$+ 2\lambda[\langle 01|H_1|01\rangle + 2\langle 00|H_1|02\rangle] + \dots \tag{1.23}$$

where the first index in Ψ_{ij} refers to orders in λ and the second refers to orders in $\mathcal{E}$. When Ψ_0 is taken as the independent-particle function of equation (1.2) the leading term of both of these expressions is given in terms of the orbitals of Ψ_0. Thus

$$\mu_z^0 = \sum_a \langle a|z|a\rangle \tag{1.24}$$

and

$$\alpha_{zz}^0 = \sum_a \langle a|z|a_1\rangle \tag{1.25}$$

where the a_1 are the solutions to the p.d.e.s

$$[h_0(i) - \varepsilon_a]a_1(i) = [-z_i + \langle a|z|a\rangle]a(i). \tag{1.26}$$

In the Hartree–Fock description the first-order corrections μ_z^1 vanish†; and a general description of one-electron perturbations has been given by Sandars[49] and one in terms of localised orbitals by Malrieu[49a]. The first-order corrections α_{zz}^1 for the HF description are particularly simple, as they

---

*The differences in energies found by Viers et al.[45] between calculations using symmetry-adapted pairs and those with non-symmetrised pairs are due to the p.t. third-order terms[46]. As mentioned below, cluster-CI expansions do not give unique results precisely due to these third-order terms which make different contributions in the different cluster procedures.

†Perhaps the first detailed study of correlation effects on single-particle operators is that of Sternheimer[48] who calculated the polarisation of inner shells by outer p-electrons and the effect on the electric field gradient at the nucleus. A qualitative discussion had been previously given by Abragam and Pryce[48a] for the case of spin-density at the nucleus as observed in the Hartree–Fock functions. When true Hartree–Fock calculations are carried out – an example of this without l^2-restrictions but still with l_z and s_z restrictions has been carried out by Larsson[48b] – then these terms appear in zeroth order, and are not part of 'electron correlation'.

only require the same functions, $a_1(i)$, calculated for α_{zz}^1 as discussed by Tuan, Epstein and Hirschfelder[50]; the α_{zz}^1 value in the general case has been given by Hameed[51].

There have been very few calculations which go beyond the leading term of these series other than the coupled-HF method[48, 52, 53] which is accurate through order λ[54, 55]. The first calculations are those of Epstein and Johnson[56] and Amos and Musher[57]; and other calculations have been performed by Tuan[58], Amos and Roberts[59], Hameed[51, 60], Schulman and Kaufman[61], and Caves and Karplus[55]. As shown by Hameed[51], and as is implicit in the computations of Langhoff, Karplus and Hurst[62], the convergence is critically dependent on the choice of H_0 as will be mentioned below.

An expansion similar to equation (1.22) is obtained when correlation effects are included in transition probabilities[63] except that order-parameters for each state must be used if the H_0 values are different. Recently Nicolaides and Sinanoğlu[64] have calculated a number of these transition probabilities both for first-order allowed and for first-order forbidden states with relatively high accuracy in the cases that can be compared with experiment.

It is worth noting again[54] that the coupled-HF theory cannot be used as a starting point for a perturbative expansion of electromagnetic properties: as I have shown elsewhere[54] a set of the correction terms to the coupled-HF result cancels *identically* against *all* the coupling terms of the coupled-HF theory. Of course, if the coupling terms are themselves evaluated perturbatively (or iteratively) they can be associated term by term with perturbative corrections to the uncoupled-HF result — as I have demonstrated for the term linear in λ — just as the perturbative expansion of the Hartree–Fock energy of an atom reduces to a Z-expansion. The perturbative description of coupled-HF as in the diagrammatic analysis of Caves and Karplus[55] is none other than a partial summation[12] of perturbative corrections to the *uncoupled*-HF result, in accord with my argument that the uncoupled-HF expansion provides the *unique* specification of correction terms. This partial summation does not seem to offer much insight into the physics of the problem* and, as with all such summation procedures, it leaves one in the rather unsatisfactory position of calculating some second and higher order terms while neglecting others. When all is said and done, the error given by the coupled-HF theory — the 'best' orbital procedure available — remains[58] non-negligible and of order λ^2.

1.3 CHOICE OF HAMILTONIAN, H_0

The arbitrary nature of the term electron-correlation is seen immediately when the one-electron Hamiltonian, $h_0(i)$, is written in its most general form

$$h_0(i) = -\tfrac{1}{2}\nabla_i^2 - \frac{Z}{r_i} + V(i) \tag{1.27}$$

---

*For a differing opinion see, however, the recent papers of Schneider, Csanak, Taylor and Yaris[190, 191].

with $V(i)$ an operator which is perhaps non-local and spin dependent and serves as an effective potential in which the independent electrons are assumed to move in zeroth-order. The two-electron correlation potential $g(ij)$ is not simply $1/r_{ij}$ but rather

$$g(ij) = \frac{1}{r_{ij}} - \frac{1}{(N-1)}[V(i) + V(j)] \qquad (1.28)$$

and depends explicitly on the choice of the $V(i)^*$. There are clearly a large number of obvious choices of h_0 [65] although only recently has one been able to compare accurate perturbation theoretical calculations based on different choices of h_0.

In the present section we examine some of the possible one-electron Hamiltonians†. Any discussion on h_0s invariably begins with the Hartree–Fock h_0, but it is almost always found convenient to modify this Hamiltonian in one way or another. But once the HF description is departed from (in any manner whatsoever), there is no longer even any physical picture to justify attempts to simulate the HF description with these modified h_0s. In addition, there is no obvious reason why the HF expansion should be considered the most rapidly convergent expansion — it was dropped by Kelly[69] immediately after he had made use of it for the first time[30] — and furthermore it is not clear why there should be any particular virtue in using the HF orbitals in a perturbation scheme[70-72]; other sets of orbitals might well provide more rapidly convergent expansions. Recently, in fact, Dalgarno[73], whose philosophy in having used the Z-expansion extensively is similar to the one argued here, has shown that a supposedly non-physical exchangeless h_0 introduced by Schulman and myself[37] converges more rapidly for He than does the physical Hartree Hamiltonian.

1.3.1   The Hartree–Fock Hamiltonian, H_0

The HF (or UHF[1]) h_0 can be written as a simple spin-dependent operator

$$h_0^{HF} = -\tfrac{1}{2}\nabla^2 - \frac{Z}{r} + \sum_{\rho=1}^{N}[-a_\rho \mid -a_\rho] - \sum_{\rho=1}^{N}[-a_\rho \mid a_\rho -] \qquad (1.29)$$

in which the spin-orbitals $a_1, a_2, \ldots$ replace $a, b, \ldots$ of equation (1.2) and where the exchange operators include integration over spin. Prejudice has generally favoured the use of this h_0 on the grounds of its providing the best value of $E_0 + E_1$, assuming that the absolute minimum of the variational procedure has been found by the iteration process used[74]. In open-shell systems this is rarely used since the orbitals would then not be doubly-occupied, and would

---

*As these are usually chosen to average the effect of the true pair potential, $1/r_{ij}$, some authors such as Daudel, consider electron correlation to play a role even in zeroth order.

†I have given a similar discussion for molecules previously[66]. The choice of h_0 for molecules is rather limited, unless non-symmetric schemes are used, and many-body molecular calculations have only been carried out using single-centre orbitals[67] and bare nuclear orbitals[68], except for the study on the hydrogen molecule[61] for which a finite Gaussian basis was used to construct different sets of SCF orbitals.

usually not be eigenfunctions of l^2 and l_z, both of which make for great inconveniences in computation. In fact, the most general choice of spin orbitals

$$a(r, \Omega, \sigma) = \sum_{l, m} [a_+^{lm}(r) Y_l^m(\Omega)\alpha + a_-^{lm}(r) Y_l^m(\Omega)\beta]$$

which would give rise to the true HF solution has never been used, even for He atom. If double occupancy is assumed then h_0^{HF} reduces to the restricted Hartree–Fock operator

$$h_0^{RHF} = -\tfrac{1}{2}\nabla^2 - \frac{Z}{r} + \sum_{\rho=1}^{N} [-a_\rho | -a_\rho] - \tfrac{1}{2}\sum_{\rho=1}^{N} [-a_\rho | a_\rho -] \qquad (1.30)$$

where the exchange operator no longer includes spin and for open shells a similar operator with either some average of the open-shell potentials or some effective interactions in the form of non-local potentials are used[75].

1.3.2 Thomas–Fermi exchange and Herman–Skillman (HS) Hamiltonians H_0

In view of the difficulties of working with the non-local exchange operator, particularly for problems of band-structure, Slater[76], Gáspár[77] and later Kohn and Sham[78] proposed the use of the operator

$$h_0^{TF} = -\tfrac{1}{2}\nabla^2 - \frac{Z}{r} + \sum_\rho [-a_\rho | -a_\rho] - k[\sum_\rho | a_\rho |^2]^{\tfrac{1}{3}} \qquad (1.31)$$

to give a Thomas–Fermi effective exchange potential. The Slater choice of k with the inclusion of a cut off in the exchange term at $r = r_0$ and the addition of a single Coulomb tail beyond r_0 leads to the Herman–Skillman Hamiltonian[79]

$$h_0^{HS} = -\tfrac{1}{2}\nabla^2 - \frac{Z}{r} + \sum_\rho [-a_\rho | -a_\rho] - k[\sum_\rho | a_\rho |^2]_{r_0}^{\tfrac{1}{3}} - \frac{1}{r}\Big|_{r_0}^{\infty} \qquad (1.32)$$

although workers using other choices of k also introduce the Coulomb tail. In these different self-consistent procedures, the a_ρ are, of course, all different.

All of the Hamiltonians generate orbital wave functions and total energies which are extremely close to the UHF or RHF values[80, 80a, 80b, 81, 82]. As such it might be assumed that they provide slightly simpler starting points for p.t. expansions and that in their convergence they should well approximate that of the HF expansion. Actually, despite the similarity of the occupied orbitals the one-electron Hamiltonians are quite different due to the different asymptotic potential; the potentials of h_0^{HF} and h_0^{HS} when acting on unoccupied orbitals go as zero and $-(1/r)$ respectively. Hence these operators have different spectra and, as it turns out, entirely different rates of convergence*. Kelly first noted this and proposed the use of the V^{N-1} potential to accelerate convergence although he could just as well have used h_0^{HS}

---

*In this sense, the Herman–Skillman h_0 without the tail would show convergence much like the h_0^{HF} as long as it possesses enough bound states to construct Ψ_0. For a detailed discussion of this see Wilson et al.[82].

whose potential has the same asymptotic behaviour. The convergence obtained by Kelly[69] for the energy of Be was indeed more rapid but that for the polarisability was not particularly satisfactory[69, 51] as shown in the Appendix E. Kelly's argument for the more rapid convergence of the V^{N-1} case is based on the analysis of the perturbative solution to the secular equation as in equation (1.9) above, wherein the more closely the basis functions approximate the exact excited states of the problem, the smaller the off-diagonal matrix elements and the more rapid the convergence*.

1.3.3 V^{N-1} potential Hamiltonians, H_0

The potential most favoured in the many-body calculations is the V^{N-1} potential introduced by Kelly[69]. An h_0 is defined for each angular momentum whose orbitals are occupied in Ψ_0 in which the outside electron of that symmetry is removed from the HF h_0 and in which the resultant potentials are spherically averaged, if necessary, in order that

$$[h_0^l, l] = 0.$$

Thus

$$
\begin{aligned}
h_0^l &= -\tfrac{1}{2}\nabla^2 - \frac{Z}{r} + \sum_{av}{}^l[-a_\rho^0 \mid -a_\rho^0] - \sum_{av}{}^l[-a_\rho^0 \mid a_\rho^0 -] \\
&= -\tfrac{1}{2}\nabla^2 - \frac{Z}{r} + V_l.
\end{aligned}
\tag{1.33}
$$

Usually the HF orbitals are used, and as these are not the SCF orbitals of this h_0 — except for the outer orbital itself — they are denoted as a_ρ^0. Such an l-dependent operator cannot, of course, lead to an H_0 of the form of equation (1.3) unless modified to

$$h_0^{N-1} = -\tfrac{1}{2}\nabla^2 - \frac{Z}{r} + \sum_l \sum_{k_l} \mid k_l \rangle \langle k_l \mid V_l \mid k_l \rangle \langle k_l \mid \tag{1.34}$$

where the k_l specify the complete set of eigenfunctions of h_0^l for *each* l; and notice that V_l must be defined even for the l-values unoccupied in Ψ_0 or implicitly set equal to zero. It appears that the calculations in the literature are not different from calculations that would use the h_0 of equation (1.34) explicitly, but caution must be exercised that this is the case, particularly when higher-order terms are computed.

Kelly and co-workers[72, 84] have recently introduced a modified form of h_0^{N-1} which permits the occupied orbitals to remain the HF orbitals yet allows the excited orbitals to be calculated in a V^{N-1} potential. The analogue to equation (1.33) is thus written as

$$h_0^{SH, l} = -\tfrac{1}{2}\nabla^2 - \frac{Z}{r} + V_{HF} + Q\Omega_l Q \tag{1.35}$$

---

*Amos and I[83] have used this argument to discuss the disadvantage in using Lödwin–Shull orbitals in calculations on intermolecular forces. Although the functions aid in obtaining rapid convergence of the contributions to a given order in p.t., they cause the p.t. expansion itself to converge more slowly.

and is called the Silverstone[70]–Huzinaga[71] one-electron Hamiltonian h_0; V_{HF} is the HF potential, $Q = 1 - \Sigma |a_\rho\rangle\langle a_\rho|$ with the a_ρ the HF orbitals and Ω_l is chosen to give the $(N-1)$-electron shielding desired. Such non-local terms in h_0 can introduce an undesirable arbitrariness as has been emphasised by Epstein[65] and as has recently been illustrated by Amos and myself[38]. While it might well make little difference, it is worth remarking that the use of such a potential is rather deceptive in that it assigns a certain 'goodness' to the new h_0, i.e. an improvement over the h_0^{N-1} used previously, which does not really seem to be warranted.

1.3.4 Exchangeless potential Hamiltonians, H_o

One of the many other possible Hamiltonians is the one I have called the 'exchangeless' Hamiltonian with

$$h_0^{EL}(\alpha) = -\tfrac{1}{2}\nabla^2 - \frac{Z}{r} + \alpha\sum_\rho [-a_\rho^0| - a_\rho^0] \tag{1.36}$$

where α is a constant to be chosen appropriately and a_ρ^0 are any radial functions, perhaps being derived from the zeroth iteration of h_0^{EL} itself. This h_0 was first introduced because the change of the factor of $\tfrac{3}{2}$ in k of the h_0^{TF} values seemed to have negligible effect on the eigenfunctions[80, 81].

The choice of α which gives an h_0 similar to the HS and V^{N-1} h_0s is $\alpha = (N-1)/N$ so that each electron 'sees' an effective field of $N-1$ electrons, except that the effective field is derived from *all* the electrons, including for each electron a fraction of its own self field. An intuitive advantage of this α is that its h_0^{EL} for two 1s-electrons with fully-iterated orbitals reduces to the Hartree Hamiltonian

$$h_0^H = -\tfrac{1}{2}\nabla^2 - \frac{Z}{r} + [-1s| - 1s] \tag{1.37}$$

It is also possible to choose α such that the pair-correction terms $g(ij)$ are small, as this is expected to play a direct role in the convergence of the p.t. expansion whereas the role of the asymptotic behaviour of h_0 itself is expected to be a secondary one. The $g(ij)$ corresponding to $h_0^{EL}(\alpha)$ is

$$g(ij;\alpha) = \frac{1}{r_{ij}} - \frac{\alpha}{(N-1)}\sum_\rho \{[-a_\rho^0| - a_\rho^0]_i + [-a_\rho^0| - a_\rho^0]_j\} \tag{1.38}$$

and asymptotically, for large r_i and r_j this becomes

$$g(ij;\alpha) \to \frac{1}{r_{ij}} - \frac{N\alpha}{(N-1)}\left(\frac{1}{r_i} + \frac{1}{r_j}\right) \tag{1.39}$$

The choice of α which minimises this asymptotic form along the line $r_1 = r_2$ is $\alpha = (N-1)/2N$, or half the value needed to simulate the h_0^{HF}, i.e. such that

$$g(ij;\alpha) \to \frac{1}{r_{ij}} - \frac{1}{2r_i} - \frac{1}{2r_j} \tag{1.40}$$

where the s-part is zero along the line $r_1 = r_2$. For the ground-state of He, the 'significant region' of space is $\tfrac{1}{4} \leqslant r_1, r_2 \leqslant \tfrac{3}{4}$, and for this region, the absolute value of the s-part of equation (1.40) is considerably smaller than the s-part of the asymptotic $g^{HF} = -(1/r_<)$ or that of the hydrogenic $g = 1/r_>$. In simple physical terms we have included as much asymptotic Coulomb attraction in the screening terms as there is repulsion in $1/r_{ij}$ so that the $g(ij)$ looks similar to that obtained in the van der Waals interaction between two neutral atoms, although we have not justified the use of the asymptotic limit in the introduction of this screening.

Thus the exchangeless Hamiltonian

$$h_0^{EL}[(N-1)/2N] = -\tfrac{1}{2}\nabla^2 - \frac{Z}{r} + \frac{N-1}{2N}\sum_\rho [-a_\rho^0| - a_\rho^0] \qquad (1.41)$$

has been introduced in order to include the screened Coulomb potential which has the asymptotic limit of (1.40). The reduced screening of this potential ought to increase the convergence since the zeroth-order electrons are more tightly bound, the orbital energies are larger in magnitude, and the ratio $|E_1/E_0|$ will be smaller than the typical value, 0.5–0.7 found for the HF, TF, V^{N-1} and EL $[(N-1)/N]$ procedures. McNaughton, Henry and Smith[85, 86] have discussed the HF-analogue of this and shown that with self-consistent orbitals $E_1 = 0$; notice that on intuitive grounds one would like the $|E_n|$ to converge rapidly but also monotonically, so that this potential could, in principle, cause undesirable oscillation. The smallness of E_1 is, however, very promising, and Riley and Dalgarno[73] have illustrated explicitly the rapid convergence of this series as shown below.

1.3.5 Hydrogenic or Z-expansion Hamiltonian, H_0

The most natural perturbation expansion is the Z-expansion which, with Z the nuclear charge as a simple scaling factor, gives $\lambda H_1 = (Z\rho_{12})^{-1}$ so that λ can be taken as Z^{-1}. The h_0 is

$$h_0^Z = -\tfrac{1}{2}\nabla^2 - \frac{Z}{r} \qquad (1.42)$$

and the energy is given as a series

$$E = Z^2 E_0 + Z E_1 + E_2 + \ldots \qquad (1.43)$$

This series has been used since the days of Hylleraas[22] by many authors but the most extensive work has been by Scherr and co-workers[25] and Midtdal[26] who have carried out these calculations to very high order and by Dalgarno[47a] who has calculated numerous energies and electromagnetic properties in this expansion. In the past few years Tolmachev and Safronova and their co-workers have made extensive use of this series in calculating total energies and relativistic and other corrections for many-electron atoms[87–90a].

The Z-expansion is, however, not very convergent for $Z \geqslant 3$; for example, the hfs of Li atom is given as the series[91, 92]

$$\text{hfs} = \frac{Z^3}{2} - \frac{4Z^2}{3} + \mathcal{O}(Z) = 13.5 - 12.0 + \mathcal{O}(3). \qquad (1.44)$$

An attempt to improve this using screening, by doing what amounts to a $[1, 0]$ Padé approximant gives

$$\text{hfs} = \frac{(Z-\sigma)^3}{2} + \mathcal{O}(Z-\sigma) = 4.7 + \mathcal{O}(2.69) \tag{1.45}$$

but this is no closer to the experimental value of $2.90a_0{}^3$. Notice that the entire screened series as in equation (1.45) can be constructed from the series equation (1.44), and also that the two series summed through the Z^0 term are equal for all values of σ. Such an observation might be expected to invalidate the use of screening approximations until higher-order terms are included.

1.4 COMPUTATIONAL PROCEDURES

The various procedures that have been used for many-electron calculations are discussed in this section. The first four procedures are mathematical techniques used to evaluate electron correlation within a perturbation theory framework; the fifth utilises experimental values as well as computations and is thus an empirical method; the sixth and seventh are Configuration Interaction methods; the eighth and ninth are relatively untried new p.t. techniques which use Hamiltonians which are not of the form of equation (1.3); and the tenth set of methods are (mostly) orbital methods which in general cannot approach the exact solution. The Section concludes with a brief discussion of the artistry involved in performing calculations.

It is worth observing that, ideally, a complicated many-electron problem would be solved by using different methods for different parts of the problem according to the precision desired and the convenience of the computation. In practice, however, the computational apparatus for any of the methods has, in the first instance, been sufficiently complex that each research group has chosen one such method and has performed all its computations with this method alone. Besides being somewhat intellectually unsatisfying, this has had the unfortunate result that no two calculations for the same number, e.g. the same term of a given p.t. expansion, have been carried out using two different computational techniques, neither within a single research group nor with the two calculations carried out in separate groups.

1.4.1 Analytical method

The use of analytical techniques is perforce limited to the hydrogenic or Z-expansion h_0 and the single-centre expansion for molecules.*†. These methods were first used by Layzer[95, 95a] and his co-workers who only calculated the single-excitation part† of E_2, and were later used extensively

---

*In principle it applies as well to the bare nucleus expansion for diatomics[93, 94a, 194].

†In this discussion I exclude the finite series analytical solution of the p.d.e. as this applies only to the single-excitation part [35, 36, 92], the two-particle part being inaccessible other than by the infinite series, or 'sum-over-states' method discussed here.

by Tolmachev and Safronova and their co-workers for total energies and other properties[87-90a]. The numerical integrations involved are rather difficult so that the use of the analytical hydrogenic functions is not particularly simple. It is, however, no more difficult to integrate over functions known analytically than over functions which have been generated numerically as in the many-body methods. In fact, as diagrammatic methods apply to any sum-over-states procedures – and the application to the Z-expansion has been discussed most thoroughly by Tolmachev[96] – the calculations could have been discussed below together with diagrammatic techniques. The fact that the basis functions and integrals are known analytically served as a justification to consider these calculations separately, although the separation is surely arbitrary.

A calculation of historic importance, both for the physics and for the use of these analytic techniques, is the evaluation of the R^{-6} term in the London–van der Waals energy by Eisenschitz and London in 1930 [97]. This calculation, which gave the leading term of the asymptotic formula for interatomic correlation energy in the limit of large atomic separations, involved integrations over single and double continua without the use of electronic computers and was accurate to within 0.5%.

1.4.2 Variational method

Hylleraas[22] introduced the use of variational procedures to solve the p.d.e.s of perturbation theory, although only scattered examples[7], such as that of Veselov and Adamov[98] appeared in the literature until the revival of interest in the p.d.e.s due to the work of Dalgarno, followed by the Schwartz article on the variational solutions[36].

As is well known from the variational calculus[47, 99], an inhomogeneous partial differential equation

$$D[\phi] = f \tag{1.46}$$

possessing certain boundary conditions can be solved by finding the extremum of a corresponding variational equation

$$\delta L[\xi] = 0 \tag{1.47}$$

where L is chosen such that the ξ which satisfies equation (1.47) equals the ϕ of equation (1.46); in other words equation (1.47) is the variational equivalent of equation (1.46). If the solution to equation (1.46) is unique, then there will be no extrema of equation (1.46) other than the extremum which satisfies equation (1.46), which is often called the corresponding Euler equation. For example, the variational equivalent to equation (1.15) for the pair function u_{ab} writes[100]

$$L = \langle \xi \mid h_0(1) + h_0(2) - \varepsilon_a - \varepsilon_b \mid \xi \rangle + 2 \langle \xi \mid g(12) - J \mid a(1)b(2) \rangle$$
$$- 2K \langle \xi \mid b(1)a(2) \rangle \tag{1.48}$$

where $J = \langle ab \mid g \mid ab \rangle$, $K = \langle ab \mid g \mid ba \rangle$ and no spatial degeneracy is assumed, i.e. a and b are both spatially s-orbitals. Similarly the variational equivalent to equation (1.26) for the electric-field perturbed orbital, a_1, writes

$$L = \langle \xi \mid h_0 - \varepsilon_a \mid \xi \rangle + 2 \langle \xi \mid z - \varepsilon_a^1 \mid a \rangle \qquad (1.49)$$

with $\varepsilon_a^1 = \langle a \mid z \mid a \rangle$.

Variational procedures were probably first applied to p.t. problems by Hassé[101] and by Slater and Kirkwood[102] who studied electric polarisabilities and van der Waals interactions. These authors did not, however, use the analogues of equation (1.48) and (1.49) but rather wrote down the Rayleigh–Ritz procedure, or its equivalent

$$\delta \langle \Psi \mid H - E \mid \Psi \rangle = 0 \qquad (1.50)$$

and then, used

$$\Psi = (1 + c\lambda H_1)\Psi_0 \qquad (1.51)$$

as a trial function and solved for Ψ by minimising with respect to c.* It was this same procedure that was used with

$$\Psi = (1 + \lambda f)\Psi_0 \qquad (1.52)$$

where f contains free parameters, in the now classic work of Guy and Tillieu[103, 103a], Adamov[104] and Stephen[107] which stimulated the revival of variational calculations on electromagnetic properties in the 1950's[105, 106, 108]. The significance of the fact that these calculations used the Rayleigh–Ritz principle, rather than the variational principle appropriate to the perturbation theory equation of the appropriate order in λ, and the disadvantages in so doing, become apparent (only) when many-electron problems are treated. Thus, when Dalgarno derived his uncoupled-HF procedure for many-electron atoms[109] and when Karplus and Kolker[110–112] derived similar equations for many-electron molecules, they found extra coupling terms which, despite their actual appearance[110], can be shown to vanish[100] and are therefore redundant. Notice that in the matrix form of coupled-HF theory[113] these redundancies do not appear and the energy expressions are quite simple as shown explicitly by Lipscomb[114]. The implication of this discussion is that the p.t. equations should be derived first, and reduced to the simplest form possible, and only then should the variational principle of equation (1.47) be derived. The 'perturbation–variation' procedure which applies to the use of the Rayleigh–Ritz procedure in a perturbative manner should therefore not be used for more than two electrons and the terminology should presumably be dropped.

The lack of real distinction between variational and many-body or sum-of-states methods† — and hence the rather arbitrary division of computational

---

*This is equivalent to having collected terms to a given order in λ since, by symmetry, all terms in λ^3 vanish for the perturbations involved. The later authors[101–108] did collect terms to the appropriate order in the perturbation parameters whence the method was called the 'perturbation-variation' procedure.

†Notice that the numerical procedure discussed below is actually a sum-over-states in a finite basis, the basis being δ-functions at the points on the grid. The unique advantage of this finite basis set is that it permits extrapolation to zero-mesh or infinite basis set.

techniques in the present section – is made particularly clear by the recent calculations of Riley and Dalgarno[73] for higher-order energies of He. As is well known, the variational equivalent to the p.t. equations, using linear parameters, provides *exactly* the same set of linear equations as does the expansion of the perturbed function in a finite basis. Thus if $\Sigma c_n \phi_n$ is substituted into the first-order p.d.e. we obtain

$$\sum_n (H_0 - E_0)c_n \phi_n = (E_1 - H_1)\Psi_0 \qquad (1.53)$$

which when multiplied on the left by ϕ_n^* and integrated gives the linear equations for the c_n constants of the variational technique. Riley and Dalgarno utilised the fact that for higher-order corrections the left hand side of equation (1.53) remains the same, so that the labour of solution is reduced if the l.h.s. is diagonalised once and for all. They, therefore, obtained a set of functions ξ_n for which

$$\sum_n (H_0 - E_0)d_n \xi_n = (E_1 - H_1)\Psi_0 \qquad (1.54)$$

implies that

$$d_n = (\varepsilon_0 - \varepsilon_n)^{-1}\langle \xi_n | H_1 | \Psi_0 \rangle \qquad (1.55)$$

exactly as occurs in the usual sum-of-states procedure; this is then easily used in obtaining the higher-order corrections.

The variational procedure has been used by Knight, Sanders and Scherr[25,115] for first-row atoms, Seung(Hui) and Wilson[115a] for ^{2}S Li and Chisholm, *et al.*[115b, c] for ^{2}S and ^{2}P Li, all of whom utilised the Z-expansion, by Byron and Joachain[42] for He and Be in the HF expansion (although the third- and higher-order terms of Be are approximate due to the neglect of three-particle terms) and by Tuan and Sinanoğlu[116] for Be in the HF expansion, and Geller, Taylor and Levine[117] have performed an 'exact-pairs' calculation on Be. In all of these examples either the variational function was restricted to a small number of r_{ij}-type functions, or the partial wave expansions were cut off at some l-value. Reasonable convergence was obtained in these calculations.

1.4.3 Numerical integration method

Direct numerical integration* of the inhomogeneous pair-equations seems to be the most natural and straightforward way to solve for the pair functions, although it is not necessarily the simplest procedure to carry out. Direct six-dimensional integration is not possible at present but the introduction of a partial-wave expansion leads to a set of *uncoupled* two-dimensional problems in atoms (and coupled two-dimensional problems in molecules). These are then easily solved using standard techniques and extrapolation using different mesh sizes increases the precision. Calculations using two-

---

*The advantages in using numerical integration for the one-dimensional radial HF equations can be appreciated by examining the comprehensive treatment of hfs in atoms of Bagus, Liu and Schaeffer[117a].

dimensional integration on atoms were first carried out by Temkin and Sullivan[118] in 1963 and were later used by Musher and Schulman [37] for He using several one-electron Hamiltonians, by Schulman and Lee[119] for Li using an h_0^{EL}, by McKoy and Winter[120, 121] for He ground and excited states using the Z- and HF-expansions, and by Barraclough and Mooney[122] who studied the s-limit. The major drawback in using numerical methods is the importance of different regions of space for different pair corrections; this necessitates the use of different meshes for different parts of the problem which then requires careful interpolation techniques so that integration can be carried out over identical mesh points[123]. Square-root[121] (and lower-root) meshes have also proven convenient.

1.4.4 Diagrammatic many-body method

Goldstone introduced the diagrammatic expansion into the many-particle problem in order to prove Brueckner's linked-cluster expansion to infinite order and in order to keep track of the various terms of perturbation theory. However, electrons in atoms behave rather differently from electrons in an electron gas or from nucleons in infinite nuclear matter; single-particle functions cannot be simply taken as plane waves, there are no possible divergences as N is always finite, and the convergence of p.t. is sufficiently rapid and basis dependent as well, to make the physical interpretation of the perturbative corrections somewhat of an academic exercise. It is thus not altogether clear why this relatively intricate apparatus — at least for calculations which are beyond second order with shifted denominators — has been applied to atoms without significant modification appropriate to its new application. For example, one possible modification would treat the (symmetric) total two-particle perturbations $g(ij)$ rather than divide them into their one- and two-particle parts $V(ij)$, $V(i)$ and $V(j)$. This should serve to simplify the algebra and, by eliminating some of the diagrams, would help reduce the physical interpretation of these contributions to the oblivion which I feel they deserve.

There are two techniques of summation of higher-order diagrams wherein the diagrammatic procedure possesses advantages over the differential equation methods. These are:

(i) the *exact* summation of diagonal higher-order diagrams which gives rise to a shifted energy denominator as in the Feenberg perturbation scheme[39, 40] to lowest order. I have given elsewhere[124] a particularly simple example of this summation in discussing atomic polarisabilities which might help the reader follow the necessarily more intricate summations treated by Kelly[30, 69]; and

(ii) the *approximate* summation of higher-order diagrams in which certain matrix elements divided by denominators are approximated by their values for certain k, k'. This permits, for example, a certain third-order diagram to be approximated by a constant multiplied by a known second-order diagram. In investigating the validity of such an approximation for the accurately soluble problem of the HF hydrogen atom polarisability, Schulman and I[125] found the remarkable result that the higher-order diagrams are indeed given by a constant multiplied by the leading diagram to better than the error of the computation*. In this example, which is the only one for which this approximate procedure has been checked, Kelly's choice[126] of multiplicative constant is within 1% of ours and his procedure is therefore considerably more accurate than one might have otherwise expected.

The diagrammatic description has been used most skillfully and successfully, first by Kelly[30, 69, 12, 13] and more recently by Das and co-workers[127-131] and by Schulman and Kaufman[61] in a variety of calculations on atoms and small molecules. As different h_0s have been used to generate the different orbitals, there is no uniqueness in any of the individual terms involved – and hence my aversion to their physical significance – but it seems that *all* of the schemes using h_0^{HF}, h_0^{N-1}, h_0^{SH}, h_0^{H} and h_0 (bare nucleus) are reasonably convergent; some examples of these are given in the Appendices.

One possible check on the precision of the computation has been introduced by Schulman and Kaufman who used single-particle sum rules to test the completeness of their finite Gaussian basis. The recent studies of Kelly[72] have used sum rules to test the procedure for the numerical generation of the orbitals – Kelly had previously used many-electron sum rules which, implicitly, served the same purpose – and it would be of interest to apply such tests to the molecular functions of Lee, Dutta and Das[68] which are more likely to be subject to error.

It is worth observing that in the calculations of Ron and Kelly for the (formidable) iron atom[132] the authors felt obliged to use empirical values for the inner-shell pair-correlation energies, anything else being beyond the range of practicality; and it should be again noted that all many-body calculations have truncated the partial-wave expansions at low l-values, which is of particular importance in view of the significance of higher l-values in Ne atom as discussed later.

The diagrammatic methods actually provide a pictorial description of what could, in a more mundane terminology, be called 'perturbative C.I.' Thus, in the C.I. subspace restricted to excitations from a, b to k, k', the second-order energy contribution is

$$\sum'' \frac{|\langle \mathscr{A} abcd... | \Sigma g(ij) | kk'cd...\rangle|^2}{\varepsilon_a + \varepsilon_b - \varepsilon_k - \varepsilon_{k'}} \tag{1.56}$$

where the double prime again means $k, k' \neq c, d, ...$ and k, k' not simultaneously equal to a and b, and where the orbital energies are assumed to be known as in equation (1.55); and the result is analogous to equation (1.17). The two-electron part of this is

$$\sum'' \frac{|\langle ab - ba | g | kk'\rangle|^2}{\varepsilon_a + \varepsilon_b - \varepsilon_k - \varepsilon_{k'}} \tag{1.57}$$

and corresponds to the leading term of equation (1.17) or to the sum of all two-particle diagrams in second order. The diagonal higher-order two-

---

*The phenomenon showed some generality for HF Ψ_0 values[58] which is not actually understood despite several rather tautological attempts to do so[18]. For an illuminating discussion see the recent article by Amos[192].

particle diagrams can then be summed exactly to give

$$\sum{}'' \frac{|\langle ab-ba \mid g \mid kk'\rangle|^2}{\varepsilon_a + \varepsilon_b + \langle ab-ba \mid g \mid ab\rangle - \varepsilon_k - \varepsilon_{k'} - \langle kk'-k'k \mid g \mid kk'\rangle}. \qquad (1.58)$$

All the diagonal third- and higher-order diagrams, not only the two-particle ones, can be summed when the three-particle terms from equation (1.56)

$$|\langle P^{(3)}abc \mid g(13)+g(23) \mid kk'c\rangle|^2 \qquad c \ne a, b$$

are added to the numerator, and give simply

$$\sum{}'' \frac{\langle \mathscr{A}abcd... \mid \Sigma g \mid kk'cd...\rangle}{H_{00} - H_{ab, kk'}} \qquad (1.59)$$

(where H_{00} and $H_{ab, kk'}$ are the expectation values of H over the functions being mixed) i.e. precisely the result of perturbative C.I.

1.4.5   Empirical and semi-empirical methods

Skutnik et al.[133, 134] have utilised Sinanoğlu's 'non-closed-shell theory' to obtain the correlation energy and energy differences for all states of atoms B through Na and their ions which belong to the set of configurations $1s^2 2s^n 2p^m$. This is done by dividing the pair energies, ε_{ab}, of equation (1.17) into parts which arise from different choices of k, k' in the summation of equation (1.18). The three parts are
(i) 'internal' in which both k and k' are within the valence shell, defined to include both the 2s and the 2p orbitals;
(ii) 'semi-internal' in which either k or k' is within the valence shell and the other is without; and
(iii) 'all-external' in which neither k nor k' is in the valence shell.
The first two parts are calculated explicitly and the third is estimated empirically.
The working hypothesis is that given the additivity* of all these contributions, an explicit calculation of the internal and semi-internal terms for each pair for a whole set of states would leave only the all-external contributions to be determined, and these might be relatively constant for all states and even for all Z; they might therefore be subject to easy parametrisation.
The internal pair energies — symmetric pairs are used — are calculated by 2×2 C.I. and are almost exactly linear in Z as expected since the square of the matrix elements should be $\sim Z^2$ (for hydrogenic orbitals $\langle 1s1s \mid r_{12}^{-1} \mid 1s1s\rangle = 5Z/8$) and the denominators $\sim Z$[134a]. The semi-internal pair energies are calculated using a small C.I. basis, the resulting sum for each state of a given Z is compared with experiment and the all-external energies for each pair are obtained from a least-squares fit. The results seem remarkably good in that for a given Z the data are fitted with small mean-square errors, and, what is more, all the empirical all-external pair energies except the $2s^2$

---

*This additivity is obvious from the second-order formulation. Actually these authors use a formulation which includes some higher-order terms which are not strictly additive, but, of course are additive through at least second order.

energy are relatively insensitive to Z such that the 45 states which arise from $Z = 7, 8$ can be fitted with *six* parameters to within the error limits of the data.

The state-independence of the all-external pair energies could have been expected since the additional Coulomb and exchange operators resulting from different numbers of valence electrons should have little effect on a non-valence-shell orbital due to its diffuse character. The apparent Z-independence of the all-external pair-correlation occurs since the square of the matrix elements are again proportional to Z^2 but the denominator for excitation outside of the valence shell is also proportional to Z^2.

The importance of these results for the interpretation of many-electron correlation energies is clear: at the level of all-external correlation, all the orbital pair energies are completely independent of each other and, in lowest order, transferable. Conversely, the non-negligible contribution of internal and semi-internal pairs makes any attempt to consider total pair-energies as independent and transferable doomed to failure, except between systems having the same effective or local set of occupied orbitals, i.e. the same so-called 'Hartree–Fock sea.' This then provides the most definitive realisation of the pair-correlation description of atoms – no more precise general statement is possible – even though it is within the somewhat arbitrary definition of the valence shell.

The CNDO semi-empirical technique has been applied by Diner, Malrieu and Claverie[41, 135, 136] to the calculation of correlation energy in a series of molecules using diagrammatic methods and a localised orbital approximation. Their Perturbed Configuration Interaction for Localised Orbitals procedure looks to be most promising because of the ease with which it can be applied to large molecules despite the crudeness of the approximations involved.

Pamuk and Sinanoğlu[137, 138] have developed a generalisation of the CNDO procedure for electron-correlation in (closed-shell) molecules. They find that the total energies of a large series of molecules can be fitted well assuming the transferability of the CNDO parameters so that only a small number of such parameters are required.

1.4.6   Nesbet's cluster method

The behaviour of *any* many-particle system can always be described in a cluster expansion as was first considered in detail by J. E. Mayer[139]. The utility of a cluster expansion requires (a) an explicit recipe for the calculation of the terms corresponding to each cluster; and (b) rapid convergence of the resultant series. The rapid convergence of cluster series for quantum mechanical calculations on atoms and molecules is assured by the empirical accuracy of independent-particle descriptions as discussed above in justifying the perturbation expansion.

Cluster expansions for the many electron wave function were first described by Sinanoğlu[34] and by Szasz[34a] utilizing the intuition applied by K. A. Brueckner and others to infinite nuclear matter and the electron gas. Sinanoğlu took the critical step of introducing 'approximations' to make the clusters independent[183, 184] (first for electron-pairs and later, in a schematic way, for higher-order clusters) and this is what distinguishes his description from that of Szasz[34a, 141, 142] and the original Bethe-Goldstone[140] description for nuclear matter. Notice that only when the clusters are uncoupled is the cluster expansion different from what is obtained in an ordered-restricted C.I. expansion, as can be seen by comparing the Čižek-Paldus[185] Coupled-Pair-Many-Electron-Theory with the Total-Pair-Excitation-Block method discussed by Barr and Davidson[178].

An explicit and clear recipe for computing all-order uncoupled clusters and the realisation thereof has been given by Nesbet[14, 15]. Despite the formidable computational difficulties, the Nesbet procedure has been among the most fruitful methods used for atoms[143-147]. The procedure defines a hierarchy of configurations indicated schematically as

$$\Psi = \Psi_0 + \Sigma(i) + \Sigma(ij) + \Sigma(ijk) + \dots \qquad (1.60)$$

with Ψ_0 the RHF function. This function is then truncated at any point and each cluster, *without* the summation, is used in an uncoupled variational calculation as a trial function. The energy contribution in this order is the sum of the contributions of these independent clusters minus those of lower-order clusters previously calculated.

It should be remembered that cluster expansions are not unique, and, for example, the cluster expansion defined by Nesbet gives different pair-energies when canonical RHF or unitary-transformed localised orbitals are used. This is a consequence[46] of the different inclusion of third-order terms when viewed within the framework of perturbation theory for different choices of the actual orbitals. For this reason it is not at all clear what is meant by statements in the literature to the effect that for 'exact pairs' the sum of pair energies is a meaningful quantity and independent of unitary transformations among the orbitals.

1.4.7 Configuration interaction and modified configuration interaction methods

The traditional C.I. method has been used in a highly successful way by Harris, Schaeffer and co-workers[148-152], Weiss[187] and other authors in recent years, and although it has always been assumed that the integral problem precluded extensive computations, in actual fact, calculations using bases sufficiently large to give most of the correlation energy are indeed possible.

Modified C.I. methods, which use the Multi-configuration–SCF procedures of Hartree[153] and Jucys[154, 154a], as revived by Wahl[16] and by Clementi and Veillard[155], and then add additional configurations, seem also to be of potential utility. However, the difficulty of performing such calculations is so great that the consummate artistry of its few practitioners is required to obtain results.

1.4.8 Kirtman's method

Kirtman[156-158] has proposed a method for calculating accurate wave functions which includes correlation perturbatively into a non-symmetric wave function which is then antisymmetrised, a procedure analogous to that proposed by Hirschfelder and Silbey[159] for intermolecular forces and later modified to the HAV method[160]; a good discussion of the derivation of such perturbation expansions has been given by Amos[161], from which it is recognised that this is only one of many possible choices of procedures to symmetrise non-symmetric schemes. Kirtman's procedure, or the Distinguished Electron Method, has been used with rather good results for He, Li and Be using only very small basis sets. However, the antisymmetrisation step — which, incidentally, could be written in a more compact form using the symmetry properties of the spatial eigenfunctions of H — seems to be cumbersome for $N > 5$ as there are $N!$ permutations to be taken unless closed shells are used. Kirtman and Mowery[157] also use a shifted h_0 for Slater 2s-orbitals in Be in order to avoid an unpleasant singularity; one must, however, view these results with caution, especially given the extreme sensitivity of the Be polarisability to the choice of α in the exchangeless expansion[51, 193].

1.4.9 Musher–Silbey method

Silbey and I have proposed a method[162] which calculates the *spatial* eigenfunctions[163–167] of H perturbatively in such a way that the zeroth iterate is not an exact eigenfunction of the appropriate permutation symmetry operator. In other words, the spatial function $\Phi_i^{[\alpha]}$ which is to be an eigenfunction of H and of $D_{ii}^{[\alpha]}$, the projector onto the Young tableau corresponding to the ith row of the irreducible representation α of $\mathscr{S}_N$, is expanded as

$$\Phi_i^{[\alpha]} = \Phi_{i,0}^{[\alpha]} + \lambda \Phi_{i,1}^{[\alpha]} + \dots \qquad (1.61)$$

where

$$H_0 \Phi_{i,0}^{[\alpha]} = E_0 \Phi_{i,0}^{[\alpha]} \qquad (1.62)$$

but where

$$D_{ii}^{[\alpha]} \Phi_{i,0}^{[\alpha]} \neq \Phi_{i,0}^{[\alpha]} \qquad (1.63)$$

or, equivalently

$$[H_0, P] \neq 0 \qquad \text{for all } P \text{ contained in } \mathscr{S}_N. \qquad (1.64)$$

This then permits solution in terms of an initial Hartree-approximation rather than HF so that the V^{N-1} potential appears naturally; but the major advantage of the method is the simplicity it affords the computations in that antisymmetry of the functions does not have to be put in explicitly. The argument for the validity of the procedure is rather simple:

(i) all eigenfunctions of a symmetric H possess certain symmetries and therefore a perturbation theory expansion starting from a non-symmetric Φ_0 will either converge to a symmetric solution Φ or it will not converge; and

(ii) if it does not converge it may oscillate among solutions which are so closely spaced that the expansion is too crude to distinguish among them, in which case even the non-convergent solutions should be highly accurate.

Thus the perturbation expansion is expected to converge to an eigenfunction of the symmetry operator in the same way that it converges to an eigenfunction of the Hamiltonian, although it might have to go to an extra-ordinarily high power of λ in order to do so. As for the physics of the 'oscillating' non-convergent cases, the argument is merely an extension of that given by Messiah[168] applied to the antisymmetrisation of the wave function for the system of two He atoms, one on the Earth and the other on the Moon.

Recent calculations on the excited states of He and on the ground state of Li have shown that accurate results can be obtained using this procedure[188,189], although no oscillating solutions have been observed.

1.4.10 Approximate (mostly orbital) methods

There are a number of methods for calculating many-electron wave-functions which are widely discussed in the literature even though they are incapable of providing the exact wave function. These are:
(i) The Boys–Handy procedure[169–171] which writes

$$\Psi = \prod_{i<j} f_{ij}(r_i, r_j)\Psi_0 \tag{1.65}$$

where f is a function of interelectronic coordinates and Ψ_0 is a single determinant both of which contain variational coefficients, although iteration for the Ψ_0 starts with the HF function.
(ii) Various so-called unrestricted HF procedures such as Spin–UHF, Orbital–UHF, etc., which only serve as reminders of the fact that the RHF usually used is not necessarily the HF solution in the sense of the definition of the best single determinant[1].
(iii) Various projected methods which take a 'primitive' function

$$\xi = \phi_1(1)\phi_2(2)\phi_3(3)...\sigma_1(1)\sigma_2(2)\sigma_3(3)... \tag{1.66}$$

and project on it with an antisymmetriser and then with spin or orbital projectors to provide eigenfunctions of L^2 and S^2. When the spin-projector is the $\mathcal{O}^S$ of Löwdin[2] then these methods correspond to Löwdin's Extended Hartree–Fock procedure and are equivalent with the GF procedure of Goddard and when the spin projectors are the remaining f_α projectors of the representation α of $\mathscr{S}_N$[166] then these methods correspond to the remaining GI procedures, all of which could be called extended-HF but with the extended definition

$$\Psi = \mathscr{A}D_{ii}^{[\alpha]}\xi \tag{1.67}$$

or

$$\Psi = \mathcal{O}^L\mathscr{A}D_{ii}^{[\alpha]}\xi \tag{1.68}$$

if orbital extension is included as well. These procedures have been used almost exclusively by Goddard[3], although there have been extensive calculations on aromatic molecules* using the Alternant Molecular Orbital method[172]. The only orbital-extended calculations are those on He by

*See, however, my discussion in Reference 27 which shows that the complications introduced by having both $\mathscr{A}$ and $D_{ii}^{[\alpha]}$ in equation (1.67) are not necessary if non-symmetric spatial functions are permitted. For $N \geqslant 6$ the results are almost identical and the algebra enormously simplified.

Auffray and Percus[173] and by Lefebvre and Smeyers[174] who obtain almost all the correlation energy with the restriction that $l \leqslant 2$ in the expansion

$$\phi_{\pm} = a_0^{\pm} Y_0 + a_1^{\pm}(Y_1^1 + Y_1^0 + Y_1^{-1}) + \dots \tag{1.69}$$

$$\Psi(12) = \mathcal{O}^S \mathcal{O}^L \mathcal{A} \phi_+(1)\alpha(1)\phi_-(2)\beta(2). \tag{1.70}$$

A description which is 'extended' still further and is capable of giving the exact result for He is the Clifford algebra description of Lefebvre and Prat[175] which, has, however, not been used in calculation.

The best orbital description[4, 4a], which permits of a description in terms of N spatial orbitals writes

$$\Psi = \mathcal{O}^L \mathcal{A} \sum_j \delta_j D_{1j}^{[\alpha]} \xi \tag{1.71}$$

where the $D_{1j}^{[\alpha]}$ operate on the spatial variables of ξ and the δ_j values are constants. The total wave function, of course, gives no more information than does the first spatial part[162–167] (with $i = 1$ arbitrary)

$$\Phi_1 = \mathcal{O}^L \sum_j \delta_j D_{1j}^{[\alpha]} \phi_1(1)\phi_2(2)\phi_3(3)\dots. \tag{1.72}$$

When the ϕ values are restricted to be eigenfunctions of l^2 and l_z, and degenerate for different m_l so that the $\mathcal{O}^L$ is unnecessary for closed shells or is a simple projector among the open-shell orbitals for open-shells, this has been named in 1968 the Best Radial N-Orbital method[4] after a town in Czechoslovakia[176]. The use of equation (1.72) is equivalent not only to equation (1.71) but also to the equally more complicated

$$\Psi = \mathcal{O}^L \mathcal{A} \sum_j \delta_j D_{1j}^{[\bar{\alpha}]} \xi \tag{1.73}$$

with the $D_{1j}^{[\bar{\alpha}]}$ operating on the spin-function. In this form Goddard has called this the Spin-Optimised-GI method[5] and Kaldor and Harris have called it the Spin-Extended-HF method[6], although the latter seems to be poorly chosen as it would rather imply the use of

$$\xi = \phi_1(1)(\alpha + \lambda_1 \beta)(1)\phi_2(2)(\alpha + \lambda_2 \beta)(2)\dots \tag{1.74}$$

as was discussed by Löwdin and Lunell[177]. It is easy to see that equation (1.74) is not as flexible as equations (1.71)–(1.73) as it only permits the use of N linear parameters whereas the others permit f_α [4].

The more general function of the form of equation (1.67) with

$$\xi = (\phi_1^+ \alpha + \phi_1^- \beta)(1)(\phi_2^+ \alpha + \phi_2^- \beta)(2)\dots \tag{1.75}$$

contains $2N$ spatial orbitals but its properties have not been considered in the literature. Carrying this further, an analogy of equations (1.71)–(1.73) is obtained by taking each simple product of the ϕs, $\phi_1^a \phi_2^b \phi_3^c \dots$, with $a, b, c, \dots$ referring to the signs, for which the sum of the associated spin-components is less than or equal to S, and writing this in the form of equation (1.72). The results of an investigation of such a function are likely to be exceedingly complicated and not very interesting, which serves as a reminder of the relative futility in searching for orbital methods which are improvements over the simplest HF and G1 (or Valence-Bond) orbital methods.

1.4.11   Artistry and conclusion

The artistry involved in the performance of calculations can be appreciated by the following list of approximations or arbitrary procedures which are used repeatedly and, often, of necessity in numerical work:

(i) limited basis sets or the equivalent of finite, if extrapolated, mesh sizes in numerical methods;

(ii) truncation of partial wave expansions in orbitals and in integrals;

(iii) choice of h_0s in all perturbation theoretical calculations, the choice of the Ψ_n in perturbative C.I. calculations, and the choice of Ψ_0 in Nesbet's cluster method (i.e. the unitary transformation among the occupied orbitals based on which the cluster expansion is defined);

(iv) approximate summation of certain diagrams and the neglect of others in many-body calculations; and

(v) the choice of which parameters are to be calculated and which are to be estimated empirically in the empirical methods.

Barr and Davidson[178] have recently presented a careful analysis of correlation calculations on the neon atom based on a HF Ψ_0 and their discussion is most illustrative. They find the best *calculated* values of the second-order p.t. energy, E_2, the Bethe–Goldstone–Nesbet energy $\Sigma\varepsilon_{ij}$, and the total C.I. with pair excitations, or Total Pair Excitation Block, E_{TPEB}, to be

$$E_2 = 123\%; \ \Sigma\varepsilon_{ij} = 104\%; \ E_{TPEB} = 90\%$$

in percentage of the exact correlation energy. The difference between E_2 and $\Sigma\varepsilon_{ij}$ which corresponds to particle–particle and hole–hole ladder diagrams summed to infinite order amounts to 19%; while the difference between $\Sigma\varepsilon_{ij}$ and E_{TPEB}, which corresponds to pair–pair interactions, amounts to 14%. This, of course, means that E_2 is about 30% 'too large' which is not at all unreasonable; this also indicates that it does not make much sense to calculate one set of third-order corrections without calculating all significant third-order corrections. Barr and Davidson are careful to point out that $\Sigma\varepsilon_{ij}$ is not invariant under a unitary transformation of the orbitals* while E_2 [46] and E_{TPEB} are so invariant. Susceptible to the usual prejudice, these authors are not concerned with the strong dependence of these values on the choice of Ψ_0 when it is permitted to be other than Hartree–Fock.

The last step of the Barr and Davidson argument attempts to account for the error involved in using a finite basis set by estimating the three-particle effects and working backwards. They perform a calculation for the three-particle terms in a small basis set from which they estimate the total con-

---

*The difference is actually negligible in this example with the tetrahedral 'localisation' used. An example in which the effect is most significant has been given earlier by Bender and Davidson[179].

tribution to be $\sim 2\%$*. This leads them to extrapolate the best calculated values upwards by 8–9% to give

$$E_2^{\text{extrap}} = 132\%; \quad \Sigma\varepsilon_{ij}^{\text{extrap}} = 113\%; \quad E_{\text{TPEB}}^{\text{extrap}} = 98\%.$$

Notice that three-particle corrections first appear in fourth-order p.t.[37] and since third-order p.t. terms contribute $\sim 30\%$, the total fourth-order energy could well contribute $\sim 10\%$. Whether the 2% estimate of Barr and Davidson is accurate so that the remainder of the $\sim 10\%$ comes from fourth-order pair terms remains to be demonstrated more conclusively. It is clear, however, that results accurate to better than 10% of the 'correlation energy' from even the best of calculations presently considered feasible are more likely than not to be fortuitous. As these authors observe, the '10% error…will be difficult to remove.'

This argument provides a quantitative basis for the fundamental role I have argued here for the perturbation theoretical description of electron correlation: the independent particle Ψ_0 provides an accurate description of a many-electron system if, and only if, higher-order perturbation theoretical corrections are of rapidly decreasing magnitudes.

The best way to perform calculations thus seems to be within some sort of perturbation theoretical framework in which one watches the convergence of the terms, order by order, without having assumed initially that any terms, e.g. three-particle fourth-order terms, are necessarily negligible. In this spirit there should be little advantage in diagonalising matrices as in the Nesbet, MET and TPEB procedures, over determining the correction terms perturbatively.

1.5 APPENDICES

1.5.1 Appendix A Helium energy

$$E_{\text{exact}} = -2.90372 \text{ a.u.}$$

	h_0^Z	h_0^{HF}	$h_0^H = h_0^{EL}(\alpha = \frac{1}{2})$	$h_0^{EL}(\alpha = \frac{1}{4})$	h_0^{NS}§
E_0	-4.0*	-1.836†	-1.836‡	-2.843‡	-2.5§
E_1	1.25	-1.026	-1.026	0.067	-0.185
E_2	-0.167	-0.037	-0.048	-0.086	-0.132
E_4	-2×10^{-4}	-8×10^{-4}	-1.3×10^{-3}	-4.5×10^{-4}	-0.284
E_6	-4×10^{-5}	—	-1×10^{-4}	8×10^{-6}	-4×10^{-4}
E_{10}	-4×10^{-7}	—	-3×10^{-6}	7×10^{-9}	-5×10^{-5}
E_{14}	-1×10^{-8}	—	-2×10^{-7}	1×10^{-11}	1×10^{-7}
E_{18}	-2×10^{-10}	—	$\sim 1 \times 10^{-8}$	$\sim 2 \times 10^{-14}$	1×10^{-7}

*From Midtdal[26].
†From Byron and Joachain[42]. Calculation carried out only as far as E_5.
‡From Riley and Dalgarno[73].
§Hydrogen-like Non-Symmetric with $Z_1 = 1$ and $Z_2 = 2$. Riley and Musher, unpublished.

---

*Similar estimates have been given by Bunge and Peixoto[180] and Das and co-workers[131] who also argue that $\sim 8\%$ of the correlation arises from orbitals of $l \geqslant 4$. For three-electron term calculations see Micha[181], as well as Kelly and Ron[132].

Comments: (i) Visual inspection shows that convergence is in the order $h_0^{EL}(\alpha = \frac{1}{4}) > h_0^Z > h_0^H$ with h_0^{HF} perhaps close to h_0^H. As far as fifth order both h_0^H and h_0^{HF} give -2.90370.

(ii) In the Riley–Dalgarno calculation for $h_0^{EL}(\alpha = \frac{1}{4})$ tabulated here HF orbitals were used. When self-consistent orbitals are used then $E_0 = -2.839$ and $E_1 = 0$ [85]. Musher and Schulman[37] iterated once towards self-consistency from hydrogenic functions and obtained $E_0 = -2.828$, $E_1 = -0.006$, $E_2 = -0.077$, $E_3 = 0.010$ which gives more rapid convergence as far as it goes than does the Riley–Dalgarno calculation.

(iii) The terms of h_0^Z series are all negative after E_3 while those of h_0^H alternate in sign (the odd energies are positive). The terms of h_0^{HF} are negative from E_2 to E_5, as far as the series has been calculated, while those of $h_0^{EL}(\alpha = \frac{1}{4})$ are of irregular sign.

(iv) A table of partial sums of the energy terms, along with the errors involved, are given for the lower orders of these series in Reference 37.

(v) The h_0^{NS} expansion is slowly convergent after E_{14}. When $Z_1 = 1.9$ and $Z_2 = 2$ as in Reference 188, $E_{18} = -1 \times 10^{-10}$ so that this series converges essentially like the h_0^Z series and perhaps even a bit faster. The total energy in the 100-element basis of this calculation is -2.90367002 both variationally and perturbatively where this value is reached by 15th order.

1.5.2   Appendix B Lithium energy

$$E_{exact} = -7.47807 \text{ a.u.}$$

	h_0^Z	h_0^{HF}	h_0^{N-1}	$h_0^{EL}(\alpha = \frac{2}{3})$	h_0^{NS}
$E^{(0)*}$	-10.1250† (35)‡	-5.0062§ (33)	$-5.7823\|$ (22)	-4.3314¶ (42)	-5.161** (31)
$E^{(1)}$	-7.0565 (6)	-7.4327 (0.6)	-7.4322 (0.6)	-7.4132 (1)	-7.412 (0.9)
$E^{(2)}$	-7.4649 (0.2)	—	-7.474 (0.05)	-7.4884 (0.1)	-7.467 (0.2)
$E^{(3)}$	-7.4726 (0.07)	—	-7.478 (0)	-7.4793 (0.02)	-7.461 (0.2)

$*E^{(n)} = \sum\limits_{i=0}^{n} E_i$

†From Seung (Hui) and Wilson[115a].
‡Errors in percent given in parentheses.
§From Roothaan C. C. J., Sachs L. M. and Weiss A. W., (1960) *Rev. Mod. Phys.* 32, 186.
‖From Chang, Pu (Poe) and Das[127].
¶From Schulman and Lee[118].
**Lee, Schulman, (Seung) Hui and Musher[189], based on the Hartree H_0.

Comment: Convergence is reasonably rapid in all schemes and the pattern does not differ significantly from that of the Helium energy described in Appendix A.

1.5.3 Appendix C Lithium hfs

$$f_{expt} = 2.9096\ a_0^3$$

	h_0^Z	h_0^{RHF}	h_0^{N-1}	$h_0^{EL}\ (\alpha = \frac{2}{3})$	Nesbet-cluster
$f^{(0)}$	13.5†(300)‡	2.065§(29)	2.065‖(29)	4.497¶(54)	2.067**(29)
$f^{(1)}$	1.5 (48)	2.695 (7)	2.578 (11)	3.149 (9)	2.699 (7)
$f^{(2)}$	—	—	2.848 (2)	2.850 (2)	2.978 (2)
$f^{(3)}$	—	—	2.89 (0.3)	—	2.896 (0.4)

$*f^{(n)} = \sum_{i=1}^{n} f_i$

†From Cohen and Dalgarno[91].
‡Errors (%) given in parentheses.
§Cohen M. H., Goodings D. A. and Heine V. (1959). *Proc. Phys. Soc. (London)* 73, 811.
‖From Chang, Pu (Poe) and Das[28], breakdown in orders given in Reference 118.
¶From Schulman and Lee[118].
**From Nesbet R. K. (1966). *Proc. Int. Colloq. No. 164, C.N.R.S., Paris* p. 87.

Comment: Except for the h_0^Z expansion which has not converged at all, the various expansions converge at almost identical rates.

1.5.4 Appendix D Nitrogen hfs

$$f_{expt} = 1.22099\ a_0^3$$

	h_0^{N-1}	Nesbet-cluster	C.I.
$f^{(0)*}$	0†(100)‡	0§(100)	0‖(100)
$f^{(1)}$	0.980 (20)	0.232 (82)	0.986 (26)
$f^{(2)}$	1.223 (0.3)	2.508 (105)	0.939 (23)
$f^{(3)}$	—	1.267 (4)	—

$*f^{(n)}$ are partial sums; order is defined differently for each method.
†From Dutta N. C., Matsubara (Dutta) C., Pu (Poe) R. T. and Das T. P. (1969). *Phys. Rev.* 177, 33. f_2 includes approximate values of higher-order terms. f_1 derives from a strong cancellation between a 2s-hole term of 6.56 and a 1s-hole term of −5.58.
‡Errors in percent are given in parentheses.
§From Nesbet R. K., (1966). *Proc. Int. Colloq. No. 164, C.N.R.S., Paris* p. 87.
‖From Platas O. R. and Schaeffer III H. F. (1971). *Phys. Rev.* A4, 33. Here order is defined relative to RHF, so that the 'first-order function' of these authors gives f_1 and the two-particle terms give f_2. The first-order function is essentially a C.I.-obtained UHF function for which its $f_0 = 0.896$ and $f_1 = 0$ by Brillouin's theorem, so that pair terms would only contribute to f_2.

Comments: (i) The convergence of the h_0^{N-1} series seems to be excellent, but in view of the difficulties with the other expansions, this calculation should be very sensitive to numerical error from the various possible sources.

(ii) The convergence of the Nesbet-cluster series is difficult to comment upon, other than to consider it as surprising, particularly in view of the C.I. results which provide such a significantly larger first-order term. It would be amusing to know if the cluster expansion defined with some unitary-transformed RHF orbitals would provide more rapid convergence.

(iii) The convergence of the C.I. series is not particularly good. If f_3 were the same size as f_2, then the result would still be 20% in error, while if it were larger then the convergence would not be so convincing.

1.5.5 Appendix E Beryllium electric polarisability

Experimental value unknown
Approximate bounds* $36.2 < \alpha < 44.7$ a_0^3

	h_0^Z	h_0^{HF}	h_0^{HS}	h_0^{N-1}	h_0^{EL} $(\alpha = \frac{3}{4})$	h_0^{EL} $(\alpha = \frac{3}{8})$
$\alpha^{(0)}$	3.2†	30.4‡	78.5§	82.1‖	73.6§	41§
$\alpha^{(1)}$	9.0	40.3	17.5	37.6**	2.7	−8
(α_{geom})¶	(45.6)	(45.5)	(43.9)	(53.3)	(37.5)	(19)
$\alpha^{(2)}$	—	—	—	52.0**	—	—
$\alpha^{(3)}$	—	—	—	46.8**	—	—

*From Cohen M., (1967). *Can. J. Phys.* 45, 3387.

†From Drake G. W. F. and Cohen M. (1968) *J. Chem. Phys.* 48, 1168. Notice the goodness of α_{geom} even though clearly one must go to high $\alpha^{(n)}$ before convergence is obtained. For example, in the geometric expansion which is used by these authors rather than the two term perturbative one, the approximate $\alpha^{(7)} = 3.2$, the same as the leading term $\alpha^{(0)}$!

‡From Caves and Karplus[55]. Earlier values of 28.9 and 37.4 were obtained for $\alpha^{(0)}$ and $\alpha^{(1)}$ respectively by Langhoff, Karplus and Hurst[62] and Epstein and Johnson[56] using the previous workers' orbitals. Notice the importance of using flexible enough variational functions in such calculations.

§From Hameed[60].

‖From Kelly[69].

¶Computed as in the text, although not restricted to the HF value and hence less likely to be meaningful.

**Order takes on a special meaning here: both $\alpha^{(1)}$ and $\alpha^{(2)}$ include the summation of some higher-order terms. $\alpha^{(3)}$ is the renormalized value of $\alpha^{(2)}$ and hence is dominated by third-order corrections.

Comments: Convergence is very difficult to assess with only two terms available, although the h_0^{HF} series seems to converge reasonably well. For the remaining series, α_1 is at least a large fraction of α_0 so convergence in all cases is suspect, even though α_{geom} gives reasonable values with one exception. Only the h_0^{N-1} value of $\alpha^{(1)}$ is within the error bounds, and even here, the normalisation (third-order) correction is 15%. Both the h_0^{HS} (close to the so-called Method d, an approximate HF method, which uses HF orbitals, but local approximations to their h_0) and the h_0^{EL} series give poor results through first order. None of this slow convergence can be attributed to the low lying $1s^2 2p^2$ ^1S state as can be seen from the h_0^Z series. The goodness of α_0 for h_0^{EL} $(\alpha = \frac{3}{8})$ is wiped out by the large *negative* α_1. Notice that this choice of α is the one corresponding to equation (1.41) of the text which gave rise to the more rapid convergence in the He atom energy. The rapidly convergent two-term series $\alpha^{(0)} = 51.6$, $\alpha^{(1)} = 51.0$ has been obtained by Scott and Kirtman[193] using Kirtman's procedure described in the text, although it seems likely that when α_2 is calculated it will turn out to be larger in absolute value than α_1.

1.5.6 Appendix F HD molecule spin–spin coupling constant

$J_{expt} \approx 42.7$ Hz*

	h_0^{HF}	h_0^H	h_0 (bare nucleus)
$J^{(0)}$	12.4† (71)‡	20.9† (51)	73.02§ (72)
$J^{(1)}$	21.8 (49)	34.2 (20)	37.73 (12)
$J^{(2)}$	—	∼38.1 (11)	42.57 (0.3)

*Approximate value since orbital and dipolar terms are generally assumed to be small, *c.* 1 Hz, but have not been calculated. The calculations reported here are only for the Fermi contact term.

†Schulman and Kaufman, Reference 61.

‡Errors in percent are given in parentheses.

§Dutta, C. M., Dutta, N. C. and Das, T. P. (1970). *Phys. Rev. Lett.* 25, 1965. $J^{(2)}$ is the sum of higher-order diagrams evaluated by the same authors.

Comments: (i) The $J^{(0)}$ in the HF series is very poor. Thus all orbital calculations in the literature which, by their stated goals, were trying to converge to this value, were highly inadequate if they agreed at all with experiment.

(ii) The convergence of the h_0^H series is relatively good although slow. For many-electron systems this cannot be used, but the result should augur well for an h_0^{N-1} series. Again notice the importance of including at least J_1 and not drawing any conclusion based on J_0 alone.

(iii) The bare nucleus series shows good convergence and an unbelievably good final result were J_{expt} solely due to the Fermi term. However, the bare nucleus expansion has been recently calculated independently in a finite basis set by J. M. Schulman and D. N. Kaufman (to be published) through first order, the values obtained being $J_0 = 46$ and $J_1 = 3$ cps with errors of 0.5 cps, and thus placing some doubt on the values of Dutta, Dutta and Das. The same value of $J_0 = 46$ cps has been independently obtained also by R. M. Pitzer and J. D. Power (unpublished).

(iv) With such poorly convergent series, there should be significant advantage in performing perturbation theory with $H_{00} - H_{nn}$ denominators rather than the usual $\varepsilon_0 - \varepsilon_n$.

Acknowledgements

This manuscript was completed while I was visiting the Theoretical Chemistry Department of the Orta Doğu Teknik Üniversitesi, Ankara, Turkey, and I would like to express my gratitude to the members of that department for their kind hospitality. The present research has been generously supported by the National Science Foundation.

References

1. Musher, J. I. (1970). *Chem. Phys. Lett.,* **7,** 397
2. Löwdin, P. O. (1955). *Phys. Rev.,* **97,** 1509
3. Goddard III, W. A. (1967). *Phys. Rev.,* **157,** 73, 81, 93
4. Hameed, S., (Seung) Hui, S., Musher, J. I. and Schulman, J. M. (1969). *J. Chem. Phys.,* **51,** 502
4a. Heiker, L. G. and Gallup, G. A. (1970). *J. Chem. Phys.,* **52,** 888
5. Ladner, R. C. and Goddard III, W. A. (1969). *J. Chem. Phys.,* **51,** 1073
6. Kaldor, U. and Harris, F. E. (1969). *Phys. Rev.,* **183,** 1
7. Bethe, H. A. and Salpeter, E. E. (1957). *Handbuch der Physik,* ed. by S. Flugge, vol. 35 (Berlin: Springer Verlag)
8. Pitzer, K. S. (1959). *Advan. Chem. Phys.,* **2,** 59
9. Löwdin, P. O. (1959). *Advan. Chem. Phys.,* **2,** 207
10. Sinanoğlu, O. (1964). *Advan. Chem. Phys.,* **6,** 315
11. Sinanoğlu, O. (1969). *Advan. Chem. Phys.,* **14,** 237
12. Kelly, H. P. (1968). *Advan. Theor. Phys.,* **2,** 75
13. Kelly, H. P. (1969). *Advan. Chem. Phys.,* **14,** 129
14. Nesbet, R. K. (1965). *Advan. Chem. Phys.,* **9,** 321
15. Nesbet, R. K. (1969). *Advan. Chem. Phys.,* **14,** 1
16. Wahl, A. C. and Das, G. (1970). *Advan. Quantum. Chem.,* **5,** 261
17. McWeeny, R. and Steiner, E. (1965). *Advan. Quantum. Chem.,* **2,** 93
18. Freed, K. F. *Ann. Rev. Phys. Chem.,* in the press

19. Sinanoğlu, O. and Brueckner, K. A. (1970). *Three Approaches to Electron Correlation in Atoms* (New Haven: Yale University Press)
20. Condon, E. U. (1968). *Rev. Mod. Phys.*, **40**, 872
21. Fock, V., Veselov, M. G. and Petrashen, M. (1940). *Zh. Eksperim. Teor. Fiz.*, **10**, 723
22. Hylleraas, E. A. (1929). *Z. Physik*, **54**, 347; (1930). **65**, 209
23. Frankowski, K. and Pekeris, C. L. (1966). *Phys. Rev.*, **146**, 46
24. Schwartz, C. (1962). *Phys. Rev.*, **126**, 1015; **128**, 1146
25. Sanders, F. C. and Scherr, C. W. (1969). *Phys. Rev.*, **181**, 84
26. Midtdal, J. (1965). *Phys. Rev.*, **138**, A1010
27. Musher, J. I. (1969). *J. Chem. Phys.*, **50**, 3741
28. Sinanoğlu, O. (1961). *Phys. Rev.*, **122**, 493
29. Sinanoğlu, O. (1961). *Proc. Roy. Soc. (London)*, **A260**, 379
30. Kelly, H. P. (1963). *Phys. Rev.*, **131**, 684
31. Møller, C. and Plesset, M. S. (1934). *Phys. Rev.*, **46**, 618
32. Brillouin, L. (1934). *Actualites Sci. Ind.* No. 159
33. Nesbet, R. K. (1955). *Proc. Roy. Soc. (London)*, **A230**, 312
34. Sinanoğlu, O. (1962). *J. Chem. Phys.*, **36**, 706, 3198
34a.Szasz, L. (1962). *Phys. Rev.*, **126**, 169
35. Dalgarno, A. and Lewis, J. T. (1955). *Proc. Roy. Soc. (London)*, **A233**, 70
36. Schwartz, C. (1959). *Ann. Phys.(N.Y.)*, **6**, 156, 170
37. Musher, J. I. and Schulman, J. M. (1968). *Phys. Rev.*, **173**, 93
38. Amos, A. T. and Musher, J. I. (1971). *J. Chem. Phys.*, **54**, 2380
39. Morse, P. M. and Feshbach, H. (1953). *Methods of Theoretical Physics*, Vol. 2, Ch. 9, (New York: McGraw-Hill)
40. Feenberg, E. (1948). *Phys. Rev.*, **74**, 206
41. Diner, S., Malrieu, J. P., Claverie, P. and Jordan, F. (1968). *Chem. Phys. Lett.*, **2**, 319
42. Byron, F. and Joachain, C. J. (1967). *Phys. Rev.*, **157**, 1, 7
43. McKoy, V. (1965). *J. Chem. Phys.*, **43**, 1605
44. King, H. F. (1967). *J. Chem. Phys.*, **46**, 705
45. Viers, J. W., Harris, F. E. and Schaeffer III, H. F. (1970). *Phys. Rev.*, **A1**, 24
46. Amos, A. T., Roberts, H. G. Ff. and Musher, J. I. (1969). *Chem. Phys. Lett.*, **4**, 93
47. Hirschfelder, J. O., Byers Brown, W. and Epstein, S. T. (1964). *Advan. Quantum Chem.*, **1**, 255
47a.Dalgarno, A. (1962). *Advan. in Physics*, **11**, 281
48. Sternheimer, R. M. (1950). *Phys. Rev.*, **80**, 102 *et seq.*
48a.Abragam, A. and Pryce, M. H. L. (1951). *Proc. Roy. Soc. (London)*, **A205**, 135
48b.Larsson, S. (1970). *Phys. Rev.*, **A2**, 1248
49. Sanders, P. G. H. (1969). *Advan. Chem. Phys.*, **14**, 365
49a.Malrieu, J. P. (1967). *J. Chem. Phys.*, **47**, 4555
50. Tuan, D. F., Epstein, S. T. and Hirschfelder, J. O. (1966). *J. Chem. Phys.*, **44**, 431
51. Hameed, S. (1971). *Phys. Rev.*, **A4**, 543
52. Peng., H. (1941). *Proc. Roy. Soc. (London)*, **A178**, 499
53. Allen, L. C. (1955). *Quart. Progr. Rept. No. 18, Solid-state and Molecular Theory Group, MIT*, 15 October 1955, p.16
54. Musher, J. I. (1967). *J. Chem. Phys.*, **46**, 369
55. Caves, T. C. and Karplus, M. (1969). *J. Chem. Phys.*, **50**, 3649
56. Epstein, S. T. and Johnson, R. E. (1967). *J. Chem. Phys.*, **47**, 2275
57. Amos, A. T. and Musher, J. I. (1967). *Mol. Phys.*, **13**, 509
58. Tuan, D. F. (1970). *Chem. Phys. Lett.*, **7**, 115
59. Amos, A. T. and Roberts, H. G. Ff. (1969). *J. Chem. Phys.*, **50**, 2375
60. Hameed, S. (1971). *J. Phys.*, **B4**, 728
61. Schulman, J. M. and Kaufman, D. N. (1970). *J. Chem. Phys.*, **53**, 477
62. Langhoff, P. W., Karplus, M. and Hurst, R. P. (1966). *J. Chem. Phys.*, **44**, 505
63. Westhaus, P. and Sinanoğlu, O. (1969). *Phys. Rev.*, **183**, 56
64. Sinanoğlu, O. and Nicolaides, C. (1970). *J. Physique*, **31**, C4-117
65. Epstein, S. T. (1964). *J. Chem. Phys.*, **41**, 1045
66. Musher, J. I. (1967). *Rev. Mod. Phys.*, **39**, 203
67. Kelly, H. P. (1969). *Phys. Rev. Lett.*, **23**, 455
67a.Miller, J. H. and Kelly, H. P. (1971). *Phys. Rev.*, **A4**, 480
68. Lee, T., Dutta, N. C. and Das, T. P. (1970). *Phys. Rev. Lett.*, **25**, 204

69. Kelly, H. P. (1964). *Phys. Rev.*, **136**, B896
70. Silverstone, H. J. and Yin, Y. L. (1968). *J. Chem. Phys.*, **49**, 2026
71. Huzinaga, S. and Arnau, C. (1970). *Phys. Rev.*, **A1**, 1285
72. Kelly, H. P. and Miller, J. H. (1971). *Phys. Rev.*, **A3**, 578
73. Riley, M. E. and Dalgarno, A. (1971). *Chem. Phys. Lett.*, **9**, 382
74. Bonačič, V. and Koutecký, J. (1971). *Chem. Phys. Lett.*, **10**, 401
75. Roothaan, C. C. J. (1951). *Rev. Mod. Phys.*, **23**, 69; (1960). **32**, 179
76. Slater, J. C. (1951). *Phys. Rev.*, **81**, 385
77. Gáspár, R. (1954). *Acta Phys. Hung.*, **3**, 263
78. Kohn, W. and Sham, L. J. (1965). *Phys. Rev.*, **140**, A1133
79. Herman, F. and Skillman, S. (1963). *Atomic Structure Calculations* (Englewood Cliffs, N.J.: Prentice-Hall)
80. Tong, B. Y. and Sham, L. J. (1966). *Phys. Rev.*, **144**, 1
80a.Cowan, R. D., Larson, A. C., Liberman, D., Mann, J. B. and Waber, J. (1966). *Phys. Rev.*, **144**, 5
80b.Liberman, D. A. (1968). *Phys. Rev.*, **171**, 1
81. Slater, J. C., Wilson, T. M. and Wood, J. H. (1969). *Phys. Rev.*, **179**, 28
82. Wilson, T. M., Wood, J. H. and Slater, J. C. (1970). *Phys. Rev.*, **2A**, 620
83. Amos, A. T. and Musher, J. I. (1967). *Phys. Rev.*, **164**, 31
84. Chase, R. L., Kohler, H. S. and Kelly, H. P. (1971). *Phys. Rev.*, **A3**, 1550
85. McNaughton, D. J. and Henry, W. J. (1969). *J. Phys.*, **B2**, 1131
86. McNaughton, D. J. and Smith Jr., V. H. (1969). *J. Phys.*, **B2**, 1138
87. Matulis, A., Safronova, U. and Tolmachev, V. (1964). *Lietuvos Fiz. Rinkinys*, **4**, 331
88. Ivanova, A. I., Safronova, U. I. and Tolmachev, V. V. (1967). *Lietuvos Fiz. Rinkinys*, **7**, 35
89. Safronova, U. I. and Tolmachev, V. V. (1967). *Lietuvos Fiz. Rinkinys*, **7**, 53
89a.Safronova, U. I. and Tolmachev, V. V. (1967). *Teor. i Eksper. Khim.*, **3**, 571, 579
90. Safronova, U. I. and Kharitonova, V. N. (1970). *Opt. i Spektroskopiya*, **28**, 1039
90a.Safronova, U. I. (1970). *Opt. i Spektroskopiya*, **28**, 1050
91. Cohen, M. and Dalgarno, A. (1963). *Proc. Roy. Soc. (London)*, **A275**, 492
92. Hall, G. G., Jones, L. L. and Rees, D. (1965). *Proc. Roy. Soc. (London)*, **A283**, 194
93. Kirtman, B. and Decious, D. R. (1968). *J. Chem. Phys.*, **48**, 3133
93a.Matcha, R. L. and Byers Brown, W. (1968). *J. Chem. Phys.*, **48**, 74
94. Goodisman, J. (1967). *J. Chem. Phys.*, **47**, 1256
94a.Goodisman, J. (1968). *J. Chem. Phys.*, **48**, 2981
95. Layzer, D. (1964). *Ann. Phys. (N.Y.)*, **29**, 101
95a.Layzer, D., Horak, Z., Lewis, M. and Thompson, D. (1964). *Ann. Phys. (N.Y.)*, **29**, 101
96. Tolmachev, V. V. (1969). *Advan. Chem. Phys.*, **14**, 421, 471
97. Eisenschitz, R. and London, F. (1930). *Z. Physik*, **60**, 491
98. Veselov, M. G. and Adamov, M. N. (1947). *Dokl. Akad. Nauk SSSR*, **57**, 235
99. Epstein, S. T. (1971). 'The variational method', *WIS-TCI reports* 336, 338, 339
100. Musher, J. I. (1965). *Ann. Phys. (N.Y.)*, **32**, 416
101. Hassé, H. R. (1931). *Proc. Camb. Phil. Soc.*, **27**, 66
102. Slater, J. C. and Kirkwood, J. G. (1931). *Phys. Rev.*, **37**, 682
103. Guy, J. and Harrand, M. (1952). *Compt. Rend.*, **234**, 616, 716
103a.Tillieu, J. and Guy, J. (1955). *Compt. Rend.*, **239**, 1203, 1283
104. Adamov, M. N. and Milevskaya, I. S. (1957). *Opt. i Spektroskopiya*, **2**, 399
105. Das, T. P. and Bersohn, R. (1956). *Phys. Rev.*, **104**, 849
106. Das, T. P. and Bersohn, R. (1959). *Phys. Rev.*, **115**, 897
107. Stephen, M. J. (1957). *Proc. Roy. Soc. (London)*, **A243**, 264, 274
108. Marshall, T. W. and Pople, J. A. (1960). *Mol. Phys.*, **3**, 339
109. Dalgarno, A. (1959). *Proc. Roy. Soc. (London)*, **A251**, 282
110. Karplus, M. and Kolker, H. J. (1963). *J. Chem. Phys.*, **38**, 1263
111. Karplus, M. and Kolker, H. J. (1963). *J. Chem. Phys.*, **39**, 2011
112. Karplus, M. and Kolker, H. J. (1964). *J. Chem. Phys.*, **41**, 1259
113. Stevens, R. M., Kern, W. and Lipscomb, W. N. (1963). *J. Chem. Phys.*, **38**, 550
114. Lipscomb, W. N. (1966). *Advan. Magnetic Resonance*, **2**, 137, Appendix B.
115. Knight, R. E. (1969). *Phys. Rev.*, **183**, 45
115a.Seung (Hui), S. and Wilson,Jr., E. B. (1967). *J. Chem. Phys.*, **47**, 5343
115b.Chisholm, C. D. H. and Dalgarno, A. (1966). *Proc. Roy. Soc. (London)*, **A292**, 264

115c. Chisholm, C. D. H., Dalgarno, A. and Innes, F. R. (1968). *Phys. Rev.*, **167**, 60
116. Tuan, D. F. and Sinanoğlu, O. (1964). *J. Chem. Phys.*, **41**, 2677
117. Geller, M., Taylor, H. S. and Levine, H. B. (1965). *J. Chem. Phys.*, **43**, 1727
117a.Bagus, P. S., Liu, B. and Schaeffer III, H. F. (1970). *Phys. Rev.*, **A2**, 555
118. Schulman, J. M. and Lee, W. S. (1972). *Phys. Rev.*, **A5**, 13
119. Temkin, A. and Sullivan, E. (1963). *Phys. Rev.*, **129**, 1250
120. McKoy, V. and Winter, N. W. (1968). *J. Chem. Phys.*, **48**, 5514
121. Winter, N. W. and McKoy, V. (1970). *Phys. Rev.*, **A2**, 2219
122. Barraclough, C. G. and Mooney, J. R. (1971). *J. Chem. Phys.*, **54**, 35
123. Lee, W. S. (1971). *Ph.D. thesis, Polytechnic Institute of Brooklyn*
124. Musher, J. I. (1970). *Chem. Phys. Lett.*, **6**, 33
125. Schulman, J. M. and Musher, J. I. (1968). *J. Chem. Phys.*, **49**, 4845
126. Kelly, H. P. (1967). *Phys. Lett.*, **25A**, 6
127. Chang, E. S., Pu (Poe), R. T. and Das, T. P. (1968). *Phys. Rev.*, **174**, 1
128. Dutta, N. C., Matsubara (Dutta), C., Pu (Poe), R. T. and Das, T. P. (1969). *Phys. Rev.*, **177**, 33
129. Lyons, J. D., Pu (Poe), R. T. and Das , T. P. (1968). *Phys. Rev.*, **178**, 103
130. Lyons, J. D., Pu (Poe), R. T. and Das, T. P. (1969). *Phys. Rev.*, **186**, 266
131. Lee, T., Dutta, N. C. and Das, T. P., *Phys. Rev.*, in the press
132. Ron, A. and Kelly, H. P. (1971). *Phys. Rev.*, **A4**, 11
133. Öksüz, I. and Sinanoğlu, O. (1969). *Phys. Rev.*, **181**, 42, 54
134. Skutnik, B., Öksüz, I. and Sinanoğlu, O. (1968). *Int. J. Quant. Chem.*, **S2**, 1
134a.Linderberg, J. and Shull, H. (1960). *J. Mol. Spectrosc.*, **5**, 1
135. Malrieu, J. P., Claverie, P. and Diner, S. (1969). *Theoret. Chim. Acta*, **13**, 18
136. Diner, S., Malrieu, P. and Claverie, P. (1969). *Theoret. Chim. Acta*, **13**, 1
137. Pamuk, H. Ö and Sinanoğlu, O., to be published
138. Pamuk, H. Ö (1968). *Ph.D. thesis, Yale University*
139. Mayer, J. E. and Mayer, M. G. (1940). *Statistical Mechanics*, (New York: Wiley)
140. Bethe, H. A. and Goldstone, J. (1957). *Proc. Roy. Soc. (London)*, **A238**, 551
141. Szasz, L. (1959). *Z. Naturforsch.* **14a**, 1014
142. Szasz, L. (1960). *Z. Naturforsch.* **15a**, 909
143. Nesbet, R. K. (1967). *Phys. Rev.*, **155**, 51, 56
144. Nesbet, R. K. (1968). *Phys. Rev.*, **175**, 2
145. Nesbet, R. K. (1970). *Phys. Rev.*, **A2**, 661, 1208
146. Nesbet, R. K. (1970). *Phys. Rev.*, **A3**, 87
147. Moser, C. M. and Nesbet, R. K. (1971). *Phys. Rev.*, **A4**, 1336
148. Harris, F. E., Kaldor, U. and Schaeffer III, H. F. (1968). *Int. J. Quantum Chem.*, **S2**, 13
149. Schaeffer III, H. F. and Harris, F. E. (1968). *Phys. Rev.*, **167**, 67
150. Schaeffer III, H. F., Klemm, R. A. and Harris, F. E. (1968). *Phys. Rev.*, **176**, 49
151. Schaeffer III, H. F., Klemm, R. A. and Harris, F. E. (1969). *Phys. Rev.*, **181**, 137
152. Schaeffer III, H. F., Klemm, R. A. and Harris, F. E. (1969). *J. Chem. Phys.*, **51**, 4643
153. Hartree, D. R., Hartree, W. and Swirles, B. (1939). *Phil. Trans. Roy. Soc. (London)*, **A238**, 229
154. Jucys, A. P. (1952). *Zh. Eksperim. Teor. Fiz.*, **23**, 129
154a.Jucys, A. P. (1969). *Advan. Chem. Phys.*, **14**, 191
155. Clementi, E. and Veillard, A. (1966). *J. Chem. Phys.*, **44**, 3050
156. Kirtman, B. (1968). *Chem. Phys. Lett.*, **1**, 631
157. Kirtman, B. and Mowery, R. L. (1971). *J. Chem. Phys.*, **55**, 1447
158. Kirtman, B. and Kaldor, U. (1971). *Phys. Rev.*, **A3**, 1295
159. Hirschfelder, J. O. and Silbey, R. J. (1966). *J. Chem. Phys.*, **45**, 2188
160. Hirschfelder, J. O. (1967). *Chem. Phys. Lett.*, **1**, 325, 363
161. Amos, A. T. (1970). *Chem. Phys. Lett.*, **5**, 587
162. Musher, J. I. and Silbey, R. (1968). *Phys. Rev.*, **174**, 94
163. Matsen, F. A. (1964). *Advan. Quantum Chem.*, **1**, 60
164. Gallup, G. A. (1968). *J. Chem. Phys.*, **48**, 1752
165. Gallup, G. A. (1969). *J. Chem. Phys.*, **50**, 1206
166. Musher, J. I. (1970). *J. Physique*, **31**, C4–51
167. Kaplan, I. G. (1969). *Symmetry of Many-Electron Systems*, (Moscow: Nauka Press), English translation, Academic Press, in the press
168. Messiah, A. (1962). *Quantum Mechanics*, (Amsterdam: North-Holland), Vol. 2, 600

169. Boys, S. F. and Handy, N. C. (1969). *Proc. Roy. Soc. (London)*, **A309**, 209
170. Boys, S. F. and Handy, N. C. (1969). *Proc. Roy. Soc. (London)*, **A310**, 43, 63
171. Boys, S. F. and Handy, N. C. (1969). *Proc. Roy. Soc. (London)*, **A311**, 309
172. Pauncz, R. (1967). *Alternant Molecular Orbital Method*, (Philadelphia: W. B. Saunders Co.)
173. Lefebvre, R. and Smeyers, Y. G. (1967). *Int. J. Quantum Chem.*, **1**, 403
174. Auffray, J. P. and Parcus, J. K. (1962). *Compt. Rend.*, **254**, 3170
175. Prat, R. and Lefebvre, R. (1969). *Int. J. Quantum Chem.*, **3**, 503
176. Cizek, J. and Paldus, J. (1969). *Chem. Phys. Lett.*, **3**, 1, Acknowledgment
177. Lunell, S. (1968). *Phys. Rev.*, **173**, 85
178. Barr, T. L. and Davidson, E. R. (1970). *Phys. Rev.*, **A1**, 644
179. Bender, C. F. and Davidson, E. R. (1967). *J. Chem. Phys.*, **47**, 360
180. Bunge, C. F. and Peixoto, E. M. A. (1970). *Phys. Rev.*, **A1**, 1277
181. Micha, D. A. (1970). *Phys. Rev.*, **A1**, 755
182. Nesbet, R. K. (1971). *Internat. J. Quantum Chem.*, **4**, 117
183. Stanton, R. E. (1965). *J. Chem. Phys.*, **42**, 3630
184. Dmitriev, Yu. Yu., Bochkariov, V. B., Labzovsky, I. N. and Zashikhin, M. N. (1971). *Chem. Phys. Lett.*, **11**, 620
185. Čižek, J. and Paldus, J. (1971). *Int. J. Quantum Chem.*, **5**, 359
186. Weiss, A. W. (1971). *Phys. Rev.*, **3A**, 126
187. Weiss, A. W. (1970). *Nuclear Instr. & Methods*, **90**, 121
188. Riley, M. E., Schulman, J. M. and Musher, J. I. *Phys. Rev.*, **A** (in the press)
189. Schulman, J. M., Lee, W. S., Seung Hui, S. and Musher, J. I. *Phys. Rev.*, **A** (in the press)
190. Schneider, B., Taylor, H. S. and Yaris, R. (1970). *Phys. Rev.*, **A1**, 855
191. Csanak, Gy., Taylor, H. S. and Yaris, R. *Adv. Atomic. Mol. Phys.* (in the press)
192. Amos, A. T. (1972). *Int. J. Quantum Chem.*, **6**, 125
193. Scott, W. R. and Kirtman, B. (1972). *J. Chem. Phys.*, **56**, 1685
194. Montgomery, H. E., Bruner, B. L. and Knight, R. E. (1972). *J. Chem. Phys.*, **56**, 1449

2
The Calculation of Energy Quantities for Diatomic Molecules

A. C. WAHL

Argonne National Laboratory, Argonne, Illinois

2.1 INTRODUCTION

The purpose of this article is to survey the progress made over the past several years and to assess the 'state of the art' in the *a priori* computation of 'energy quantities' for diatomic molecules.

It is thus limited to a discussion of various methods used for computing the total electronic energy of diatomic molecules and their ions. The discussion of each method will be illustrated by representative calculations employing that method. This section on methods will be followed by a survey of the calculation of ground-state energies, excited-state energies, potential curves and dissociation energies, ionisation potentials and electron affinities. Since the recent calculational literature concerning diatomic molecules has leaned heavily towards application of traditional theory to systems *larger* than LiH, so will this survey. In the opinion of this reviewer the major advances which have been recently made have occurred primarily through the gradual improvement of *traditional* techniques leading to either *true chemical* accuracy in energy quantities of interest or to a firm understanding of the power and limitations of less accurate methods.

However, the significance of this seemingly adiabatic advance is quite dramatic. The *accuracy* now attainable and which promises to become routine in *a priori* computations of diatomic energies and potential curves has stimulated greatly increased contact with the experimentalist particularly in the areas of collisions, scattering, spectroscopy and transport processes. This illustrates the rising confidence among experimentalists in the reliability of *a priori* computations and has encouraged the application of successful techniques to larger diatomic systems of practical interest*.

Furthermore our increased understanding of the results of less accurate *a priori* calculations has greatly enhanced their utility as a tool when appropriately used. In addition to these major advances based primarily on traditional approaches, two recent techniques have been applied to diatomic molecules; the method of transcorrelated pairs and the application of many-body perturbation theory to the strong chemical interaction region. Neither of these approaches has yet been implemented well enough technically to be fairly compared or to be competitive with the power of more traditional methods on larger diatomic molecules. However their application is new, they do work, and as their technology improves they will deserve our continued attention. Within the more traditional framework the ideas of the *independent electron-pair approach* are being increasingly employed with considerable success.

There have been a number of recent reviews and compendia related to the theoretical aspects of the electronic structure of diatomic molecules. These are by Matsen and Browne[1] with emphasis on ground and excited states of He_2 and LiH, a compendium of results by Krauss[2], a brief survey of methods by Wahl and co-workers[3], the annual review of quantum chemistry and dynamics by Allen[4], a survey of calculations by Schaefer[5] and a bibliography of hydride calculations by Cade and Huo[6]. A number of specific method reviews have also been recently published, namely for the Hartree–Fock method[7,8] many-body perturbation theory[9–11], the independent electron-pair approach[12], multiconfiguration SCF[13], the direct calculation of natural orbitals[14], and for the relationship between various calculations methods[15].

---

*An annotated bibliography of recent calculations on diatomic molecules prepared for this review by the author giving states calculated, method and basis set used, and name (not value) of quantities computed is available as ANL-7936 from National Technical Information Service, U.S. Department of Commerce, 5285 Port Royal Road, Springfield, Virginia 22151.

2.2 METHODS

The basic theoretical approach to calculating electronic energies and studying and interpreting the electronic structure of diatomic molecules is to solve approximately the electronic Schrödinger equation at a fixed internuclear distance. It is primarily in our ability to solve this equation well enough to yield the accurate molecular energy differences required that advances have occurred over the past several years. In this section we will outline some of the most commonly used methods, recent contributions to them, and the two new methods which have recently been applied to diatomic molecules.

Due to the presence of the two-electron interaction term the Schrödinger equation has only been solved *approximately* for many electron systems. The approximation consists in the form chosen for the wave function. In the following discussion we shall outline various methods for achieving an approximate solution to the electronic Schrödinger equation and present recent illustrations of each. These methods differ primarily in the chosen form of the wave function. All the major methods used on molecules to date rely on expanding, at some point, the wave function in terms of some finite set of functions called basis functions. The two most commonly used basis functions are Slater-type or Gaussian-type orbitals. When a sufficient number of either type is used the true physical properties of the wave-function form are achieved from the calculation. When small basis sets are used many properties of the wave function remain dependent on the basis set while some properties are reliable.

The first method to be discussed (Section 2.2.1) will be the SCF method, which, if an infinite basis set were used, yields the Hartree–Fock molecular orbitals. As we shall see, by a judicious choice of an *extended* basis set, SCF results very close to the Hartree–Fock limit can be achieved. These Hartree–Fock results contain the correlation error which is defined as the difference between a dynamical variable computed from the Hartree–Fock wave function and the value which would be obtained from an exact non-relativistic wave function.

The other methods discussed (Section 2.2.2–2.2.8) can eliminate most of this correlation error and it is our increased understanding of how to use these methods to achieve reliable energetic quantities which will be stressed. This discussion is somewhat complicated by the fact that due to the use of a smaller basis set a calculation which includes electron correlation may have a higher energy than an SCF calculation (which of course contains no correlation but utilised a more extensive basis set). This *will not usually mean* that the SCF calculation will yield a better potential curve or excitation energy. This is because small expansion errors often cancel as a function of internuclear distance or electronic configuration while correlation energy differences in general do not.

2.2.1 Self-consistent field molecular-orbital method

In this method, which continues to be the most widely applied, the wave function is expanded as a product of one-electron spin orbitals Φ:

$$\psi = A\prod_i \Phi_i$$

where A is an antisymmetrising operator required for the identical electrons. For molecules the spatial part ϕ_i of the spin orbital Φ_i is expanded in terms of some known set of functions x_p thus

$$\phi_i = \sum_p c_{ip} x_p$$

and the best (SCF) orbitals for the basis expansion x_p, is obtained by varying the c_{ip} values to reach a minimum energy for the system. Evidence continues to accumulate as more and more molecular Hartree–Fock calculations are performed that[7,8]:

(a) Near-Hartree–Fock wave functions (within 0.01 a.u.) can readily be obtained by starting from relatively small optimised basis sets for the separated atoms and augmenting with one or two carefully chosen basis functions of higher atomic symmetry for each molecular symmetry type. The atomic basis sets need only be of nominal accuracy, i.e. two Slater basis functions for each s orbital and two or three basis functions for each p and d orbital, provided that the exponents are carefully optimised. Gaussian basis sets which yield similar accuracy for the atoms have also been shown to be equally effective[74]. In order to approach true Hartree–Fock accuracy (expansion errors of less than 1 in 10^7 of the total energy), one must either augment with several basis functions in each molecular symmetry, and again vary the exponents of *all* basis functions until a minimum value of the energy is obtained or use a very large saturation set. The process of re-optimisation can use up many tens of hours of machine time, even for small first-row molecules.

(b) Correlation energy is, in general, not constant as a function of internuclear distance. The improper dissociation forced by the MO picture seriously affects the Hartree–Fock solution even at R_e. Thus the accuracy attainable in the Hartree–Fock values is in general poor for D_e, fair for ω_e; however R_e is fortuitously good. There are notable exceptions. (See (d) below.)

(c) One-electron properties come out fairly well when the wave function is close to the Hartree–Fock limit. However, minimal basis set SCF one-electron properties are quite unreliable.

(d) Hartree–Fock calculations give potential curves of useful accuracy for diatomic rare-gas systems[18], highly ionic[19] diatomic molecules, hydrides[20], and molecules or molecule-ions consisting of a closed-shell atom or ion interacting with a half-open shell one (He_2^+, Fe_2^-, Ne_2^+, LiHe, NaNe, etc.)[21]. In fact quantitative potential curves are obtained for three classes of systems. (i) closed shell–closed shell interactions, (ii) half-open–closed shell interactions, (iii) open shell–open shell interactions of the highest multiplicity.

This point will be elaborated upon further in the section dealing with potential curves and dissociation energies.

(e) Although *in general* in error by several electron volts even excitation energies can be quantitatively computed in the Hartree–Fock model if the system is appropriate, for example for a closed-shell atom, ion, or proton interaction with various orientations of another atom[22].

(f) Ionisation potentials and electron affinities are in general in error by several electron volts, that is, by the correlation error.

(g) By *systematic semi-empirical* correction the correlation error in dissociation energies, excitation energies, ionisation potentials and electron affinities can be estimated within a few tenths of an electron volt. (This usually requires a *series* of computations on related molecules with similar electronic structure, one member of which is well documented experimentally[23].)

(h) Properly shaped *full* and *continuous* potential curves require a model explicit including electron correlation. (Unless the system is a member of class (d) above.)

Open-shell SCF theory has received particular attention. Several methods for eliminating the off-diagonal Lagrangian multipliers in open-shell SCF equations, one by basis set transformation, are given[24]. In related work some aspects of orbital invariance[25] for the single-eigenvalue solution of the open-shell problem is discussed and comparisons are made between various open-shell SCF techniques[26].

In the area of Hartree–Fock perturbation theory a time-dependent form[27] has been applied to polarisabilities and a comparison made with the Karplus–Kolker method[28]. Upper bounds are discussed for second-order Hartree–Fock energies.

Other contributions to the Hartree–Fock method include work on the symmetry properties of spin orbitals[29], null spaces in the Roothaan procedure[30], a new extrapolation method[31], and a quadratically convergent technique[31], generalisation of Brillouin's theorem[93], and a discussion of saddle point characteristics[94].

Several simplified or approximate SCF schemes[32] have been proposed and the $X\alpha$ method which utilises an approximation to the exchange contribution has been compared with the rigorous SCF scheme[33, 95].

Typical recent Hartree–Fock calculations which illustrate the state of the art are those on KrF and KrF^+ [34], NaHe and LiHe [74], the triplet sigma state of NaLi[35] and several states of CaO[36]. Such calculations involving fairly large atoms are now routine. The potential curves calculated for KrF, NaHe and LiHe can be expected to be quite realistic since they fall in class (b) of the Hartree–Fock model, for, since no new electron-pairs are formed, and the system dissociates properly the correlation energy can be expected to be relatively constant as a function of internuclear distance. These calculations predicted a repulsive potential curve indicating the KrF would not exist in the gas phase. In the work on NaHe and LiHe, potential curves were computed for the $^2\Sigma$ ground state and the first $^2\Sigma$ and $^2\Pi$ excited states. Due to the fact that these are appropriate Hartree–Fock systems, the potential curves and their spacings are quite accurate, probably to better than 0.1 eV.

As is found for KrF^+, the Hartree–Fock model for CaO does not dissociate properly and the four states computed[36] for this molecule could not be expected to lie in the correct order or have the correct spacing due to correlation energy differences between them. Thus these calculations were incapable of predicting the ground state with confidence, although recent experiments indicate that the $^2\Pi$ state, predicted to be the ground state by the Hartree–Fock calculation, is in fact the true ground state. This again points up the fact that the ability of the Hartree–Fock model to predict in general quantitative energy differences is not high.

The well-known deficiencies of the Hartree–Fock model may thus be associated with (i) the increase in correlation energy caused by forming new electron pairs and bringing more electrons closer together and (ii) the inability of the MO model to describe the dissociation of the molecule into its asymptotic fragments.

A great qualitative improvement over the Hartree–Fock model is obtained by merely correcting defect (ii) by choosing a wave function form which can continuously describe the atoms as they form the molecule. All forms of configuration interaction (CI) discussed in Section 2.2.3 are capable of achieving this. However, they have not been applied in general with this philosophy in mind.

Once the asymptotic difficulties are remedied additional configurations must be added to achieve additional extra molecular correlation arising from sources mentioned in (i).

Sections 2.2.2–2.2.6 represent various methods for removing the correlation error in binding energies, potential curves, excitation and ionisation energies and electron affinities. It should be kept in mind that if only part of the correlation energy is evaluated for a molecular system the error in energy differences will only be eliminated if the neglected correlation energy is constant as a function of internuclear distance, in the case of potential curves, or the same for different electronic configurations, in the case of ionisation and excitation energies and electron affinities.

A growing understanding of how to neglect those invariant parts of the correlation energy has characterised recent advances and a large part of this understanding has come from the independent electron-pair approach in which the correlation energy is evaluated as the sum of that arising from the correlation between the $N(N-1)/2$ electron pairs in the N-electron system. This concept and its validity has guided the choice of configurations, the neglect of certain excitations, and has greatly simplified the calculation of correlation energies.

2.2.2 Electron correlation and the independent electron-pair approach (IEPA)

The IEPA has been reviewed recently in a number of articles[9, 11]. Many of the calculations utilising the methods discussed in Sections 2.2.3. and 2.2.4 have used the separability of electron-pair correlation to greatly simplify the computational procedure. Gradually an increased knowledge of the correlation effects important in describing molecular formation, excitation and ionisation is being built up by analysing large-scale computations in terms of individual pair energies. In particular, inner-shell correlation energies have been shown to be relatively constant during molecular formation and changes in *intershell* correlation are clearly very important in accurately describing chemical bonding. In addition it appears that there exist correlation effects which are *uniquely molecular* in nature and that such effects are the most important ones to include as a *first step* in attacking the *molecular* correlation problem[58].

In its most simplified form the IEPA assumes that the total correlation energy in an N-electron system may be expressed as the simple sum of the correlation energy of each of the $N(N-1)/2$ electron pairs evaluated in separate calculations involving a single pair. This idea has been widely applied to atoms[10] utilising various technologies and it appears that correlation energies within 10% of the total are obtained and energy differences are evaluated to between 1 and 10% precision. The application of similar ideas to diatomic molecules is now yielding similar precision. Two good examples of the application of the independent electron-pair model to diatomic molecules are afforded by calculations on BH [37] and CO [38].

The BH calculation utilised the method of direct determination of approximate natural orbitals to evaluate the pair energies and investigated the intra- and inter-pair correlation energies at several internuclear distances. The major results and conclusions of this work were that both the interpair and intrapair correlation energy involving the K shell were independent of internuclear distance and that the interpair correlation energy varies much more strongly with internuclear distance than the intrapair energies. A total energy of 25.262 55 Hartrees was obtained by summing the SCF energy and the pair energies. This total energy can be compared with the experimental energy of 25.290 Hartrees. The calculated internuclear distance was 2.316 a.u. and the calculated force constant was 3.190 (mdyn $Å^{-1}$) as compared with experimental values of 2.336 a.u. and 3.028 (mdyn $Å^{-1}$).

In the pair-energy study[38] of CO the pair energies were defined as the energy improvements obtained from CI calculation in which single and double excitations from one pair of SCF space functions are made. These independent pair energies added up to 93% of the correlation energy of -0.525 a.u. while the full variational CI calculations involving all pairs simultaneously gave 70% of the correlation energy. No potential curve was computed in this study. Similar results were obtained in a more comprehensive study [39] of MgH in which the sum of the independent pair energies yielded 67% of the correlation energy while the full variational calculation involving all configurations simultaneously yielded 57%.

2.2.3 Configuration-interaction methods without orbital optimisation

One of the most widely applied and successful methods for including electron correlation has been the configuration-interaction technique. In the configuration–interaction method the wave function is expressed as,

$$\psi(1,2,...N) = \sum_K A_K \Phi_K(1,2,...N)$$

that is, the wave function is expanded in terms of a series of configurations which differ from each other in the assignment of the N electrons among some number $M > N$ of spin orbitals. This scheme has been applied to a variety of diatomic systems and a number of different choices have been made for the M spin orbitals making up the configurations. These choices have been atomic orbitals (giving rise to a valence bond wave function), occupied and virtual molecular orbitals obtained from Hartree–Fock calculations, or some arbitrary choice of orbitals permitted by the basis set.

A major problem which continues to challenge configuration-interaction calculations is that if one does not evaluate *all* the correlation energy one must devise a computational scheme which evaluates it *to the same extent* in both the molecule and atoms (if potential curves are desired), in different molecular states (if term values are desired), or in the ion and parent system (if electron affinities or ionisation potentials are desired). Thus an unthinking brute force CI attack on the correlation problem for molecules is seldom successful unless it goes all the way, and the most successful evaluation of energy properties for *large* diatomics has occurred when an effort has been made to separate the molecular correlation problem from the atomic one[54, 55, 58] If one chooses the CI approach one is also faced with the problem of the slow convergence of the multiconfiguration expansion of a many-electron wave function. At this point one must make a choice between trying to cope with very large numbers of configurations or improving the convergence. Although the former choice is now feasible with large computers (and one can always extract simple quantities from a complicated wave function and leave everything else out of sight in the machine), it is possible that one would be more likely to be able to comprehend the steps of the calculation and find unexpected relations and regularities by working directly with the most convergent expansion obtainable. If one constructs all configurations from a common set of orthonormal orbitals, then this reasoning leads immediately to variational equations for a finite set of orbitals which must be solved simultaneously with the secular equation for the configuration expansion coefficients. Such methods which determine optimal orbitals will be discussed in Section 2.2.4.

Recent formal and computational contributions to the CI method have included numerical integration directly over complete orbitals rather than over basis functions[40], perturbation theory for CI calculations[41], calculation of matrix elements for valence-bond wave functions[42], partitioning techniques using the Padé approximation[43], natural orbital interactions[44], symmetry behaviour of natural orbitals[45], and relation of cluster expansion to correlated wave functions[46].

Some typical recent calculations involving the straight-forward CI approach are valence-bond calculations[47] on F_2 and Cl_2 utilising accurate Hartree–Fock atomic orbitals as the valence-bond functions. This study, which involved some 300 configurations, yields only half the binding energy and emphasised the need to provide for considerable atomic distortion during bond formation and for a type of truly molecular correlation not achievable by utilising only ground-state atomic orbitals.

Another typical valence-bond study[48] was carried out on eight low-lying states of BH utilising Gaussian lobe functions of 'double zeta' quality. In this study the 1s shell of boron was kept doubly occupied and all configuration away from permutations between the 2s and 2p atomic orbitals on boron and the 1s orbital on hydrogen considered. This study yielded potential curves for the eight states in the correct order. However predicted equilibrium distances were all too large and a stable $^1\Pi$ state was not predicted by the calculation (this was improved by varying the hydrogen exponent). For the $^1\Sigma$ ground state the binding energy predicted by the calculation was 2.45 eV which comprises c. 70% of the experimental value of 3.57 eV. The calculated vertical transition energies agreed within about 10% of the experimental values. There have also been a number of interesting CI studies on CO[49],

N_2 [50] and O_2 [51] utilising all configurations possible from a minimal basis set of STO's. These studies yield a very large number of states (c. 100) and, in general, qualitative features are correct. Thus the states are properly ordered and if theoretically bound do in fact exist. Also the general shape, i.e. equilibrium distances and maxima, are qualitatively reflected. However, as one would expect, if states are very close together in energy at significantly different internuclear distances their order can be reversed.

Very high precision CI calculations have been performed on excited states of He_2 [52], BeO [53] and O_2 [54] in which a large basis set and a large number of configurations were included. These studies are representative of the true power of the CI method.

2.2.4 Configuration-interaction methods with orbital optimisation

It was well known by specialists and by now has become generally obvious that the reason for the slow convergence of CI calculations utilising arbitrary orbitals is that the 'excited' orbitals were not in the correct physical space. In particular they should occupy the *same physical space* as the orbital for which they are intended to provide correlation.

The improvement of the convergence of the CI expansion can and has been achieved in a number of ways discussed in the following sub-section. All these methods provide a means of placing the correlating orbital 'on top of' the orbital being correlated. Thus, although the excited orbital may be *nodally orthogonal* to the ground-state orbital it does in fact occupy the same physical space so that the two orbitals overlap each other. This tends to maximise the exchange integral between the orbitals being correlated and the correlation orbitals, which in turn leads to a strong coupling between the dominant and excited configurations, and a significant lowering of energy. All the variational methods for forming effective correlation orbitals result in orbitals similar to the natural orbitals of the system. In several methods the actual numerical techniques are also similar. For example, the scheme for the direct calculation of natural orbitals[14] leads to pseudo-eigenvalue equations very similar to those resulting in the MCSCF process[55]. In the following section we will discuss various methods for obtaining and using optimal orbitals in CI calculations.

2.2.4.1 Multiconfiguration self-consistent field method

(a) *Form of wave function*—The functional form of the wave function is the same as in the straight CI approach, that is

$$\Psi = A_0\Phi_0 + \sum_k A_k\Phi_k$$

where the A values are mixing coefficients, the Φ values are Slater determinants formed from a common set of orbitals $\phi_{\mu l}$, and Φ_0 is the dominant configuration.

(b) *Optimisation of the wave function*—The basic variational conditions[55] for an optimum wave function are:

(i) $\delta E_{A_k} = 0$ for a given set of orbitals. This condition yields secular equations for best-mixing coefficients as in the CI approach.

(ii) $\delta E_{\phi_{\mu l}} = 0$ for a given set of expansion coefficients.

This condition yields a set of Fock-type equations for best orbitals; in general, one equation for *each* orbital of the wave function. This pseudo-eigenvalue problem is solved iteratively as in the conventional SCF procedure. An alternative procedure utilising a generalisation of Brillouin's theorem has recently received attention[98].

When this doubly iterative process converges we have the optimum orbitals *and* mixing coefficients for a given basis set. Recent theoretical and computational[55] contributions to this method have been an improved iterative scheme leading to faster convergence, a means of selecting initial starting orbitals by maximising their exchange integral with the orbital or orbitals being correlated and a means of choosing the important configurations for describing the extra molecular correlation. These are those that are required for the proper dissociation of the molecule into two Hartree–Fock atoms and another class which contributes to the molecule but *not* to the atoms.

The most recent work[58] on F_2 provides a good example of the state and precision of multiconfiguration self-consistent field wave functions for diatomic molecules. The major difficulties previously quoted in earlier applications of the MCSCF method, namely slow convergence of the orbitals and the arbitrary choice of starting orbitals, have been solved and a highly accurate potential curve evaluated for F_2. The spectroscopic constants obtained from the curve compare quantitatively with experiment and are given in Table 2.5.

This result was achieved by evaluating the major part of the correlation energy *changing* with internuclear distance through a six-configuration MCSCF base-function which included those configurations which contribute to the molecule but formally vanish in the atom, and the one additional configuration required to formally dissociate the molecule to two Hartree–Fock atoms. This simple six-configuration wave function yielded the results given in Table 2.6. When additional non-molecular correlation effects were included, which were primarily atomic in nature, the values improved.

The same MCSCF scheme has recently been applied to a number of other molecular systems with similar success. These include NaLi[59], LiH[60] and O_2^- [61]. The advantages of this scheme are the relative simplicity of the resulting wave function and the insurance that each configuration is the most effective of its kind possible within the framework of the basis set used.

The GI method[89a], which is a restricted form of a general MCSCF scheme, achieves part of the molecular correlation by projecting from the unrestricted Hartree–Fock determinant and then varying the resulting wave function. This method, which does correct the asymptotic difficulties of the Hartree–Fock model, has been recently applied to diatomic molecules (Table 2.1).

2.2.4.2 Approximate natural orbital method

The natural orbital concept has had tremendous impact on the calculation of configuration-interaction wave functions. The basic feature of natural orbitals is that they are the set of orbitals which lead to the most rapid convergence of the first-order density matrix (though this does not necessarily lead to most rapid convergence of a CI expansion). The main problem with the evaluation of exact natural orbitals is that to compute them by diagonalization of the first-order density matrix one requires an exact wave function from which to construct the first-order density matrix.

A variety of simplifying schemes for evaluating approximate natural orbitals has thus been developed. In the *iterative* natural orbital method one performs a large CI calculation, obtains natural orbitals and retains only those having large coefficients, redetermines the natural orbitals from a new CI, and so on. Although one arrives at a small set of effective natural orbitals the process still seems to be quite costly. A typical recent calculation employing this technique was performed[38] on CO in which 70% of the correlation energy was obtained. The same technique has been widely applied to hydrides[62] with similar success in the total energy and has been incorporated within the computational framework of many configuration-interaction calculations[54].

2.2.4.3 Pseudo-natural orbital method

This computational scheme goes one step further in simplifying the calculation of natural orbitals. In this technique the natural orbitals are obtained not from a complete CI but from a CI calculation involving only selected pairs of electrons. The resulting pseudo-natural orbitals are then used in a larger CI calculation involving excitations from several pairs simultaneously.

This scheme, which may be viewed as the computational implementation of the independent electron-pair model, has been incorporated to some extent within most of the computational schemes discussed in Section 2.2.4. A good example of the use of the technique is provided by the recent work on KrF^+ [34]. In essence both the MCSCF scheme[58] and the direct calculation of approximate natural orbitals[14] are evaluating pseudo-natural orbitals since both schemes as applied to date have concentrated on correlation between selected pairs of electrons.

2.2.4.4 Direct calculation of natural orbitals

In many ways this technique is quite similar in its application and techniques to the multiconfigurational self-consistent field method, in that the natural orbitals are determined by the solution of a secular equation and a pseudo-eigenvalue problem by a double iterative procedure. In the main applications of this method further simplications have been made, namely, the introduction of the concepts of the independent electron-pair model so that approximate natural orbitals and correlation energies are determined for a given pair of electrons one at a time. The individual pair correlation energies are then summed to yield the total correlation energy. This scheme of course contains the already-mentioned shortcomings of the IEPA together with its significant simplifications. An example of the application[37] of this technique to BH was given in Section 2.2.2 within the IEPA approach.

2.2.5 Geminals

In the geminal approach the wave function is expressed as

$$\Psi(1,2,\ldots 2N) = A \prod_{k=1}^{N} \Psi(2K-1,2K)$$

where the $\Psi(i,j)$ is a spin geminal for the electron pair i,j which is expressed as the product of a spatial $\Lambda(i,j)$ and a spin $\theta(i,j)$ part. The simplest spin geminal is a doubly occupied orbital, which would yield the familiar Hartree–Fock orbitals; however, the point of the method is for the geminal $\psi(i,j)$ to include electron correlation between the electron pairs. In most geminal calculations the spatial part of $\Psi(1,2)$, the spin geminal, is expressed as

$$\Lambda(1,2) = \sum_{i} d_i X_i(1) X_i(2)$$

where the X_i values are the natural (the most rapidly convergent) orbitals of the geminal pair, and the d_i values the occupation coefficients. This method has been applied to a number of first-row diatomic hydrides and reasonably accurate results have been obtained. However, the difficulty of including intershell correlation with a single product of separated geminals appears to be a serious shortcoming of the method. The largest diatomic system to which the method has been applied is NH [64].

The current status of geminal calculation is well illustrated by extensive work on BH [63]. In this calculation an antisymmetrised product of separated geminals was used. The major conclusion and results of this work were that the 1s shell geminal in B was very similar to the innermost geminal in BH, leading to the transferability of this geminal from the atom into the molecule. The wave function yielded 25% of the correlation energy in B and 46.7% of the correlation energy in BH, which thus resulted in a binding energy of 3.858 eV as compared with an experimental binding energy of 3.592 eV. The fact that only half the correlation energy was recovered was attributed to the inability of a single product of separated geminals to handle intershell correlation effects as well as the need for using the *same* orbitals to describe different intrashell correlation effects. A potential curve which dissociates properly to B and H using the geminal wave function was constructed as a function of internuclear distance; this process of course requires the same degree of accuracy in the wave function at each value of the internuclear distance. The potential curve thus obtained was a good *qualitative* accuracy as evidenced by the spectroscopic constants when compared with experiment namely $B_e = 12.085$ (12.016), $\omega_e = 2928$ (2367), $\omega_e \chi_e = 45.4$ (49), $\alpha_e = 0.4887$ (0.408) (in cm^{-1}), $R_e = 1.23$ a.u. (1.236 a.u.). These results are in general representative of similar results obtained on LiH [63] and NH [64] by the same authors although the LiH results are somewhat better due to the unimportance of intershell effects.

2.2.6 Direct correlation, method of moments and transcorrelated pairs

In the direct correlation method the interparticle coordinates are included explicitly in the wave function. Some new applications to more than two electrons have been accomplished on atoms by a combination of a few interelectronic distances with conventional wave functions[68]. However it still appears that, although effective, direct correlation is computationally prohibitive for polyelectronic molecular systems.

One of the newer methods, that of transcorrelated pairs[69], bypasses such computational difficulties by utilising the method of moments[70] to greatly simplify the integrals arising from the use of direct correlation functions. This method has been applied to the neon atom, the LiH molecule at its equilibrium distance, and to the beryllium atom, yielding promising results[71]. The energy results for LiH were good, namely, within 0.007 a.u. of the experimental value, and this method will probably become an alternative to conventional variational procedures.

2.2.7 Perturbation theory

Application of the many-body perturbation theory approach to atoms have been numerous and encouraging. However only a very few diatomic molecules have been attempted, all using single-centre wave functions, making it difficult to judge the meaning of the results or the effectiveness of the method.

These calculations, on H_2 [65] and HF [66], both used single-centre wave functions and the results were good but not impressive in comparison with traditional variational calculations. The use of single-centre wave functions also makes them difficult to compare with other methods whose techniques are better implemented. A special perturbation theory of molecular energies at short range has been developed[67].

2.2.8 Equations of motion method

Recently the equation of motion method[96] has been applied to a number of diatomic molecules[97]. This scheme is developed in terms of many-body theoretical formalism or excitation language and attempts to evaluate excitation energies directly without solving the Schrodinger equation for each state by taking advantage of common correlation terms between various states. Excitation energies obtained to date are within 10% of experimental values and calculated oscillator strengths show fair agreement.

2.3 APPLICATIONS

In a survey of the evaluation of the techniques discussed in the preceeding section several points should be kept in mind.

All the methods demonstrated the capability of achieving quantitative accuracy when well implemented and thoroughly applied.

The efficiency and effectiveness of a method when applied is a synthesis of theory, numerical analysis, computer programming and available computer time.

Four distinct models and their associated levels of accuracy can be identified and the applications will usually be discussed in terms of these four levels; minimal basis set SCF, near Hartree–Fock SCF, minimal basis set correlated wave functions, extended basis set correlated wave functions.

As stressed earlier we have learned a great deal about when it is appropriate to use each model and what quality of result to expect from it. In the following section the application of various techniques to the study of energetic quantities of chemical interest will be illustrated by means of specific diatomic systems.

2.3.1 Ground-state energies

It is of course aesthetically satisfying to obtain the lowest possible energy, and therefore approach as closely as possible the true molecular energy. This virtuous quality of low total energies however must be balanced by the fact that attainment of the exact wave function is not yet practical and all energetic quantities of chemical interest depend on *small* ($<1\%$ of the total energy) *energy differences* and not on absolute energy. Thus more important than seeking the lowest possible total energy, is the calculation of the energy within a comprehensive model which properly reflects the *changes* taking place in the energy with the molecular geometry, the electronic state or the number of electrons. Thus it is usually not the calculations which yield the lowest total molecular energy at one internuclear distance which have provided the most useful evaluation of the energy differences involved in chemistry. In fact it has been the computational schemes which identified the 'structure dependent' part of the correlation energy and do not put a great effort into correlating the invariant parts, which have proved most successful in yielding accurate energy differences on 'interesting' diatomic systems.

It is worth while to take stock of the progress made in obtaining total

Table 2.1   Comparison of energetic quantities for BH from various types of calculation

Method	Total energy/a.u.	R_e/Bohr	ω_e/cm^{-1}	D_e/eV
Min. SCF*	25.0570			
Min. CI†	25.0903			
SCF (near Hartree–Fock)[20]	25.1314	2.305		2.78
Geminals[63]	25.2053	2.329	2928	
VB [48]	25.1456	2.536		2.45
IEPA (with natural orbitals) [37]	25.1262	2.316		
CI (extended CI) [62]	25.2621	2.336		c. 3.33
SOGI‡	25.1801	2.336		3.23
Experimental	25.290	2.336	2367	c. 3.57

*Sahni, R. C. (1956). *J. Chem. Phys.*, **25**, 332
†Fraga, S. and Ransil, B. J. (1962). *J. Chem. Phys.*, **36**, 1127
‡Blent, R. J., Goddard, W. A. and Ladner, R. C. (1970). *Chem. Phys. Lett.*, **5**, 302, and preprint by Blint and Goddard (1972).

ground-state energies by looking at three different molecules of varying complexity namely BH, F_2 and O_2. By inspecting Table 2.1 for BH, we see that the minimal basis set SCF solution yields 99% of the total energy, the Hartree–Fock solution 99.4% of the total energy, the minimal basis set CI result yields 25%, and the extended basis set CI yields 81% of the correlation energy. Note that the error in the minimal basis set SCF result

Table 2.2 Summary of calculations on F_2 including electron correlation

Form of wave function	Total energy/a.u.	Binding energy obtained/eV
Fraga and Ransil: 14 configurations obtained by promoting all possible electron pairs plus the single excitations from the $1\sigma_u$ and $2\sigma_u$ into the $3\sigma_u$ orbital (minimal basis set)*	− 197.9558	2.05
Harris and Michels: Same 14 configurations as above using *simple symmetry* MOs constructed from the basis set (minimal basis set)†	− 197.9560	2.04
Harris and Michels: 25 *valence bond* configurations constructed by all possible occupations of a $1s_a$, $1s_b$, $2s_a$, $2s_b$, $2p_{\pi a}$, $2p_{\pi b}$, $2p'_{\pi a}$, $2p'_{\pi b}$, $2p_{\sigma a}$ and $2p_{\sigma b}$ atomic orbital (minimal basis set)†	− 197.9560	2.04
Harris and Michels: Same 25 configuration as above in which 'double zeta' Hartree–Fock atomic orbitals were used‡	− 198.8179	0.41
Allen: Valence bond calculation utilising Gaussian lobe basis set§	− 198.7780	0.35
Schaefer: 25 configuration *valence-bond* calculation using *accurate* Hartree–Fock atomic orbitals¶	− 198.8303	0.69
318 configurations constructed from the $1\sigma_g$, $2\sigma_g$, $3\sigma_g$, $1\sigma_u$, $2\sigma_u$, $3\sigma_u$, $1\pi_u$, $1\pi_g$, $4\sigma_g$, $4\sigma_u$, $2\pi_g$, $2\pi_u$, $1\delta_g$, $1\delta_u$ molecular orbitals. The $4\sigma_g$, $4\sigma_u$, $2\pi_g$, $2\pi_u$, $1\delta_g$ and $1\delta_u$ were constructed from a single d Slater-type orbital of each fluorine atom and the other orbitals from Hartree–Fock AOs¶	− 198.9619	0.69
198 configurations constructed from the Hartree–Fock AO molecular orbitals plus a $4\sigma_g$, $4\sigma_u$, $2\pi_g$, $2\pi_u$, orbital constructed from an additional p orbital on each fluorine atom¶	− 198.9467	0.80
Das and Wahl: ODC two partially optimised configurations $1\sigma_g^2$, $1\sigma_u^2$, $2\sigma_g^2$, $2\sigma_u^2$, $1\pi_u^2$, $1\pi_g^2$, $3\sigma_g^2$, $1\sigma_g^2$, $1\sigma_u^2$, $2\sigma_g^2$, $2\sigma_u^2$, $1\pi_u^2$, $1\pi_g^2$, $3\sigma_u^2$ in which a partial simultaneous optimisation of both configurations was achieved via the MCSCF technique‖	− 198.8377	0.54
ODC: The same two configurations used above were fully optimised via the MCSCF technique[8]	− 198.8485	0.81
OVC: MCSCF wave function which included those six configurations obtained by doubly exciting the $2\sigma_g$, $2\sigma_u$, $3\sigma_g$, $1\pi_u$, and $1\pi_g$ into the $3\sigma_u$ orbital[9]	− 198.8546	0.95
OVC: A six-configuration MCSCF wave function obtained by promoting the $3\sigma_g$, $1\pi_u$, $1\pi_g$, electron pairs into the $3\sigma_u$ orbital and including the two split shell excitations $3\sigma_g$, $1\pi_u \rightarrow 3\sigma_u$, $2\pi_g$; $3\sigma_g$, $1\pi_g \rightarrow 3\sigma_u$, $2\pi_u$ then atomic correlation was added[58]	− 198.9809	1.67 ± 0.06

*Fraga, S. and Ransil, B. J. (1962). *J. Chem. Phys.*, **36**, 1127
†Harris, F. E. and Michels, H. H. (1967). *Int. J. Quantum Chem.*, **1S**, 329
‡Harris, F. E. and Michels, H. H. (1969). *Int. J. Quantum Chem.*, **3S**, 329
§Allen, L. C. (1966). *Quantum Theory of Atoms, Molecules and the Solid State* (London: Academic Press)
¶Schaefer, H. F. (1970). *J. Chem. Phys.*, **52**, 6241
‖Das, G. and Wahl, A. C. (1966). *J. Chem. Phys.*, **44**, 87

is not recovered in the CI calculation. This is an important point and indicates that basis set inadequacies at the SCF level persist after introducing correlation. For the case of F_2 (Table 2.2) a similar pattern emerges, but in this case the correlation energy changes taking place during molecular formation completely obscure the chemical binding. This will be discussed at greater length in the section on dissociation energies. It is also important to note that as the basis set was improved the value of the binding energy decreased in both the SCF and VB model, pointing to the importance of extended basis sets for obtaining *quantitative* results connected with energy differences.

As noted previously very low total energies have been obtained for CO [38] and MgH [39] (70% and 60% of the correlation energy respectively). However, for O_2 and F_2 an effort was made to identify and include the 'structurally dependent' part of the correlation, and although only 23% of the correlation energy was achieved, the calculated interaction potentials are among the most accurate so far obtained (Table 2.6)[54, 58].

2.3.2 Excited-state energies

One of the encouraging new developments in quantum-mechanical calculations on diatomic molecules and their ions has been the increased emphasis on excited states. Excitation energies are calculated as the difference between the calculated total energies for two states, $\Delta E = E(AB^*) - E(AB)$. The calculated energies, $E(AB^*)$ and $E(AB)$, for the two states may be derived from two *different variational* wave functions or from a frozen-orbital approximation in which electrons are redistributed among the fixed orbitals.

The calculation of excited-state energies $E(AB^*)$ which are the lowest of their symmetry species may be analysed in the same terms as ground-state energies. However since the transition energy between two states is the difference between the two absolute energies we are faced with the problem of making the errors in the calculated energy the same for the two states involved.

It is instructive to analyse the excitation energy in terms of the Hartree–Fock contribution and the correlation contribution. Thus we have

$$\Delta E = E_1^{HF} - E_2^{HF} + E_1^{corr}(R) - E_2^{corr}(R')$$

and we can express the excitation energy as the sum of the Hartree–Fock contribution and the correlation energy contribution

$$\Delta E = \Delta E_{12}^{HF} - \Delta E_{12}^{corr}$$

between states 1 and 2.

Hartree–Fock excitation energies are thus in error by the difference between the correlation energy for the states involved. This quantity is typically of the order of several electron volts, and is often sufficient to reverse the order of excited states. There are notable cases however where the correlation error in the HF calculation can be expected to be similar for the two states; for example a proton interacting with an atom such as HF^+

or HCl^+ [22], or a closed-shell system interacting with an atom such as NaHe, LiHe [74] and NaO [75] (essentially Na^+O^-). All such systems undergo only the small correlation energy changes (c. 0.1 eV) associated with the reorientation of one atom in relation to the other. Thus they do not differ in the *number* or *nature* of electron pairs. (Figure 2.1.)

The extended basis-set studies on HF^+ [22] and NaO [75] are illustrative of such cases in which the Hartree–Fock model yields such quantitative results. For HF^+ the $^2\Pi \rightarrow {}^2\Sigma$ separation obtained from calculation is 0.143 eV while the experimental value is 0.139 eV. The results for NaO indicate that the $^2\Pi \rightarrow {}^2\Sigma$ excitation energy is 0.2 eV and recent experiments also suggest a very small value.

Conversely, extended basis-set HF calculations on CaO^{36} provide a good example of the type of precision obtained in term values, for a system in which there are large correlation-energy changes. In this work the calculated order of the lowest-lying states was correct in the HF model but was placed in doubt by an estimated correlation difference of several electron volts.

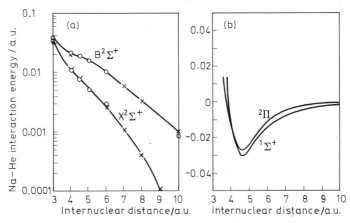

Figure 2.1 Typical potential energy curves obtained from extended basis sets (both STO and GTO) SCF calculations on a system appropriate for the Hartree–Fock model (see Section 2.2.4.1): (a) Interaction energies of the ground and excited states of NaHe, × STF, o GTF; (b) Interaction energies of the ground ion $^1\Sigma^+$ and the neutral $^2\Pi$ states of NaHe
(From Krauss, Maldonado and Wahl[74], by courtesy of the American Institute of Physics)

In general, SCF calculations point to the need for including electron correlation in the calculation of accurate excitation energies.

Going beyond the Hartree–Fock model are typical minimal basis-set CI calculations for 72 excited states of CO [49]. The major and encouraging result of this study was that, except for two very close states with a small internuclear distance, the predicted order agreed with experiment and excitation energies agreed semiquantitatively (c. 20%) with existing experimental data, although binding energies showed only qualitative agreement; namely, often only 50% of the experimental value. These results did show the minimal basis-set full-CI calculation to provide a powerful tool for sorting out the

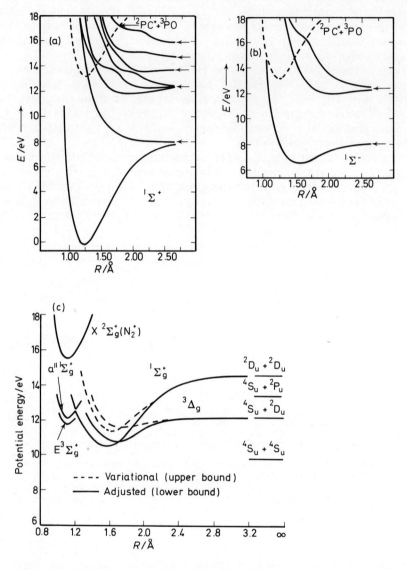

Figure 2.2 Potential curves obtained from a full CI calculation employing a *minimal* basis set. Such economical calculations yield a wealth of correct qualitative information about the order of states, the number of bound states and the existence of maxima. See Sections 2.2.3, 2.3.2 and 2.2.3. (a) Calculated potential energy curves for the $^1\Sigma^+$ states of CO. The dotted line indicates the approximate potential curve for the ground state CO^+ and the arrows near the margin indicate dissociation limits.
(From O'Neil and Schaefer[49], by courtesy of the American Institute of Physics)
(b) $^1\Sigma^-$ states for CO (see (a))
(c) Calculated potential energy curves for N_2O
(From Michels[50], by courtesy of the American Institute of Physics)

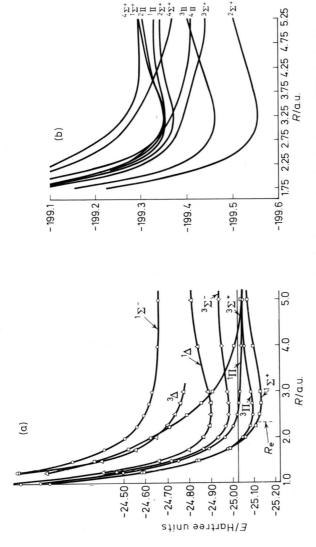

Figure 2.3 (a) Potential energy v. internuclear separation for the ground and excited states of **BH** obtained from a double-zeta basis set valence-bond calculations. The semi-quantitive results give vertical transition energies as shown in Table 2.3. Some features of the curves were basis-set dependent.
(From Harris and Allen[48], by courtesy of Academic Press)
(b) Potential energy curves for the MgH molecule obtained from NO calculations. The spacings agree to within 10% of experiment
(From Chan and Davidson[39], by courtesy of the American Institute of Physics)

qualitative complexity of the many states arising from the interaction of a few states of the constituent atoms. Similar calculations have been performed on O_2 [51], CN [76], and N_2 [50] (in which two new low-lying states were predicted for N_2) (Table 2.3 and Figure 2.2).

Table 2.3   Excitation energies (in eV)

State	BH [48] VB–CI	MOSCF	Expt.
$^1\Delta$	6.01	5.33	5.72
$^3\Sigma^-$	3.30 + 0.77	3.14 + 0.58	3.36 + x
$^1\Pi$	2.90	2.90	2.86
$^3\Pi$	0.77	0.58	x
$^1\Sigma^+$	0.00	0.00	0.00
	CO [49] minimal CI		
$^3\Sigma^+$	5.06 (5.72)*		6.92
$^3\Pi$	5.22 (5.60)*		6.04
$^1\Sigma^+$	0		

*Exponents optimised

Quantitive results (± 0.3 eV) were obtained for extended basis-set CI calculations (Table 2.3) as illustrated by the recent work on BH [48], and extended CI calculations on MgH [39] which yielded the $^2\Pi \rightarrow {}^2\Sigma$ transition energy within 5% of experiment (Figure 2.3). Other similar high-precision excited-state calculations were performed for Be_2 [77], B_2 [78] and BeO [53], namely calculations approaching better than 10% accuracy in term values; for SiH [79], which used separately obtained SCF orbitals for each contributing parent state, 3% precision in term values was claimed.

Representative excited state calculations of very high precision are those on the $^3\Pi_u$ state of He_2 in which CI calculations on the potential curve for the excited state agree with experiment to *c.* 0.02 eV over all values of the internuclear distance [52].

2.3.3   Interaction potential curves and dissociation energies

We again stress that the main problem which must be faced in applying the *ab initio* techniques discussed in Section 2.2 to the evaluation of interaction potentials is that they must be applied in such a way as to make the error in the total energy constant or nearly constant as a function of internuclear distance.

Recent progress towards achieving this has come from two directions:

(i) An increased understanding of the circumstances under which SCF interaction potentials are quantitative or semiquantitative.

(ii) Identification and evaluation of those correlation effects which are 'structure dependent', that is, which change significantly as a function of internuclear distance.

In the light of these two points it is convenient to analyse the computed interaction potential, $V(R)$, for two atoms A and B in terms of the Hartree–Fock contribution and the correlation contribution:

$$V(R) = E_{AB}^{HF}(R) - E_A^{HF} - E_B^{HF} + E_{AB}^{cor}(R) - E_A^{cor} - E_B^{cor}$$

where E^{HF} = Hartree–Fock energy computed from Hartree–Fock wave function, E^{cor} = Correlation energy defined as $E_{exact} - E^{HF}$.

As before we shall neglect the relativistic contributions in this discussion. Hartree–Fock calculations are routine and widely applied at both the minimal basis-set and the extended basis-set levels. From such calculations an SCF interaction potential is defined as

$$V_{(R)}^{HF} = E_{AB}^{HF}(R) - E_A^{HF} - E_B^{HF}.$$

For the Hartree–Fock interaction potential to be accurate the correlation energy of the system AB must be nearly equal to the sum of the correlation energies of the two atoms A and B. As discussed in Section 2.2.2.1 this fortunate near-equality does occur for three classes of systems and the resulting Hartree–Fock interaction potentials are semiquantitive (0.1– 0.5 eV errors in the interaction potential) when:

(i) A and B are both closed-shell systems, i.e. He_2, Ne_2, Ar_2, $NeAr$ [18], Li^+Ne, Na^+He [74] (Figure 2.1).

(ii) A is closed shell, B is half-open: HAr, HHe, LiNe, NaHe [74], Li_2^+, $NaLi^+$ [59], HF^+ [22], HCl^+, Ne_2^+, Cl_2^- [80] (Table 2.4), KrF, ArF, NeF, HeF [81].

(iii) A and B are half-open and AB is the highest multiplet formed from A and B: $Li_2(^3\Sigma_u)$, $NaLi(^3\Sigma)$ [59] (Figure 2.4).

In all these cases, *no new electron pairs* are formed or *old ones broken* and AB *dissociates* in the HF model to $A + B$.

If AB does not fall into one of these three classes, then:

(i) AB dissociates incorrectly in the HF model, i.e., H_2 to $\frac{1}{2}(H^+ + H^-) + \frac{1}{2}(H + H)$.

(ii) New electron pairs are formed in AB so that $E_{AB}^{cor} \neq E_A^{cor} + E_B^{cor}$. Thus errors of 1–5 eV are produced in the interaction potential.

The calculation of minimal basis-set Hartree–Fock wave functions for a large series of diatomic and polyatomic molecules has demonstrated that minimal basis-set SCF calculations yield bond lengths within a few per cent [17]. The results on O_2 (Tables 2.6) are typical, while those for F_2 (Table 2.6) represent among the worst agreement with experiment obtained from minimal basis-set SCF calculations.

A number of quantitative Hartree–Fock interaction potentials utilising large basis sets have been computed for class (i) systems, namely, the closed-shell systems interaction, and quantitative to semiquantitative results have been obtained. These systems include He_2, Ne_2, Ar_2 [18b], HeNe, ArNe, ArHe [18a], Li^+Ne, Na^+He [74] and FH^+ [22].

Hartree–Fock calculations of class (ii) interaction potentials have been performed on HHe, LiHe, NaHe [74], Li_2^+, $NaLi^+$ [59], He_2^+, Ne_2^+, Cl_2^-, Ar_2^+, F_2^- [80], ArF [81], KrF [34], NeF [81], HeF [81], yielding semiquantitative interaction potentials and dissociation energies for the bound systems (see Table 2.4).

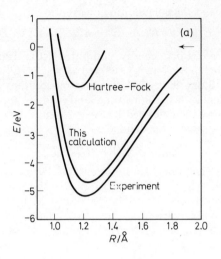

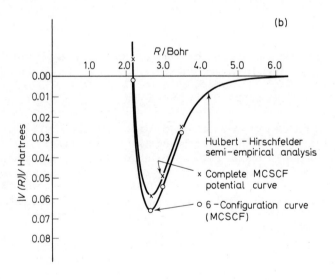

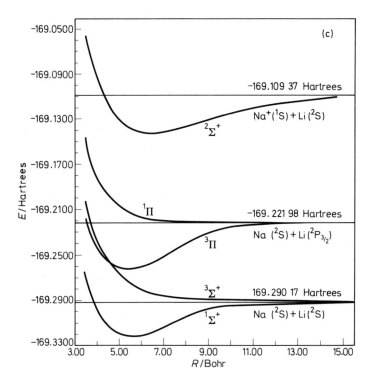

Figure 2.4 *Quantitative* accuracy is achieved in potential energy curves when *extended* basis sets are used and care is taken to include the configurations which *vary significantly* with internuclear distance in the CI calculations. (a) Potential energy curves for the $X^1\Sigma_g^-$ ground state of the oxygen molecule. The arrow on the right indicates the dissociation limit 149.620 82 Hartree which is the calculated energy, in the basis set, of two ^3P oxygen atoms. The energy of two Hartree–Fock ^3P oxygen atoms is very nearly the same, -149.618 76 Hartree. This calculation is a CI using approximate natural orbitals
(From Schaefer[54], by courtesy of the American Institute of Physics)
(b) Theoretical and experimental potential curves for the F_2 molecule obtained by the MCSCF technique
(From Das and Wahl[58], by courtesy of the American Institute of Physics)
(c) Potential energy curves for the NaLi molecule in the $^1\Sigma^+$ relaxed Φ_0, $^3\Sigma^+$, $^3\Pi$, $^1\Pi$ states. The ground-state curve was obtained by the MCSCF technique. The $^2\Sigma^+$ and $^3\Sigma^+$ curves are appropriate for the Hartree–Fock calculations. The $^3\Pi$ and $^1\Pi$ curves are also calculated by the Hartree–Fock method but they are probably not as accurate as the $^3\Sigma^+$ due to the probable mixing of a nearly-degenerate excited state
(From Bertoncini and Wahl[59], by courtesy of the American Institute of Physics)

Recent Hartree–Fock calculations on typical systems of class (iii) have been done for $Li_2(^3\Sigma)$ [92], NaLi $(^3\Sigma)$ [59].

Examples of systems which do not fall into these three categories, favourable for Hartree–Fock calculations, for which portions of the potential curve near R_e have been computed include the alkali halides[19], the fluorides of O, S, N, P, C, Si, Se [23], and NaO [75] and CaO [36]. However, spectroscopic constants obtained from these curves indicate the adequacy of the Hartree–Fock model in reflecting the potential-curve shape near R_e. In fact, for the alkali halides[19], the shape is excellent since once they have formed, they fall into class (i) systems and it is thus only the well *depth* but not the shape which is in error.

For the wide class of systems for which the Hartree–Fock model yields

Table 2.4   Ground-state potential curve characteristics[80]

	D_e/eV calc.	expt.	R_e/Bohr calc.	ω_e/cm^{-1} calc.	B_e/cm^{-1} calc.	ΔE_{LR}/eV calc.
He_2^+	2.67	2.49	2.0	1790	7.4	1.02
Ne_2^+	1.65	1.35 ± 0.07	3.2	660	0.59	1.77
Ar_2^+	1.25	1.6 ± 0.3	4.6	300	0.139	0.69
F_2^-	1.66	1.29 ± 0.1	3.6	510	0.50	1.89
Cl_2^-	1.28	1.26 ± 0.1	5.0	260	0.136	0.78

Table 2.5   Comparison of spectroscopic constants from theory and experiment for MCSCF wave function for F_2
(From Das and Wahl[58], by courtesy of the American Institute of Physics)

Wave function	R/Bohr	D_e/eV	ω_e/cm^{-1}	α_e/cm^{-1}	B_e/cm^{-1}
6-Configuration	2.64	1.82 ± 0.2	986	0.0159	0.902
Final	2.67	1.67 ± 0.06	942	0.0160	0.88
Experimental	2.68	1.68	932	0.0146	0.89

incorrect dissociation products and interaction potentials (in error by 1–5 eV) a large number of CI calculations which eliminate these difficulties have been performed. At the minimal basis-set level complete CI calculations discussed in Section 2.3.2 form a general and economical method for obtaining correct *qualitative* information on the ordering of molecular states, excitation energies, bond lengths and the general shape of the interaction potentials[49–51, 76].

When extended basis sets are used quantitative interaction potentials have been obtained. Good examples are calculations on the He_2 and BH interaction potential for which quantitative results have been achieved by most methods. Calculated interaction potentials for HF, O_2 [54], F_2 [58], NaLi [35], MgH [39] and BeO [53] are illustrative of the high-accuracy results achieved on larger systems. One of the most encouraging features emerging from these recent calculations is that it appears that by utilising *only* those configurations which lead to proper dissociation of the molecule plus *truly* molecular configurations (namely those configurations which exist in the molecule, but vanish formally in the atoms), potential curves accurate to ± 0.1 eV in the strong interaction region can be obtained routinely. This approach (Tables 2.5 and 2.6) has been advocated and developed in the optimised valence configuration method[55, 73] and applied to Li_2, $NaLi$[35], Fe_2[58] and O_2^-[61].

Similar ideas have recently been incorporated into the CI approach in the

Table 2.6 Comparison of calculated potential curves with experiment

	$O_2(^3\Sigma_g)$ Total energy	D_e/eV	ω_e/cm^{-1}	r_e/Å
Min. SCF				1.217
Near Hartree–Fock	149.6639	1.43	2000	1.152
Min. CI	149.1703	3.81	1582	1.30
Extended CI*	149.7944	4.72*	1614	1.22
Expt.		5.21	1580	1.207

	$F_2(^1\Sigma^+)$			r_e/a.u.
Min. SCF	197.8865			2.50
Near Hartree–Fock	198.7683	-1.37	1292	2.51
Min. CI	197.9560	2.04		
(MCSCF) [58]	198.9809	1.67	942	2.67
Expt.	199.67	1.68	932	2.68

*Expansion error is 0.53 eV. Thus bonding energy with better basis would be 5.24 eV, giving excellent agreement[54].

first-order wave function concept, and high precision results obtained for BeO[53], O_2[54], KrF^+[34] and BH[83]. Further encouragement comes from the apparent constancy of the correlation energy of the inner shells during molecular formation as illustrated by these and other calculations previously discussed.

Other important developments in the evaluation of interaction potentials have occurred in the variational approach to long-range dispersion forces. Again, by an appropriate selection of only those excitations which correspond to *interatomic* correlation, the full interaction potential around the van der Waals well has been obtained by the MCSCF technique[84] and by direct configuration interaction[85] for He_2. Recent similar work on the van der Waals interaction in HHe, LiHe[86] and Li_2[87] support a 10% accuracy in the computed potential curve. All these variational long-range calculations focused *only* on interatomic correlation. These new developments are discussed in greater detail in a companion article by Certain and Bruch in this volume (Chapter 4).

Among other interesting diatomic potential-curve calculations are those by the CI method on BH ($B^1\Sigma^+$ state) illustrating a double minimum, and the similar calculations on excited states of He_2 yielding interesting features, which have been reviewed recently[1].

2.3.4 Ionisation potentials and electron affinities

There are two main approaches for calculating ionisation potentials or electron affinities. The first, called the direct method, consists of calculating the total energy for the neutral and the charged species and identifying the difference as the ionisation potential or electron affinity. The second method requires a Hartree–Fock wave function for the parent species and identifies the ionisation potential with the orbital energy (Koopman's theorem); a generalisation of Koopman's theorem for multiconfiguration wave functions[88] has been recently proposed. The Koopman theorem method, which corresponds to a vertical ionisation potential, yields results which are typically 5–10% (1–2 eV) in error. Such results have been obtained and reported routinely from Hartree–Fock calculations (Table 2.7).

The direct method requires a wave function for the parent and ionised species and, of course, must therefore deal with the consistency of errors in the parent and ionised wave functions. Although there are notable cases in which the Hartree–Fock model can yield reliable ionisation energies or

Table 2.7 Ionisation potentials (in eV) from SCF wave functions[23]

System	Koopmans*	Direct†	Corrected‡ for correlation	Expt.
CF		8.7	9.2 ± 0.5	8.9–15
NF		13.1	13.2 ± 0.3	13.1 ± 0.2
OF	16.6	12.3	13.1 ± 0.5	13.1 ± 0.3
SiF		7.1	7.4 ± 0.2	7.3 ± 0.1
PF		9.6	9.7 ± 0.2	< 13.8 ± 0.6
SF	11.4	9.6	10.0 ± 0.4	Not measured
SeF	10.8	9.2	9.5 ± 0.4	Not measured
ClO	15.2	10.8	11.2 ± 0.4	11.0 ± 0.5

*Only valid for ionisation from closed shell or half-open shell systems unless corrected.
†Difference of total Hartree–Fock energies for neutral system and ion.
‡The direct value has been corrected by adding the correlation energy change associated with ionisation estimated from an isoelectronic system or interpolated from adjacent fluoride.

electron affinities, usually the correlation energy changes associated with electron removal or capture are of the order of several eV. This error is often of sufficient size to reverse the level of the ionic states and thus lead to an incorrect assignment of where the electron is 'coming from'. In order to correct this discrepancy electron correlation must be introduced in such a way that the correlation energy neglected is the same in parent and charged species. An example of how this can be achieved is provided by CI calculations on N_2^+ [89b] which gave the three lowest ionisation energies in the correct order when the Hartree–Fock model did not. Hartree–Fock ionisation potentials can also be corrected with semi-empirical estimates of the correlation correction, and results correct to ± 0.5 eV are claimed (Table 2.7).

Two new interesting developments are the high accuracy results obtained for K shell ionisation energies for NO and O_2 [90], which can be related to ESCA experiments, and the electron affinity of O_2, obtained by computing to high accuracy the dissociation energy of O_2^- [61] and the utilising the electron affinity of O. An interesting calculation of the proton affinities[91] of CO and N_2 has also been reported.

2.4 SUMMARY AND PROGNOSIS

This survey has illustrated the fact that quantitative energetic information can now be achieved for diatomic systems of chemical interest through a variety of variational techniques. The Hartree–Fock model is now routinely achievable, its capabilities are well understood, and it yields accurate energetics for certain classes of systems. The introduction of electron correlation still requires considerable skill in order to obtain correct physical results. However we do now possess the necessary skill. This statement could not have been made four years ago. We have also witnessed the maturation of the *users* of these quantum mechanical techniques. The capabilities of various levels of calculation are well understood by most users and by now we know when we are right, when we are lucky, and when we are foolish. Considerable but not prohibitive work is necessary to achieve ± 0.1 eV in interaction potentials, dissociation energies, ionisation potentials or transition energies. However, several techniques for achieving this accuracy have been successfully demonstrated on a wide variety of diatomic systems and, for the smaller systems ± 0.02 eV accuracy is being achieved. On the other hand when only qualitative information is desired minimal basis-set SCF and CI calculations are capable of yielding it for a number of molecular characteristics at a very modest effort (often less than a minute of computer time for the smaller diatomic systems).

Extrapolating from the work surveyed, it appears that the greatest possibility for the extension of these methods to even larger diatomic systems lies in the use of the configuration interaction method utilising some optimised set of orbitals (INO, PNO, or MCSCF) in which an effort is made to evaluate only the varying part of the correlation energy. Also it appears that solving the asymptotic difficulties of the MO model with a MCSCF 'base function' and then proceeding to correlate further from that level rather than from the single configuration HF level would eliminate the need for many single excitations and other complications now required in direct configuration interaction in order to achieve a full and continuous potential curve.

We have also seen that realiable *a priori* capability is no longer limited to two- or four-electron diatomic molecules but has been demonstrated for several systems containing up to 50 electrons. In predicting the future it should be remembered that most of the calculations reported were performed on *relatively* old and slow computers using established stable 'checked out' programmes (CDC6600 or IBM 360/75). Thus with no improvement in theoretical techniques, calculations which are now taking up to an hour would only take minutes on *existing* hardware (IBM 195 or CDC-7600) to which only a few have access. Computers encompassing advanced features (such as parallel processing, mathematical function macros, and extended *fast* memory) will bring both the time *and* cost down by at least another order of magnitude, without either new faster techniques or tailor-coding for the machine characteristics and will thus extend accurate techniques to larger systems. Even within this conservative view the power of *a priori* calculations must be taken seriously. For diatomic systems we are already seeing the machinery for the calculation of wave functions

and energies becoming a reliable sub-program buried within a comprehensive computational framework for computing the results of scattering
and collision processes, transport phenomena and spectroscopic studies.
This process is being accelerated as more and more experimentalists
use such calculations as routine sources of information complementary to
experiment and as the number of *valid predictions* on diatomic systems of
chemical interest increases.

References

1. Brown, J. C. and Matsen, F. A. (1972). Ab initio *Calculations on Small Molecules, Advan. Chem. Phys.,* **23**
2. Krauss, M. (1967). *National Bureau of Standards Technical Note No. 438*
3. Sutton, P., Bertoncini, P., Das, G., Gilbert, T. L., Wahl, A. C. and Sinanoglu, O. (1970). *Int. J. Quantum Chem.,* **3S,** 479
4. Allen, L. C. (1969). *Ann. Rev. of Phys. Chem.,* **20,** 315
5. Schaefer, H. F. III. (1972). *Electronic Structure of Atoms and Molecules.* (Addison Wesley)
6. See Reference 20.
7. Nesbet, R. K. (1967). *Advan. Quantum Chem.,* **3,** 3
8. Wahl, A. C., Bertoncini, P., Das, G. and Gilbert, T. (1967). *Int. J. Quantum Chem.,* **1S,** 123
9. Nesbet, R. K. (1969). *Advan. Chem. Phys.,* **14,** 237
10. Kelly, H. P. (1969). *Advan. Chem. Phys.,* **14,** 129
11. Sinanoglu, O. (1969). *Advan. Chem. Phys.,* **14,** 237
12. Silver, D. M., Mehler, E. L. and Ruedenberg, K. (1970). *J. Chem. Phys.,* **42,** 1174
13. Wahl, A. C. and Das, G. (1970). *Advan. Quantum Chem.,* **5,** 261
14. Kutzelnigg, W. (1972). *Molecular Calculations Including Electron Correlation,* preprint
15. Condon, E. U. (1968). *Rev. Mod. Phys.,* **40,** 872
16. Nesbet, R. K. (1971). *Int. J. Quantum Chem.,* **45,** 117
17. Newton, M. D., Lathan, W. A., Hehre, W. J. and Pople, J. A. (1970). *J. Chem. Phys.,* **32,** 4064
18. Matcha, R. L. and Nesbet, R. K. (1967). *Phys. Rev.,* **160,** 72; Gilbert, T. L. and Wahl, A. C. (1967). *J. Chem. Phys.,* **47,** 3425
19. Matcha, R. L. (1970). *J. Chem. Phys.,* **53,** 485; (1968), ibid., **49,** 1264; (1968). ibid., **48,** 335; (1967). ibid., **47,** 5295; (1967). ibid., **47,** 4595
20. Cade, P. E. and Huo, W. M. (1967). *J. Chem. Phys.,* **47,** 614; (1967). ibid, **47,** 649
21. See references 74 and 80.
22. Julienne, P. S., Krauss, M. and Wahl, A. C. (1971). *Chem. Phys. Lett.,* **11,** 16
23. O'Hare, P. A. G. and Wahl, A. C. (1971). *J. Chem. Phys.,* **55,** 666; (1970). ibid., **53,** 2469; (1971). ibid., **55,** 4563
24. Dahl, J. P., Johansen, H., Truax, D. R. and Ziegler, T. (1970). *Chem. Phys. Lett.,* **6,** 64; Hunt, W., Dunning, T. and Goddard, W. (1970). ibid., **3,** 606
25. Walker, T. E. H. (1971). *Chem. Phys. Lett.,* **9,** 174
26. Sleeman, D. A. (1968). *Chem. Phys. Lett.,* **11,** 135; Jungen, M. (1968). ibid, **11,** 193
27. Epstein, I. R. (1970). *J. Chem. Phys.,* **53,** 1881
28. Sadlej, A. J. (1971). *Chem. Phys. Lett.,* **8,** 100
29. Musher, J. I. (1970). *Chem. Phys. Lett.,* **7,** 397
30. Carbo, B. (1971). *Chem. Phys. Lett.,* **8,** 75
31. Winter, N. W. and Dunning, T. H. (1971). *Chem. Phys. Lett.,* **8,** 169; Hunt, W., Goddard, W. III and Dunning, T. (1970). ibid., **6,** 147
32. Hillier, I. H. and Saunders, V. R. (1970). *Int. J. Quantum Chem.,* **4,** 503; Brown, R. D., Burden, F. R. and Williams, G. R. (1970). *Theoret. Chim. Acta,* **18,** 98; Body, R. G. (1970). *Theoret. Chim. Acta,* **18,** 107
33. Swarz, K. and Connolly, J. W. D. (1971). *J. Chem. Phys.,* **55,** 4710
34. Liu, B. and Schaeffer, H. F. III (1971). *J. Chem. Phys.,* **55,** 2369
35. Bertoncini, P. J., Das, G. and Wahl, A. C. (1970). *J. Chem. Phys.,* **52,** 5112

36. Carlson, K. D., Kaiser, K., Moser, C. and Wahl, A. C. (1970). *J. Chem. Phys.*, **52**, 4678
37. Gelus, M., Ahlrichs, R., Staemmler, V. and Kutzelnigg, W. (1971). *Theoret. Chim. Acta*, **21**, 63
38. Siu, A. K. Q. and Davidson, E. R. (1970). *Int. J. Quantum Chem.*, **4**, 223
39. Chan, A. C. H. and Davidson, E. R. (1970). *J. Chem. Phys.*, **52**, 4108
40. Schaeffer, H. F. III (1970). *J. Chem. Phys.*, **52**, 6241
41. Adams, W. (1971). *Chem. Phys. Lett.*, **9**, 199
42. Shull, H. (1970). *Int. J. Quantum Chem.*, **3**, 523
43. Brandas, E. J. and Bartlett, R. J. (1971). *Chem. Phys. Lett.*, **8**, 153
44. Kouba, J. E. and Orhn, Y. (1971). *Int. J. Quantum Chem.*, **5**, 539
45. Bingel, W. A. (1970). *Theoret. Chim. Acta*, **16**, 319
46. Roby, K. R. (1971). *Int. J. Quantum Chem.*, **5**, 119
47. Heil, T. G., O'Neil, S. V. and Schaefer, H. F. III (1970). *Chem. Phys. Lett.*, **5**, 253
48. Harrison, J. F. and Allen, L. C. (1969). *J. Mol. Spectrosc.*, **29**, 432
49. O'Neil, S. V. and Schaefer, H. F. III (1970). *J. Chem. Phys.*, **53**, 3994
50. Michels, H. H. (1970). *J. Chem. Phys.*, **53**, 841
51. Schaefer, H. F. III and Harris, F. E. (1968). *J. Chem. Phys.*, **48**, 4946
52. Gupta, B. K. and Matsen, F. A. (1969). *J. Chem. Phys.*, **50**, 3797
53. Schaefer, H. F. III (1971). *J. Chem. Phys.*, **22**, 176
54. Schaefer, H. F. III (1971). *J. Chem. Phys.*, **54**, 2207
55. Wahl, A. C. and Das, G. (1970). *Advan. Quantum Chem.*, **5**, 261; (1970). *Phys. Rev. Lett.*, **24**, 440; (1966). *J. Chem. Phys.*, **44**, 87; (1972). *J. Chem. Phys.*, **56**, 1769
56. Mukherjee, N. G. and McWeeny, R. (1970). *Int. J. Quantum Chem.*, **4**, 97
57. For a discussion of errors in expectation values of MCSCF wave functions see: Tuan, D. F. T. (1970). *J. Chem. Phys.*, **52**, 5247
58. Das, G. and Wahl, A. C. (1970). *Phys. Rev. Lett.*, **24**, 440; (1970). *Advan. Quantum Chem.*, **5**, 261; (1972). *J. Chem. Phys.*, **56**, 3532
59. See Reference 35 and Bertoncini, P. J., Das, G. and Wahl, A. C. (1972). *A Theoretical Study of Na₂.*, unpublished
60. Hinze, J., *MCSCF Calculations for Ground and Excited States of LiH*. To be published in *J. Chem. Phys.*
61. Zemke, W., Das, G. and Wahl, A. C. (1972). *Chem. Phys. Lett.*, **14**, 310
62. Bender, C. F. and Davidson, E. R. (1969). *Phys. Rev.*, **183**, 23
63. Mehler, E. L., Ruedenberg, K. and Silver, D. M. (1970). *J. Chem. Phys.*, **52**, 1181
64. Silver, D. M., Ruedenberg, K. and Mehler, E. L. (1970). *J. Chem. Phys.*, **52**, 1206
65. Kelly, H. P. (1970). *Phys. Rev.*, **A2**, 1261; Schulman, J. D. and Kaufman, D. N. (1970) *J. Chem. Phys.*, **53**, 477
66. Lee, T., Dutta, N. C. and Das, T. P. (1970). *Phys. Rev. Lett.*, **25**, 304
67. Byers Brown, W. (1968). *Chem. Phys. Lett.*, **1**, 655; Byers Brown, W. and Power, J. D. (1970). *Proc. Roy. Soc. (London)*, **A317**, 545
68. Sims, J. S. and Hagstrom, S. (1971). *Phys. Rev. A*, **4**, 908
69. Boys, S. F. and Handy, N. C. (1969). *Proc. Roy. Soc. (London)*, **A309**, 209; (1969). ibid., **A310**, 43
70. Handy, N. C. and Epstein, S. J. (1970). *J. Chem. Phys.*, **53**, 1392; Hegyi, M. G., Mezei, M. and Szondy, T. (1971). *Theoret. Chim. Acta*, **21**, 168
71. Boys, S. F. and Handy, N. C. (1969). *Proc. Roy. Soc. (London)*, **A311**, 309; for a study of Li₂ see Gombas, P. and Szondy, T. (1970). *Int. J. Quantum Chem.*, **4**, 603
72. Schaefer, H. F. III and Harris, F. E. (1968). *Phys. Rev. Lett.*, **21**, 1561
73. Wahl, A. C. and Das, G. (1970). *Advan. Quantum Chem.*, **5**, 261
74. Krauss, M., Maldonado, P. and Wahl, A. C. (1971). *J. Chem. Phys.*, **54**, 4944
75. O'Hare, P. A. G. and Wahl, A. C. (1972). *J. Chem. Phys.*, **56**, 4516
76. Schaefer, H. F. and Heil, T. J. (1971). *J. Chem. Phys.*, **54**, 2573
77. Bender, C. F. and Davidson, E. R. (1967). *J. Chem. Phys.*, **47**, 4972
78. Bender, C. F. and Davidson, E. R. (1967). *J. Chem. Phys.*, **46**, 3313
79. Wirsam, B. (1971). *Chem. Phys. Lett.*, **10**, 180
80. Gilbert, T. L. and Wahl, A. C. (1971). *J. Chem. Phys.*, **55**, 5247
81. Liebman, J. F. and Allen, L. C. (1970). *J. Amer. Chem. Soc.*, **92**, 3539; (1969). *Chem. Commun.*, 1354; (1971). *Int. J. Mass. Spectrom. Ion. Phys.*, **7**, 27
82. In the papers referred to in Reference 23 the Hartree–Fock ionisation potentials were semi-theoretically corrected for this series of fluorides and predictions made.

83. Pearson, P. K., Bender, C. F. and Schaefer, H. F. III. (To be published)
84. Bertoncini, P. J. and Wahl, A. C. (1970). *Phys. Rev. Lett.*, **25**, 991
85. Schaefer, H. F. III, McLaughlin, D. R., Harris, F. E. and Alder, B. J. (1970). *Phys. Rev. Lett.*, **25**, 988
86. Das, G. and Wahl, A. C. (1971). *Phys. Rev.*, **A4**, 825
87. Kutzelnigg, W. and Gelus, M. (1970). *Chem. Phys. Lett.*, **7**, 296
88. Levy, B. and Berthier, G. (1968). *Int. J. Quantum Chem.*, **2**, 307
89a. Goddard, W. A. and Ladner, R. C. (1971). *J. Amer. Chem. Soc.*, **93**, 6750
89b. Dunning, T. H., *Generalised Valence Bond Calculations on* N_2^+ (in preparation)
90. Bagus, P. S. and Schaefer, H. F. III (1971). *J. Chem. Phys.*, **55**, 1474
91. Jansen, H. B. and Ros, P. (1971). *Theoret. Chim. Acta*, **21**, 199
92. Kutzelnigg, W., Staemmler, V. and Gelus, M. (1972). *Chem. Phys. Lett.*, **13**, 496
93. Coulson, C. A. (1971). *Mol. Phys.*, **20**, 687
94. Coulson, C. A. (1969). *Acta Phys. Acad. Sci. Hungaricae*, **27**, 345
95. The Xα method has recently been applied to several diatomic molecules and is yielding semiquantitative results. A review of this method is in the press (private communication, K. Johnson)
96. Rowe, D. J. (1968). *Rev. Mod. Phys.*, **40**, 153; Shibuya, T. and McKoy, V. (1970). *Phys. Rev. A*, **2**, 2208
97. Rose, J., Shibuya, T. and McKoy, V. (1972). *Application of the Equations of Motion Method to the Excited States of* N_2, *CO, and* C_2H_4, preprint (To be published in *J. Chem. Phys.*)
98. Grein, F. and Chang, T. C. (1971). *Chem. Phys. Lett.*, **12**, 44

3
Ab Initio
Molecular Orbital Theory of
Organic Molecules

L. RADOM and J. A. POPLE
Carnegie-Mellon University, Pittsburgh, Penn.

3.1 INTRODUCTION

For many years, molecular orbital theory has proved to be a valuable theoretical method for describing key features of the electronic structure of organic molecules. Early applications were mostly concerned with the understanding of the π-electrons of aromatic and other planar molecules, first at the independent electron level[1] following the work of Hückel[2] and later with approximate theories taking some account of the details of electron interaction[3–5]. In the last decade, the advent of electronic computers has permitted generalisation of these methods to all electrons (or, at least, all valence electrons) in a molecule and numerous techniques have been proposed for the evaluation of molecular orbital wave functions for the general three-dimensional molecule. The most widely used such methods are semi-empirical in nature and combine mathematical approximations with some parameterisation based on appropriate experimental data[6–12]. More recently, advances in mathematical technique and programming have reached a point at which organic molecules of moderate size can be studied fully by *ab initio* methods involving no empirical parameterisation. Although necessarily more expensive in application, such methods do have the advantage of eliminating some of the subjective aspects of parameterisation and permit the theory to advance in a more well-defined and systematic manner. The purpose of this article is to review the current status of *ab initio* molecular orbital theory in the area of application to organic molecules.

The earliest applications of *ab initio* methods were to molecules that were already well characterised experimentally. These studies demonstrated that such a theory gives reasonable descriptions of a variety of molecular properties. This has encouraged subsequent studies on systems and properties which are difficult to examine experimentally. Thus, for example, theory has been used to follow the paths of chemical reactions, to determine structures of intermediates, such as carbonium ions, and to test qualitative notions about electron distribution.

Molecular orbital theory is now in a position to play an important predictive role and to complement experimental techniques in a number of important research areas. We should emphasise, however, that in order that theoretical predictions on unknown systems be meaningful, they should be preceded by thorough testing of the same theoretical method on known systems.

Our coverage of the field is limited in several ways. We shall be concerned only with the predictions that have been made from single-determinant molecular orbital theory as outlined in Section 3.2. Further, we restrict our attention to molecular properties that are of most chemical interest, namely, structures, stabilities and charge distributions. The necessary background to the study of these properties is presented in Section 3.3. We have omitted many other interesting properties (electric and magnetic properties, electronic excitation, ionisation potentials and hydrogen bonding). Finally, we examine only molecules that are composed of the elements H, C, N, O and F and that include at least one carbon atom. These are discussed in detail in Section 3.4 and form the major part of the review. Within the limitations mentioned above, we have attempted to cover the literature of *ab initio* molecular orbital theory of organic molecules up to July 1971.

3.2 THEORY

3.2.1 LCAO SCF molecular orbital theory

We are only concerned in this review with calculations of molecular properties using linear combination of atomic orbitals (LCAO) self-consistent field (SCF) molecular orbital (MO) theory. In this approximation, the molecular orbitals (ψ_i) are written as linear combinations of basis functions (ϕ_μ)

$$\psi_i = \sum_\mu c_{\mu i} \phi_\mu \qquad (3.1)$$

The full molecular wave function (Ψ) is then written as a single determinant of the spatial orbitals ψ_i with appropriate spin functions α or β.

For a closed-shell system of $2n$ electrons in doubly occupied MOs, the wave function is

$$\Psi = [(2n)!]^{-\frac{1}{2}} |\psi_1(1)\alpha(1)\psi_1(2)\beta(2)\ldots\ldots\psi_n(2n-1)\alpha(2n-1)\psi_n(2n)\beta(2n)| \qquad (3.2)$$

For open-shell systems, there are two approaches that are commonly used to define the wave function, frequently described as restricted and unrestricted. In the restricted procedure, the same spatial orbitals are assigned to pairs of α and β electrons. However, if α and β electrons are unequal in number, their environments are different and giving them the same spatial orbital is a restriction. In the unrestricted procedure, different spatial orbitals are assigned to each electron.

The total energy may be calculated from the appropriate (closed- or open-shell) wave function as

$$E = \int \Psi^* H \Psi \, d\tau \qquad (3.3)$$

where H is the many-electron Hamiltonian. Minimisation of the energy leads to the Roothaan[13, 14] or Pople–Nesbet[15] self-consistent equations for the LCAO coefficients $c_{\mu i}$. As the number of basis functions ϕ_μ increases, the flexibility of the representation is improved and more accurate wave functions are obtained. The limiting description given for sufficiently large basis sets is called the *Hartree–Fock* limit.

This single-determinant theory takes no account of correlation between the motions of electrons with anti-parallel spins. To proceed beyond this level, linear combinations of determinants have to be considered (configuration interaction). This subject will not be treated in detail here since we are mainly interested in the description of electronic structure within the single determinant framework.

3.2.2 Basis functions

The self-consistent field procedure leads to a molecular energy for any nuclear arrangement once the basis functions ϕ_μ have been specified. The choice of the ϕ_μ is important and leads to differences between various *ab initio* molecular orbital calculations. An increase in the number of basis functions should lead to a better description of the electronic structure. However, this leads to an increase in computation so that, in selecting the size of basis set, a compromise between accuracy and efficiency is involved.

3.2.2.1 Location of basis functions

Three categories of molecular orbital calculations differing in the location of basis functions are in current use.

(a) Most commonly, the ϕ_μ are centred on each of the nuclei in a molecule. This was, of course, the original idea behind LCAO theory and is the method used for most of the calculations discussed in this review.

(b) In order to avoid some of the problems associated with the evaluation of multicentre integrals, many of the early calculations used basis functions centred on just one atom in a molecule. This *single-centre expansion* method is quite successful for central hydrides of the type AH_n but has found little application for other types of molecules. Much of the work that has been carried out with this method is summarised in recent review articles[16, 17] and will not be discussed in detail here.

(c) Finally, calculations have been carried out, notably by Preuss[18], Frost[19], Christoffersen[20, 21] and co-workers, in which the functions ϕ_μ are allowed to be at points in space other than the nuclei. These have been called Floating Spherical Gaussian Orbital (FSGO) calculations since the ϕ_μ are taken as spherical Gaussian functions. Such calculations use smaller basis sets than the standard LCAO procedures. However, the number of non-linear parameters is large since both the location and the size of the ϕ_μ have to be specified; this becomes troublesome in larger molecules for which some simplification is then necessary. Christoffersen has tackled this problem by using parameters from component fragments.

3.2.2.2 Minimal basis sets

The simplest basis sets ϕ_μ in LCAO calculations use the smallest number of atomic orbitals of appropriate symmetry required for the description of the atomic ground states. For the atoms considered in this review, these are

H: 1s
C,N,O,F: 1s,2s,$2p_x$,$2p_y$,$2p_z$

Such a set of ϕ_μ is usually called a *minimal basis set*.

Three main types of minimal basis set are in general use. The first uses single Slater-type exponential functions (STOs) with exponents chosen from atomic values or by optimisation in the molecule. This sort of basis set has an appealing simplicity and has been used extensively. Early applications were to diatomic molecules by Ransil[22], linear molecules by Clementi and McLean[23, 24] and other small polyatomic molecules by Pitzer, Lipscomb and co-workers[25]. The principal disadvantage of the STO basis lies in the time taken to evaluate the multicentre two-electron integrals with adequate accuracy, since the most effective procedures involve some numerical integration. Some improvement in the efficiency of STO calculations has come from the development of programmes[26] involving Gaussian transforms[27].

A closely related type of minimal basis set is one in which *contracted* (or fixed linear combinations of) Gaussian functions ($e^{-\alpha r^2}$, Gaussian type orbital, GTO) are used to simulate Slater-type orbitals. Such a procedure was first employed by Foster and Boys[28] and it has been developed by a number of other authors[29-32]. The use of Gaussian functions considerably simplifies the evaluation of the multicentre integrals as was demonstrated originally by Boys[33]. A systematic development of least-squares fits to Slater-type orbitals leads to the STO–NG basis sets with N Gaussian functions per STO[32]. The STO–NG bases have the further special feature of sharing Gaussian exponents between s- and p-functions for computational efficiency. A full study of STO–NG bases shows that convergence of equilibrium geometries, atomisation energies, dipole moments and other properties to the STO limit as $N \to \infty$ is quite rapid. Most such applications have been carried out at the STO–3G level.

The third type of minimal basis set consists of contracted Gaussian functions which simulate accurate Hartree–Fock atomic orbitals rather than the Slater-type exponential approximations to them. A contracted basis of this kind has been used by Clementi and associates in studies of pyrrole[34] and, subsequently, larger molecules. More recently, a series of such contracted basis sets (LEMAO–NG or least-energy minimal atomic orbitals at the N-Gaussian level) has been obtained[35, 36] by minimisation of atomic energies. However, convergence to limiting behaviour as $N \to \infty$ is less rapid with such bases than with the STO–NG sets.

An important feature of all minimal atomic orbital basis sets is the re-scaling of the functions on going from atoms to molecules. It was recognised many years ago that hydrogen 1s atomic orbitals should be reduced in size by a scale factor of about 1.2 in many molecules but it has only recently been realised that similar contractions apply to carbon[37]. Sets of standard or average re-scaling factors have been proposed[32, 36] for both the STO–NG and LEMAO–NG sets.

3.2.2.3 Extended basis sets

If the number of basis functions is larger than the minimal requirement, the basis set is usually described as *extended*. Such bases give increased flexibility and their use may overcome some of the inherent limitations of minimal sets.

If the ϕ_μ are taken to be individual GTOs the basis is called *uncontracted*. The use of a large number of uncontracted Gaussian functions corresponds to an extended basis. Such sets have been proposed by a number of authors[38–41], but their effectiveness is limited because of the large dimensions of the various matrices in the wave function determination. More commonly, work with extended basis sets has used either Slater-type orbitals or carefully selected contracted Gaussian functions. A number of basis sets in the latter category have been suggested[39, 42–51].

A widely used contracted Gaussian set was proposed by Whitten[43]. This is a slightly extended set with three s-type orbitals for atoms such as carbon rather than two as required in a minimal set. Whitten proposed single p-functions, but represented them as a linear combination of two off-centre s-type 'lobes' rather than as a cartesian p-type Gaussian function. Other authors have made use of Whitten s-functions, but combined them with a single contracted p-type Gaussian function obtained by Huzinaga[38].

An important step in the development of extended basis sets is the separation of the valence-shell atomic orbitals into inner and outer parts. This leads to increased flexibility in describing the variation of orbital size at any atom in a molecule, since the molecular orbitals may have variable coefficients associated with the inner and outer parts. Further, such basis sets have the important advantage of being able to describe anisotropic atomic dimensions since the inner–outer weightings may differ between the three p-functions on a given atom. (The same effect can be allowed for by independent scaling of the p-functions, but this turns out to be less practical.)

Several 'split valence bases' of this type are in common use in the literature[43, 45, 51]. A widely-used version[45] is based on Whitten's[43] three-group s-function and the Huzinaga[38] group of p-functions with the further relaxation that the outermost Gaussians in the contractions are treated as separate ϕ-functions. In this laboratory, rather simpler split valence bases have been developed for extensive application to organic molecules. The most widely used is the 4–31G set[51] which consists of 4-Gaussian inner shell atomic orbitals and valence orbitals split into 3-Gaussian inner parts and 1-Gaussian outer parts.

Further improvement in the basis sets leading to close approach to the Hartree–Fock limit follows by adding atomic functions of higher angular symmetry, such as p-functions for hydrogen and d-functions for first row atoms[52–57]. These are called *polarisation* functions. They bear little resemblance to excited state atomic orbitals; their principal function is to polarise the s- and p-functions. Such basis sets are expensive to use, but they have been applied to some of the smaller significant organic molecules.

3.3 PROPERTIES

3.3.1 Charge distributions

Because atoms have, to some extent at least, lost their identity in molecule formation, it is somewhat ambiguous to speak of an *atom in a molecule*[58-62]. In particular, the *charge* on an atom in a molecule has no precise physical meaning. However, chemists often find it useful to speak of atomic charges in discussions of molecular properties and so several definitions have been proposed.

The most commonly used method of assigning atomic charges is due to Mulliken[58] who proposed that the gross electron population associated with an atomic orbital (ϕ_μ) be given by

$$q_\mu = P_{\mu\mu} + \sum_{v(\neq \mu)} P_{\mu v} S_{\mu v} \tag{3.4}$$

where $P_{\mu v} = \sum_i c_{\mu i} c_{v i}$ and $S_{\mu v} = \int \phi_\mu \phi_v d\tau$

There are two main weaknesses in this definition. In the first place, the population $P_{\mu\mu}$ associated with an orbital ϕ_μ centred on an atom A is assigned entirely to A even though its maximum electron density may be quite distant from A and located reasonably close to an adjacent atom B. This situation reaches an extreme in one-centre calculations where all basis functions are centred on the one atom and hence all electronic charge is assigned to this atom! To overcome this difficulty, it is necessary to use a basis set which is *balanced* between atoms[59]. However, because the balance in basis sets is variable, comparison of Mulliken charge distributions calculated using different basis sets should be avoided. The second point is that the definition equation (3.4) distributes the overlap charge distribution ($2P_{\mu v} S_{\mu v}$) equally between the contributing atomic orbitals (ϕ_μ and ϕ_v). This is realistic only when ϕ_μ and ϕ_v are identical. Hence, comparison of charges on X in molecules X—A and X—B may sometimes be misleading. Comparison of charges on X in molecules X—Y—A and X—Y—B is likely to be more meaningful. However, in spite of these limitations, the Mulliken population analysis has proved useful and is currently the only widely used procedure for calculating atomic charges.

3.3.2 Dipole moments

Unlike the atomic charge, the dipole moment of a molecule is a well-defined quantity which may be precisely measured and calculated. It is given by the formula

$$\mu = \left[\sum_A Z_A r_A - \sum_{\mu,v} P_{\mu v} r_{\mu v} \right] \tag{3.5}$$

where $r_{\mu v} = \int \phi_\mu r \phi_v d\tau$, r_A and r are nuclear and electron position vectors and Z_A is the nuclear charge on atom A. An examination of dipole moments calculated with different types of basis set leads to the following conclusions.

The calculated moments are, in general,

(a) underestimated by minimal STO basis sets (including STO–NG) though this is not always the case,

(b) overestimated by energy-optimised minimal GTO basis sets,

(c) overestimated by extended sp basis sets, and

(d) reduced by adding polarisation functions to the extended basis set but still somewhat high even near the Hartree–Fock limit[59, 63].

3.3.3 Molecular geometries

The determination of molecular geometries by means of *ab initio* calculations generally involves the calculation of the total energy as a function of the geometric parameters and obtaining the structure of minimum energy by means of some type of search procedure. Studies with several different basis sets have shown that geometries in reasonable agreement with experiment may be obtained from single determinant calculations[36, 51, 64, 65]. Comparisons of geometries calculated with the STO–3G, LEMAO–4G and 4–31G basis sets (with standard exponents) have been reported[36, 66, 67]. These basis sets represent three types commonly used in geometry determinations. Overall, the 4–31G basis set is most successful, followed by STO–3G and then LEMAO–4G. Deviations between calculated and experimental values show some systematic behaviour. For the series of hydrides of first row atoms[67] (CH_4, CH_3, CH_2, CH, NH_3, ... etc.), the STO–3G bond lengths are mostly too long and 4–31G values generally too short; the mean absolute deviations between theory and experiment are 0.023 and 0.010 Å, respectively. STO–3G bond angles are systematically too small while 4–31G values are too large; the mean deviations in this case are 5.3 and 4.1 degrees respectively. Valence angles in molecules with lone-pair electrons are generally overestimated by extended sp basis sets and are diminished by addition of polarisation functions[68–70]. For positive ions, STO–3G bond lengths are often overestimated, apparently because the standard molecular scale factors are less appropriate for ions than for neutral molecules.

3.3.4 Internal rotation

Internal rotation about single bonds has been studied extensively using *ab initio* molecular orbital theory. Much of the attention has been directed towards the determination of rotational barriers (i.e. energies required to interconvert structures at local minima in the internal rotation potential function) but the related question of conformational energy differences (i.e. differences in the energies of the structures at local minima) has also been studied. Values of calculated threefold and twofold rotational barriers for neutral molecules are summarised in Table 3.1. Less-symmetrical potential functions cannot be described by a single rotational barrier and are not included. These will be discussed in the appropriate parts of Section 3.4. For those molecules for which experimental data are available, the general agreement is sufficiently encouraging to suggest that internal rotation

Table 3.1 Calculated and experimental rotational barriers (kcal mol^{-1}) for neutral molecules

	Experimental*	Calculated
Threefold barriers		
CH_3—CH_3	2.93	3.3 [25], 3.62[71], 2.52 [72], 2.88 [73], 3.45 [73], 3.5 [74], 3.07 [75], 3.3 [26], 3.33 [77], 2.9 [78], 2.8 [78], 3.26 [79]
CH_3—NH_2	1.98	2.42 [80], 2.02 [73], 2.13 [79]
CH_3—OH	1.07	1.35 [72], 1.59 [73], 1.12 [79]
CH_3—CH_2CH_3	3.33	3.12 [81], 3.45 [77], 3.69 [82], 3.70 [79]
CH_3—$NHCH_3$	3.22	3.62 [79]
CH_3—OCH_3	2.72	2.98 [79]
CH_3—CH_2F	3.30	3.63 [79]
CH_3—NHF		3.04 [79]
CH_3—OF		1.86 [79]
CH_3—$CH_2CH_2CH_3$		2.94 [83], 3.40 [77], 3.63 [82]
CH_3—CH_2CH_2OH		3.49 [82]
CH_3—CH_2CH_2F		3.46 [82]
CH_3—CH_2CH_2CN		3.64 [82]
CH_3—CH_2CHCH_2 (*cis*)	3.99	4.96 [77]
CH_3—CH_2CHCH_2 (*skew*)	3.16	3.46 [77]
CH_3—$CH(CH_3)_2$	3.90	3.88 [77]
CH_3—$CHCH_2$	1.98	1.48 [84], 1.55 [77], 3.6 [85], 1.25 [86]
CH_3—$CHCHCH_3$ (*cis*)	0.73	0.42 [77]
CH_3—$CHCHCH_3$ (*trans*)	1.95	1.54 [77]
CH_3—$C(CH_3)CH_2$	2.21	1.70 [77]
CH_3—$CHCHF$ (*cis*)	1.06	1.07 [86]
CH_3—$CHCHF$ (*trans*)	2.20	1.34 [86]
CH_3—CHO	1.16	1.09 [87]
CH_3—NO	1.10	1.8 [88], 1.05 [89]
CH_3—CO		0.38 [90]
CH_3—$CCCH_3$	<0.03	0.006 [77]
Twofold barriers		
NH_2—CHO	17–21	21.73 [91]
C_6H_5—C_2H_5†	1.3	2.43 [92]
C_6H_5—CH_2F		0.25 [92]
C_6H_5—CHF_2		0.18 [92]
C_6H_5—OH	3.36	5.15 [92]
C_6H_5—$CHCH_2$		4.42 [92]
C_6H_5—CHO	4.90	6.60 [92]
C_6H_5—$COCH_3$	3.1	4.39 [92]
C_6H_5—$COOH$		5.76 [92]
C_6H_5—NO	3.9	4.84 [92]
C_6H_5—NO_2	3±1.5	5.74 [92]

*For references to experimental barriers, see corresponding theoretical papers.
†The (CCCC) orthogonal conformation has lowest energy.

potential functions are reasonably described within the framework of the Hartree–Fock theory.

In many cases, the use of a simple (e.g. minimal) basis set and fixed (e.g. standard or experimental) bond lengths and angles is sufficient to obtain reasonable potential functions. However, a flexible rotor model is desirable when large changes in bond lengths or angles occur during rotation[26, 75, 77, 94]; this effect is particularly noticeable when steric effects are important[77]. Also, a larger basis set which includes polarisation functions has been found necessary to describe adequately the rotation in hydrogen peroxide[26, 75, 93–95]. Further calculations with polarised basis sets are desirable to determine when such basis sets are necessary in studies of internal rotation.

There have been many attempts to explain the 'origin' of rotational barriers. The earlier work has been summarised in a comprehensive review by Lowe[96]. More recently, Sovers et al.[97] have proposed from a bond-orbital analysis that the barrier in ethane is predominantly due to the Pauli repulsion between closed-shell localised C—H bond orbitals. A similar conclusion was reached by Allen[98] who suggested that barriers in general are either of a repulsive type (analogous to the repulsion of two helium atoms) or of an attractive type (analogous to the attraction of two hydrogen atoms). Allen calls the barrier attractive dominant if the change (ΔE_{attr}) in the attractive component (the nuclear–electron attraction term) of the total energy during rotation is greater in magnitude than the change (ΔE_{rep}) in the repulsive component (which is the sum of nuclear–nuclear repulsion, electron–electron repulsion and kinetic energy terms) and repulsive dominant when $\Delta E_{rep} > \Delta E_{attr}$. Although a number of workers have found it useful to describe barriers as attractive dominant or repulsive dominant, this classification has recently been criticised by Veillard[75] who points out that the qualitative conclusions about the (attractive or repulsive) nature of the cis barriers in hydrogen peroxide and hydrogen persulphide are changed by altering the bond lengths and angles or the basis set.

A procedure which has been found useful in analysing internal rotation potential functions $V(\phi)$ is to decompose $V(\phi)$ into Fourier components,

$$V_n(\phi) = \frac{V_n}{2}(1 - \cos n\phi)$$

as in equation (3.6)

$$V(\phi) = \frac{V_1}{2}(1 - \cos \phi) + \frac{V_2}{2}(1 - \cos 2\phi) + \frac{V_3}{2}(1 - \cos 3\phi) \qquad (3.6)$$

It is found[79] that such a decomposition helps to separate various effects which contribute to the potential function. An example of this decomposition is given in Section 3.4.2.1.

3.3.5 Inversion

Inversion processes of two main types may be distinguished. These are (a) *pyramidal inversion* in which invertomers containing a (tri-coordinated) pyramidal atom are interconverted via a planar (or near-planar) transition state and (b) *planar inversion* in which invertomers containing a (di-coordinated) bent atom are interconverted via a linear (or near-linear) transition state. Most examples of inversion for molecules containing the atoms H, C, N, O and F involve the nitrogen atom but inversion may also occur at the appropriately coordinated carbon and oxygen. Some aspects of this subject have been recently reviewed by Lehn[99] and by Rauk, Allen and Mislow[100].

Calculated inversion barriers for organic molecules (and ammonia) are summarised in Table 3.2. These have been obtained with a variety of basis

Table 3.2 Calculated and experimental inversion barriers (kcal mol⁻¹)

Molecule	Experimental*	Calculated
NH_3	5.8	0–26†, 5.1 [68], 5.9 [70]
$(CH_3)_2NH$	4.4	8.6 [109]
NH_2CN	2.03	1.9 [110]
NH_2CHO	1.1	0.1 [91]
$C_6H_5NH_2$	1.6	2.7 [92]
$\overline{CH_2CH_2N}H$	18–21	18.0 [111], 18.3 [112], 15.5 [113]
$\overline{CHCHN}H$		35.1 [113]
$\overline{CH_2ON}H$	31–34	32.4 [112]
CH_2NH	25–27	26.2 [114, 115]
$NHCNH$	6.7	8.4 [115]
CH_2NCO		3.2 [116]
CF_3		27.4 [117]
CH_3^-		5.6 [118], 1.2–9.0 [119, 120]
$\overline{CH_2CH_2C}H^-$		20.9 [121]
$\overline{CHCHC}H^-$		52.3 [122]
CH_3CO		29 [90]
CH_2CH		8 [118]
CH_2CH^-		38.9 [114]
CH_2OH^+		17.2 [123]

*Experimental barriers from reference 99.
†From references 64, 68, 70, 101–108.

sets and in most cases, there is satisfactory agreement with experiment. Calculations by Rauk, Allen and Clementi[68] have shown that a very large basis containing only s- and p-functions is unsuitable for calculating the inversion barrier in ammonia. These workers[68] and Stevens[70] have obtained good values of the inversion barrier using a basis set which includes d-functions on N, p-functions on H and is close to the Hartree–Fock limit.

3.3.6 Enthalpy of reaction

An important objective of quantum chemistry is the quantitative estimation of the relative stabilities of molecules. This goal has often seemed remote for single-determinant molecular orbital theory because of the difficulty in obtaining precise values of total energies due to neglect of correlation effects. In particular, the energy of the simplest reaction of the simplest type of molecule, namely, the dissociation of a diatomic molecule into atoms, is generally poorly given by the theory because of a large change in correlation energy.

This fact has undoubtedly discouraged attempts to tackle this important problem.

An important step forward was made by Snyder[124] who suggested that for reactions in which closed-shell reactants give closed-shell products, there is a reasonable prospect of cancellation of correlation corrections. An extensive set of calculations by Snyder and Basch[125] using an extended basis set has confirmed this prediction.

It has been suggested[126, 127] that even better results might be expected for formal reactions in which the number of bonds *of each type* is conserved but their relationship to one another is changed. Such a reaction is termed *isodesmic*. A particular example of an isodesmic reaction is the *bond separation reaction* in which a molecule with three or more heavy (non-hydrogen) atoms is converted into the simplest parent molecules containing two heavy atoms and the same types of bonds. For example, for acetonitrile (CH_3—$C\equiv N$) the bonds between heavy atoms are C—C and $C\equiv N$ so that the associated parent molecules are CH_3—CH_3 and $HC\equiv N$. The bond separation reaction for CH_3—$C\equiv N$ is then

$$CH_3—C\equiv N + CH_4 \longrightarrow CH_3—CH_3 + HC\equiv N$$

a molecule of methane being added to the left-hand side for stoichiometric balance. The energy change in such a reaction is called the *bond separation energy*.

Calculated bond separation energies for a large set of molecules have been found[128, 129] to be in reasonable agreement with experimental values; the mean absolute errors are c. 3 kcal mol^{-1} for 4–31G and 5 kcal mol^{-1} for STO–3G. The good results with the minimal STO–3G basis set are particularly noteworthy since this basis is most economical to use for larger molecules. Poorest results (for all basis sets) are obtained for small cyclic molecules. In all cases these molecules are predicted to be unstable relative to acyclic molecules with the same types of bonds, i.e. the bond separation energies are always too negative[126, 127].

The bond separation energies are a measure of *interactions* between bonds. (They would all be zero if a simple bond additivity hypothesis were truly valid.) These interactions are discussed in Section 3.4. Another important application of bond separation energies is in estimating total energies or enthalpies of formation of large molecules. If the energies (or enthalpies of formation) of the parent (one and two heavy atom) molecules in bond separation reactions are known either experimentally or from more sophisticated theoretical studies, a calculation of the bond separation energy suffices to predict the energy (or heat of formation) of the general molecule.

3.3.7 Potential surfaces

The results presented in Sections 3.3.3–3.3.6 suggest that under certain conditions, potential surfaces (including reaction paths) can be usefully studied with single-determinant molecular-orbital theory. One requirement is that the number of doubly-occupied molecular orbitals be conserved in the portion of the surface that is under examination. In such cases, the most attractive basis set to use (in terms of accuracy) is a large extended set since this may reliably predict both geometries and energies. Several calculations of this type with complete geometry optimisation (subject only to symmetry restrictions) have been carried out. However, the computation time involved in such calculations often discourages the study of surfaces involving larger molecules and so an alternative procedure is desirable. Caution must be exercised in using a minimal basis set alone since, although the calculated geometries are generally satisfactory, reliable relative energies may only be expected when the number of bonds of each type is conserved. In several studies[67, 78, 130, 131], a two-step process has proved very useful. Firstly, geometries at various points in the surface are calculated using a simple (e.g. STO–3G) basis and then, relative energies are obtained by carrying out single calculations with an extended (e.g. 4–31G) basis set at the previously optimised geometries. This procedure in many cases has been found to give relative energies close to those obtained from the full extended-basis treatment and from experiment. Examples of studies of potential surfaces will be discussed in appropriate parts of Section 3.4.

3.4 ORGANIC MOLECULES

3.4.1 Aliphatic hydrocarbons

3.4.1.1 Alkanes

Ab initio calculations have been reported for methane[32, 36, 51, 57, 65, 67, 76, 125, 127–129, 131, 132, 135–149], ethane[25, 26, 32, 36, 51, 65, 71–75, 77–79, 125, 127–129, 131, 136–140, 145, 146, 149–160], propane[77, 79, 81, 126–129, 131, 140, 149, 161], n-butane[77, 83, 139] and isobutane[77]. Optimised STO–3G geometries are included in Figure 3.1. Rather than quoting geometries obtained with other basis sets for the few molecules where these are available, we have consistently quoted STO–3G values in Figure 3.1 and later on in Figure 3.7 (cations) so that useful comparisons may be made.

The conformations and rotational barriers of the alkanes have received considerable attention. All molecules are predicted to have staggered conformations. The rotational barriers when calculated by a uniform method are found[77] to increase in the series CH_3—CH_3, CH_3—CH_2CH_3, CH_3—$CH(CH_3)_2$. Coupling of the rotation of the two methyl groups in propane has been considered[77, 81] and found to be relatively small. Internal rotation in n-butane is well described by the theory[77, 83].

Propane is found[126, 127] to have a small positive bond separation energy (cf. Section 3.3.6) reflecting the stabilisation of two adjacent C—C bonds. This is the simplest example of stabilisation due to branching in alkanes. In agreement with experiment, propane is found to be slightly polar with the central methylene group at the positive end of the electric dipole[131, 140].

3.4.1.2 Alkenes

Calculations have been reported for ethylene[32, 36, 51, 65, 77, 78, 125, 127–129, 131, 132, 136–140, 146, 149, 162–174], propene[77, 84–86, 126–129, 131, 140], isobutene[77], *cis*-but-2-ene[77, 175, 176], *trans*-but-2-ene[77, 175, 176], but-1-ene[77], buta-1,3-diene[77, 139, 173, 177], allene[77, 126–129, 131, 176, 178–181], buta-1,2-diene[77] and butatriene[77]. The calculations

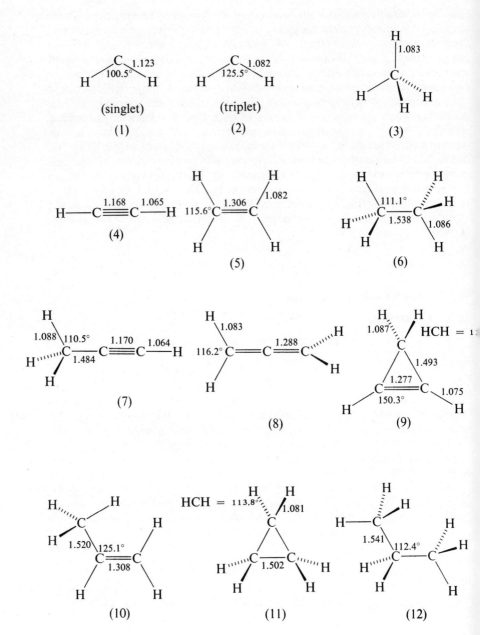

Figure 3.1 STO–3G calculated geometries[65, 78, 131] (bond lengths in Å) for hydrocarbons

confirm that barriers to rotation about the double bonds in ethylene, allene and butatriene are large and decrease with increasing number of cumulated double bonds[77, 162, 164, 178, 180, 181]. Absolute values of these barriers are greatly overestimated because structures such as the perpendicular form of ethylene are not well described by a single-determinant wave function. Much better results are obtained with a configuration-interaction treatment[164, 178].

It has been shown[77, 84-86] that the most stable conformation of a methyl group adjacent to a carbon–carbon double bond in propene, buta-1,2-diene, isobutene, and *cis-* and *trans-*but-2-ene has a methyl C—H eclipsing the double bond. Barriers to rotation are generally slightly lower than experimental values but experimental trends are well reproduced. In particular, the methyl rotational barrier in *trans-*but-2-ene is close to the propene value while the barrier in *cis-*but-2-ene is considerably smaller[77]. This is due to a relative destabilisation of the stable (HCCC both *cis*) conformation of *cis-*but-2-ene because of hydrogen–hydrogen steric repulsions. The steric interaction is also reflected in widened calculated CCC angles (128.0 degrees) in *cis-*but-2-ene compared with *trans-*but-2-ene (124.5 degrees). The gearing of methyl rotors in isobutene and *cis-* and *trans-*but-2-ene has been studied[77] but the coupling terms in the potential energy expression are found to be fairly small. The most stable forms of but-1-ene are found[77] to be *cis* (C—C eclipsing C=C) and *skew* (C—H eclipsing C=C). In agreement with experiment, the *skew* form is favoured. Buta-1,3-diene is predicted[77] to exist as *cis* and *trans* forms; the latter is more stable by 2.05 kcal mol^{-1}. However, the potential minimum at the *cis* position is very flat. The bond separation energy of propene is positive[126, 127], reflecting the hyperconjugative stabilisation of a double bond by an adjacent methyl group. Propene is found[131, 140] to have a small dipole moment which is inclined at a small angle to the double bond with polarity $\overset{+}{C}H_3$ - - - $\overset{-}{C}H_2$. The calculated moment is largely associated with a polarisation of the π-electrons in the double bond, this effect being larger than any transfer of electrons into the π-system.

3.4.1.3 Alkynes

Calculations have been reported for acetylene[23, 32, 36, 51, 65, 78, 113, 125, 127–129, 131, 135–140, 146, 150, 152–155, 163, 169, 174, 176, 182–191], propyne[65, 77, 113, 126–129, 131, 140, 146, 192, 193], but-1-yne[77], but-2-yne[77] and vinyl acetylene[77]. A point of structural interest is that the reduction in the length of the carbon–carbon single bonds in the series propane, propene and propyne that is found experimentally (1.526, 1.501, 1.459 Å) is reproduced by the theory[65, 131] (1.541, 1.520, 1.484 Å, see Figure 3.1). The calculated rotational barrier in but-2-yne (dimethylacetylene), where two methyl groups are separated by an acetylenic linkage, is close to zero[77]. As in propene, the calculated dipole moment of propyne is largely associated with polarisation of π-electrons in the triple bond[131, 140, 192]. The positive end of the dipole is on the methyl group.

3.4.1.4 Methylene, trimethylene

There have been several studies of methylene (CH_2), the simplest carbene[47, 67, 78, 113, 137, 194–200]. These all support the most recent experimental evidence for bent structures for both singlet ((1), $^1A_1$) and triplet ((2), $^3B_1$) states. Calculated STO–3G geometries are included in Figure 3.1; other theoretical geometries are in general agreement with these results but give a slightly wider HCH angle in the triplet (in better agreement with the experimental value, 136 degrees). Configuration-interaction calculations are required to obtain reasonable estimates (22 and 19 kcal mol^{-1}) of the triplet–singlet energy difference[199, 200].

Calculations have been carried out[201–204] for various possible structures of trimethylene ($CH_2CH_2CH_2$) including the (0,0) form in which the terminal methylene groups lie in the CCC plane and the (90,90) form in which the terminal methylene groups are perpendicular to the CCC plane. Stable structures for both singlet and triplet states of the (0,0) form are found[203]. These have approximately equal energies in a configuration-interaction treatment. Firm conclusions regarding the (90,90) form will have to await calculations in which both geometrical optimisation and interaction of configurations are taken into account[203].

3.4.1.5 Alicyclic hydrocarbons

Calculations have been reported for cyclopropane[113, 125–127, 131, 146, 154, 155, 176, 201, 205–212], cyclopropene[113, 125–127, 193, 207, 208, 210, 212–214], cyclobutane[215], bicyclobutane[216], cyclobutene[214, 217, 218], cyclobutadiene[219, 220], tetrahedrane[220], cyclopentane[176, 221], cyclohexane[176, 221] and cubane[176]. The two C_4H_4 isomers, tetrahedrane and cyclobutadiene have been examined with partial geometrical optimisation[219, 220]. Single-determinant calculations show cyclobutadiene to be more stable than tetrahedrane by 35 kcal mol^{-1} but because cyclobutadiene is not a closed-shell species, this estimate is probably low; a configuration-interaction calculation[220] gives an energy difference of 70 kcal mol^{-1}.

Calculations with assumed geometries on several conformations of cyclopentane and cyclohexane have produced relative stabilities in close agreement with experiment[221]. For cyclopentane, calculated relative energies (in kcal mol^{-1}) for structures of different symmetry are as follows: half chair (C_2 symmetry, 0), envelope (C_s, 0.24), and planar (D_{5h}, 8.07). For cyclohexane, the values are chair (D_{3d} symmetry, 0), twisted boat (D_2, 6.05, expt. 4.8–5.9), boat (C_{2v}, 7.19) and transition state between boat and chair (C_2, 11.22, expt. 10.8).

Bond separation energies for cyclopropane and cyclopropene have been found[126, 127] to be negative and reflect the strain energy in these small ring systems. Several authors have examined electron densities in cyclopropane and found that the maxima in the C—C bonds lie outside the internuclear axis[210, 211]. This is consistent with previous predictions of 'bent bonds' in these highly strained molecules.

Cyclobutene has been studied[217, 218] with particular emphasis on the electrocyclic transformation of cyclobutene to cis-butadiene. A limited amount of geometrical optimisation has been carried out. In agreement with the prediction from Woodward and Hoffmann's orbital symmetry rules, the preferred mode of transformation is found to be conrotatory. The non-linear variation of the individual geometric parameters during the transformation emphasises the importance of geometrical optimisation in order to determine the minimum energy path.

The dipole moment direction in cyclopropene has been predicted from several calculations to be about 0.4 D with the negative end of the dipole on the methylene group[131, 207, 213]. This is the reverse of the experimental direction[222].

3.4.2 Heteroaliphatic and heterocyclic molecules

Calculations on a large number of heteroaliphatic molecules have been reported. Many of these (containing three heavy atoms) are listed in Table 3.3

Table 3.3   Calculated (4–31G) bond-separation energies (kcal mol^{-1})*†

Molecule‡		Molecule	
$CH_3-CH_2-CH_3$§	1.2	$HO-CH_2-OH$	15.2
$CH_3-CH=CH_2$§	3.9	CH_3-O-OH	6.0
$CH_3-C\equiv CH$§	9.0	$HO-CH=O$ [123, 125, 140, 231–234]	30.1
$CH_2=C=CH_2$§	−2.5	$O=C=O$ [23, 65, 125–127, 140, 182, 186, 235–237]	52.5
$CH_3-CH_2-NH_2$	3.6	CH_3-CH_2-F [140]	6.1
$CH_3-NH-CH_3$ [126, 127, 140, 223]	2.5	$CH_2=CH-F$ [140, 172]	5.4
$CH_2=CH-NH_2$	13.3	$CH\equiv C-F$ [140, 176, 186, 191, 237]	−11.6
$CH_3-CH=NH$	8.7	$F-CH_2-F$ [65, 126, 127, 140, 141, 143, 147, 227]	11.5
$CH_2=N-CH_3$	3.1	NH_2-CH_2-OH	13.0
$CH_3-C\equiv N$ [113, 126, 127, 140, 224, 225]	13.2	$CH_3-NH-OH$	5.4
$CH\equiv C-NH_2$	13.5	CH_3-O-NH_2	3.1
$CH_2=C=NH$	−0.3	$NH_2-CH=O$ [45, 88, 91, 125–127, 140, 232, 238–241]	35.9
$NH_2-CH_2-NH_2$	8.3	$NH=CH-OH$ [88]	21.8
$CH_3-NH-NH_2$	4.1	$CH_2=N-OH$ [88]	14.6
$NH_2-CH=NH$	25.6	$CH_3-N=O$ [88, 89]	13.0
$CH_3-N=NH$	8.8	$HN=C=O$ [242]	26.8
$CH_2=N-NH_2$	7.7	$HO-C\equiv N$	−1.1
$NH_2-C\equiv N$ [110, 226]	17.8	NH_2-CH_2-F	14.7
$NH=C=NH$ [115]	5.6	CH_3-NH-F	7.1
CH_3-CH_2-OH [145]	5.4	$NH=CH-F$	12.9
CH_3-O-CH_3 [126, 127, 140, 223]	2.3	$CH_2=N-F$	14.5
$CH_3-CH=O$ [87, 123, 126, 127, 140, 160, 228]	10.4	$F-C\equiv N$ [65, 186, 224, 237, 243]	−16.9
$CH_2=CH-OH$	8.2	$HO-CH_2-F$ [244]	13.3
$CH_2=C=O$ [113, 126, 127, 140, 179, 230]	13.6	CH_3-O-F	8.8
$CH\equiv C-OH$	−0.2	$O=CH-F$ [125, 140, 232, 245]	19.2

*Taken from reference 128.
†STO–3G calculations for this set of molecules are presented in reference 129.
‡References against individual molecules are to additional calculations.
§Hydrocarbons included for completeness.

which will be discussed below. In addition, calculations have been reported for CH_3NH_2 [73, 80, 127–129, 140, 157, 223], $(CH_3)_3N$ [140, 223], CH_3OH[72, 73, 127–129, 140, 145, 157, 223, 246], $CH_3CH_2CH_2OH$[82], CH_3F [32, 36, 51, 127–129, 136, 139–141, 143, 147, 154, 157, 225, 362–364], CHF_3 [65, 140, 141, 143, 147], CF_4 [32, 65, 126, 127, 140, 143], $CH_3CH_2CH_2F$ [82], CH_3CHF_2 [92], CH_3CF_3 [92, 140], CF_3CF_3[139], CHF[199], CF_2 [199, 247], CH_2NH[114, 115, 128, 129, 248–250], CH_2O [28, 32, 36, 51, 65, 123, 125, 127–129, 136, 139, 140, 146, 167, 169, 190, 251–258], $CH_2O\cdot H_2O$ [229], CH_3COCH_3 [92, 140], CH_3COOH [92], $CHOCHO$ [259], $(HCOOH)_2$ [260], *cis*-CH_3CHCHF [86], *trans*-CH_3CHCHF [86], CH_2NHO [88], CH_3NO_2 [92, 140],

CH_3CONH_2 [92], $(NH_2CHO)_2$ [240, 262], F_2CO [140], CH_3CFO [92], HCN [32, 36, 51, 65, 113, 125, 127–129, 136, 140, 146, 150, 169, 183, 186, 237, 242, 243, 249, 252, 255, 263–268], CH_3NC [224], CH_2=CHCN [269], CH_2=CHNC [224], $CH_3CH_2CH_2CN$ [82], NCCN [183, 186, 224, 243, 264, 270], NCCCH [186, 224, 237], CH_2NN [125, 179], CO [32, 113, 125, 132, 136, 271–275], CF_3CCH [140], OCCCO [276]. Heterocyclic molecules are listed in Sections 3.4.2.4 and 3.4.2.5.

3.4.2.1   Saturated molecules

The conformational preferences of the saturated molecules in Table 3.3 have been rationalised[79, 128] in terms of *contributions* from three principal effects:

(a) Staggered arrangements of bonds are preferred.

(b) The axis of a lone-pair orbital prefers to be co-planar with an adjacent electron-withdrawing polar bond or orthogonal to the axis of an adjacent lone-pair orbital.

(c) Dipole-moment components perpendicular to the internal rotation axis prefer to be opposed.

Both (a) and (c) are well known. The second effect (b) warrants further discussion. Essentially, it reflects the tendency for lone-pair electrons to delocalise. In the case of a lone pair adjacent to an electron-withdrawing polar bond, the most effective arrangements for delocalisation are shown in Figure 3.2. Electron withdrawal in the C—X bond decreases the occupancy of the

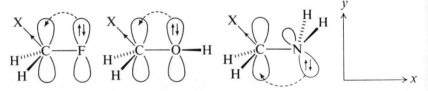

Figure 3.2   Electron donation from fluorine, hydroxyl or amino lone pairs into C—X bond

carbon $2p_y$ orbital which is then available to accept more electrons from the appropriate lone-pair orbital of F, OH or NH_2. It should be noted that for an OH substituent, electrons are most easily donated from the 2p-type lone-pair orbital on O that is perpendicular to the COH plane, hence the arrangement shown in Figure 3.2. It is easily seen that for adjacent lone-pair orbitals, delocalisation is most effective in orthogonal arrangements.

The three effects above are usefully separated by breaking down the potential function into Fourier components (cf. Section 3.3.4). An example of such a decomposition for fluoromethanol (FCH_2OH) is shown in Figure 3.3. The V_3 component reflects the tendency for staggered conformations to be preferred to eclipsed conformations (by 0.96 kcal mol^{-1}) as in (a). V_3 (ϕ) is similar to the overall potential function for methanol (where $V_3 = 1.12$ kcal mol^{-1}, $V_1 = V_2 = 0$). The large V_2 term shows a preference for an orthogonal conformation (13) over the (FCOH) *cis* (14) and *trans* (15) forms because of stabilisation through electron donation from the oxygen lone pair into the electron-withdrawing C—F bond (as in (b)). Finally, the large V_1 term shows the preference for the *cis* over the *trans* form, a result easily rationalised in terms of dipole interactions (cf. (c)). The net result of these effects is to produce a minimum in the overall $V(\phi)$ at $\phi = 55$ degrees; the expected additional *trans* minimum has disappeared due to the unfavourable effects associated with the V_1 and V_2 terms at $\phi = 180$ degrees. This result has also been obtained in independent calculations by Wolfe *et al.*[244]. An important point that

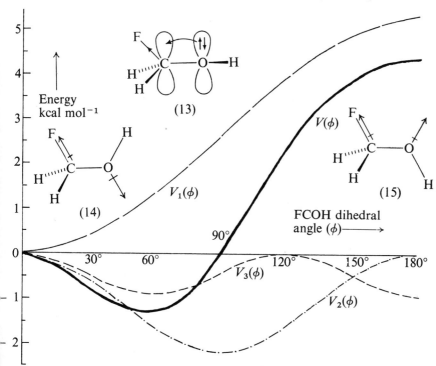

Figure 3.3 Fourier decomposition of internal rotation potential function for fluoromethanol (FCH$_2$OH)

emerges from these and related results is that the tendency for staggering of bonds (V_3 term of the potential function) is fairly small (e.g. it amounts to only *c.* 1 kcal mol^{-1} in alcohols). It is thus easily submerged in other effects

such as dipole–dipole interactions, electron delocalisation and, in larger molecules, steric effects; this is indeed the case for fluoromethanol.

Other conformational determinations[128, 129] include ethanol, where the more stable conformation has CCOH *trans* (16), ethylamine, where CCN: is *gauche* (17) and fluoromethylamine, where FCN: is *trans* (18). In methane diol, there are *gauche* arrangements about both C—O bonds (19); this result and that for fluoromethanol are relevant to the anomeric effect in carbohydrate chemistry[244].

The positive values of bond separation energies for all the saturated molecules in Table 3.3 indicate stabilising interactions between the bonds. For the molecules X—CH_2—X' where X and X' are CH_3, NH_2, OH and F, the results may be rationalised in terms of electron transfer of the type

$$X\overset{\frown}{\quad}CH_2\overset{\frown}{\quad}X'$$

proceeding via a π-type donation and a σ-type acceptance as shown in Figure 3.2. The interaction therefore depends on both the π-electron donating properties of X' and the σ-electron accepting properties of X. Thus, when X' is a strong π-electron donor (e.g. NH_2 or OH) and X a strong σ-acceptor (e.g. OH or F), the bond separation energy is large (e.g. FCH_2NH_2 or $HOCH_2OH$).

3.4.2.2   Molecules with one multiple bond

Rotational isomerism is possible in several of these molecules. For molecules HOCH=Y, the (HOCY) *cis* form (20) is favoured over the *trans* form when Y is CH_2, NH or O. In particular, the *cis* form of formic acid (where Y is O) is favoured over the *trans* in agreement with experiment; estimates[123, 128, 129, 233, 234] of the calculated energy difference range from 2.4 to 9.5 kcal mol^{-1}.

(20)   (21)   (22)   (23)

The barrier to interconversion of these forms has been calculated[234] to be 13.0 kcal mol^{-1}. Formaldoxime (CH_2=NOH) is found[88, 128, 129] to have CNOH *trans* (21) in agreement with experiment. For the substituted imines XCH=NH, the (HCNH) *syn* form (22) is more stable when X is CH_3, NH_2 or OH while the *anti* form (23) is favoured when X is F. The *trans* form of methyl di-imide is favoured[128, 129] over the *cis* form by 13.1 kcal mol^{-1}. *Trans*-glyoxal is predicted[259] to be 6.4 kcal mol^{-1} more stable than *cis*-glyoxal. Molecules with a methyl group adjacent to a double bond are all found to have most stable conformations in which a C—H bond eclipses the double bond[77, 84–89].

Calculated bond-separation energies (Table 3.3) for the molecules XCH=Y (X = CH_3,NH_2,OH,F; Y = CH_2,NH,O) may be rationalised in terms of the π-electron transfer

$$X\overset{\frown}{\quad}CH\overset{\frown}{=\!=}Y$$

the ability of X to donate π electrons and the ability of Y to accept them.

Highest values of the bond-separation energy (corresponding to greatest stabilisation) are found when X is a strong π-donor (e.g. NH_2) and Y a strong π-acceptor (e.g. O).

For triply bonded molecules $X—C{\equiv}Z$ (Z is CH or N), the σ-electron withdrawing or donating properties of X play a major role in determining the bond interactions. Electron withdrawal from the $—C{\equiv}CH$ and $—C{\equiv}N$ groups is found to be unfavourable with the result that negative bond separation energies are calculated for $HOC{\equiv}CH$, $FC{\equiv}CH$, $HOC{\equiv}N$ and $FC{\equiv}N$.

The theory may be used to estimate relative energies of isomeric species not easily determined experimentally. For example, the enol form (vinyl alcohol, $CH_2{=}CHOH$) of acetaldehyde is calculated[128] to be 12.9 kcal mol^{-1} less stable than the keto form (CH_3CHO). Calculations on the various isomers of empirical formula CH_3NO have been carried out by Robb and Csizmadia[88] who find relative stabilities in the order:

$$NH_2—CH{=}O > NH{=}CH—OH > CH_2{=}N—OH > CH_3—N{=}O >$$
$$CH_2{=}\overset{+}{N}H—\bar{O} > \overline{CH_2—NH—O}$$

Their relative energies for the first four members of the series are in reasonable agreement with other calculated values[128].

3.4.2.3 Molecules with cumulated double bonds

The calculated bond separation energies in Table 3.3 show that the interaction of cumulated double bonds in molecules $Y{=}C{=}Y'$ ranges from slightly destabilising when Y and Y' are both CH_2 to strongly stabilising when Y and Y' are both O and extensive three-centre delocalisation is possible. Several examples of the isomerisation of molecules with cumulated double bonds to molecules with single and triple bonds have been examined. The singly and triply bonded species are found[128] to be more stable except when a $C{=}O$ bond is one of the cumulated double bonds. For example, $CH_2{=}C{=}O$ is 33.4 kcal mol^{-1} more stable than $CH{\equiv}C—OH$ while $CH_3—C{\equiv}N$ is 44.6 kcal mol^{-1} more stable than $CH_2{=}C{=}NH$.

3.4.2.4 Small (three- and four-membered) heterocyclic molecules

Calculations have been reported for aziridine[111-113, 125, 208-210, 212, 248, 250], oxirane[113, 125, 208-210, 212, 277], oxaziridine[88, 112, 248, 250, 261], diaziridine[125, 209, 261], 2-azirene[113, 208], 1-azirene[113, 208], oxirene[113, 208], diazirene[207, 210, 213], difluoro-diazirene[213, 278], cyclopropenone[279], dioxiranone[280], cyclobutanone[281], β-propiolactone[281], malonic anhydride[281], 2-amino-4-methyloxetene[281] and 2-amino-4-methyleneazetine[281]. The angle (ϕ) between the N—H bond and the ring plane has been determined for aziridine[112, 113] ($\phi = 64,61$ degrees), 2-azirene[113] ($\phi = 68$ degrees) and oxaziridine[112] ($\phi = 67.5$ degrees). These high values demonstrate the increased pyramidal nature of the bonds at nitrogen in a strained ring. There is a corresponding increase in the calculated inversion barriers for these molecules (see Table 3.2).

Clark[113, 208, 279] has studied a number of the three-membered rings with particular emphasis on their aromatic or anti-aromatic nature. Examples of anti-aromatic species (four π electrons) are oxirene and planar 2-azirene. The result[113] of this anti-aromaticity in 2-azirene is that the N—H bond bends out of the ring plane by 68 degrees with a large inversion barrier of 35.4 kcal mol^{-1}. For comparison, the inversion barrier in aziridine is calculated[112, 113] to be 15.5 or 18.3 kcal mol^{-1}. 2-Azirene is found[113] to be 27.0 kcal mol^{-1} less stable than 1-azirene.

3.4.2.5   Larger heterocyclic molecules

Calculations have been reported for pyrrole (24)[34, 176, 239, 282], furan (25)[176, 283], imidazole (26)[284], pyrazole (27)[284], 1,2,4-triazole (28)[284], oxazole (29)[284], isoxazole (30)[284], urazole (31)[282], pyridine (32)[285, 286], pyrazine (33)[286–288], thymine (34)[239, 289–291], cytosine (35)[239, 289, 290], adenine (36)[239, 289, 290], guanine (37)[289] and the guanine–cytosine base pair[260].

There has been considerable interest in the π-electron distributions in these molecules and they are therefore summarised in Figure 3.4. In order to make comparison of values more meaningful, all the quoted results refer to calculations[34, 282–285, 287, 289] with the same or (in the case of furan) very similar basis sets. However, similar electron distributions are in fact obtained for cytosine, thymine and adenine with other basis sets[239, 291]. Calculated π-overlap populations are useful as a measure of the degree of double-bond character. The values for furan are therefore included in Figure 3.4 as an example. These show that there is more localisation of the double bonds (high π-overlap populations) in furan than in benzene (where there is complete delocalisation of the π-electron distribution).

Clementi[292] has studied the pyridine positive ion and finds considerable redistribution of both σ- and π-electrons compared with pyridine. He has therefore suggested that σ–π separability may not always be valid and that the π-electron approximation should therefore be used with reservation. Similar conclusions have been reached by Kramling and Wagner[282] from their calculations on pyrrole, urazole and their ions.

3.4.3   Aromatic hydrocarbons and related molecules

3.4.3.1   The C_6H_6 and C_8H_{10} isomers

Calculations have been reported for the C_6H_6 isomers benzene (38)[32, 65, 92, 139, 140, 149, 173, 176, 211, 293–300], fulvene (39)[298, 300], 2,3-dimethylenecyclobutene (40)[298, 300], trimethylenecyclopropane (41)[300] and Dewar benzene (42)[300], and C_8H_{10} isomers naphthalene (43)[300–302], azulene (44)[300, 302] and fulvalene (45)[300]. Optimised geometries have been obtained for benzene[65, 296] giving a C—C distance of 1.39 Å (experimental 1.397 Å).

Calculated orders of stability (with assumed geometries) are[298, 300] benzene > fulvene > 2,3-dimethylenecyclobutene > trimethylenecyclopropane > Dewar benzene and[300, 302] naphthalene > azulene > fulvalene. Absolute values

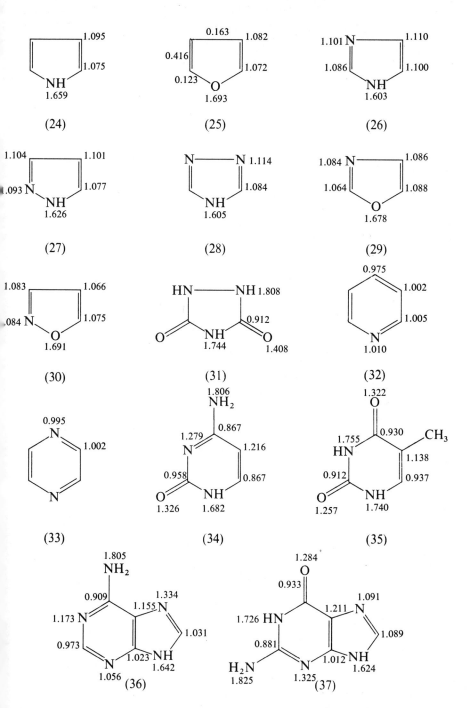

Figure 3.4 Theoretical π-electron populations and π-overlap populations (for furan, (35), only) for heterocyclic molecules

of the energy differences are over-estimated, particularly in the molecular fragment study[300] but the qualitative conclusions may be meaningful.

Fulvene has a small calculated dipole moment ($\mu_{calc} = 0.97$ D, $\mu_{expt} = 1.1$ D) with its positive end on the methylene group[298]. Dimethylene-cyclobutene also has a small calculated moment ($\mu_{calc} = 0.56$ D, $\mu_{expt} = 0.62$ D) but with its positive end in the ring[298]. The main contribution to the

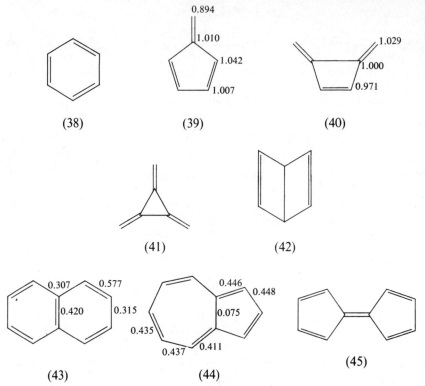

Figure 3.5 C_6H_6 and C_8H_{10} isomers. Theoretical π-electron populations shown for (39) and (40) and π-overlap populations shown for (43) and (44)

dipole moment in these molecules arises from the π-electrons (Figure 3.5). The π-overlap populations for naphthalene and azulene (included in Figure 3.5) have been reported[302] and correlate qualitatively with bond lengths.

3.4.3.2   Substituted benzenes

Calculations have been carried out[92] for a large set of monosubstituted benzenes (C_6H_5—X, X = H, CH_3 [140], CH_2CH_3, CH_2NH_2, CH_2OH, CH_2F, CHF_2, CF_3, NH_2, $NHCH_3$, $NHNH_2$, $NHOH$, NHF, OH, OCH_3, ONH_2, OOH, OF, F [140], $CHCH_2$, CHNH, CHO, $COCH_3$, $CONH_2$, COOH, COF, NCH_2, NNH, NO, NO_2, CCH, CN and NC). All the calculations have used assumed values of bond lengths and angles.

The conformations of these molecules are determined largely by conjugation effects. Thus, molecules such as phenol, benzaldehyde and nitrobenzene are found to be planar so as to maximise charge delocalisation. When the conformations most favoured by conjugation involve steric interactions (as, for example, in anisole), bond angle or torsional distortions occur. Calculated rotational barriers are higher than values for corresponding aliphatic molecules. This is consistent with the partial double-bond character in the C—X bond which is reflected in the positive π-overlap populations. Absolute values of the rotational barriers (Table 3.1) are overestimated. Aniline is found to be non-planar. The HNH plane is bent by about 48 degrees from the aromatic plane with an inversion barrier of 2.7 kcal mol^{-1}, somewhat higher than experiment.

The interactions of the side chain with the ring have been examined by calculating energy changes for the formal reaction (3.7)

$$C_6H_5X + CH_4 \longrightarrow C_6H_6 + CH_3X \qquad (3.7)$$

The energy change in reaction (3.7) compares the effect of the substituent X on the stability of benzene with its effect on the stability of methane. Positive values are obtained for the most stable conformations of all the substituted benzenes indicating that the phenyl group is stabilising compared with methyl in these cases. Large values are obtained for phenol, aniline and fluorobenzene in agreement with experiment.

Absolute values of calculated electric dipole moments of substituted benzenes are generally too low but most experimental trends are reproduced. In particular, calculated and experimental mesomeric moments are in close agreement. For example, comparison of calculated moments for nitrobenzene (4.26 D) and nitromethane (3.33 D) reveal a mesomeric component in the $\overset{+}{C}$—$\overset{-}{N}$ direction of about 0.9 D which is close to the experimental value.

The theoretical π-electron distributions of some of the molecules are shown in Figure 3.6. These results support many of the qualitative ideas of classical organic chemistry. For example, substituents such as OH and NH$_2$ lead to an increased number of π-electrons at the *ortho* and *para* positions while substituents such as NO$_2$ and CN lead to decreased π-electron densities at these positions. However, these effects are only partly due to actual *transfer* of π-electrons from the substituent into the aromatic ring (implied by the contributing structures which are commonly drawn). *Polarisation* of the π-electrons within the ring is an important additional factor.

The calculated π-electron distributions are also interesting in some of the orthogonal conformations. These may be tested experimentally with model compounds. For example, the charge effects of planar nitrosobenzene are reversed in the orthogonal form. Thus, whereas the NO substituent withdraws π-electrons from the ring in planar nitrosobenzene, there is π-charge donation from the nitrogen lone pair in the orthogonal form and the charge alternation is reversed (increased π-electron densities at the *ortho* and *para* positions).

Calculations have been carried out on the related aromatic species, the dehydrobenzenes (or benzynes)[303, 304]. However, these studies indicate[304] that a single determinant wave function does not give an adequate description of the singlet states of these molecules.

3.4.4   Cations, radicals and anions

Calculations have been reported for the hydrocarbon cations CH^+ [67, 78, 136, 305, 306], CH_2^+ [67, 78], CH_3^+ [67, 78, 118, 120, 136, 139, 176, 307−315], CH_4^+ [67, 78, 315−317], CH_5^+ [67, 78, 318−325], C_2^+ [78, 326], C_2H^+ [78, 334], $C_2H_2^+$ [78, 188], $C_2H_3^+$ [78, 314], $C_2H_4^+$ [78], $C_2H_5^+$ [78, 176, 314, 327−332], $C_2H_6^+$ [78, 156], $C_2H_6^{2+}$ [333], $C_2H_7^+$ [78, 322, 334], C_3H^+ [335], $C_3H_3^+$ [113, 122, 335], $C_3H_5^+$ [121, 336−338], $C_3H_7^+$ [82, 130, 206, 339−341], $CH_3CH_2CH_2CH_2^+$ [82], $(CH_3)_2CHCH_2^+$ [82], $(CH_3)_3CCH_2^+$ [82], $\overline{CH_2CH_2CHCH_2^+}$ [82], $\overline{CH_2CH_2C(CH_3)CH_2^+}$ [82],

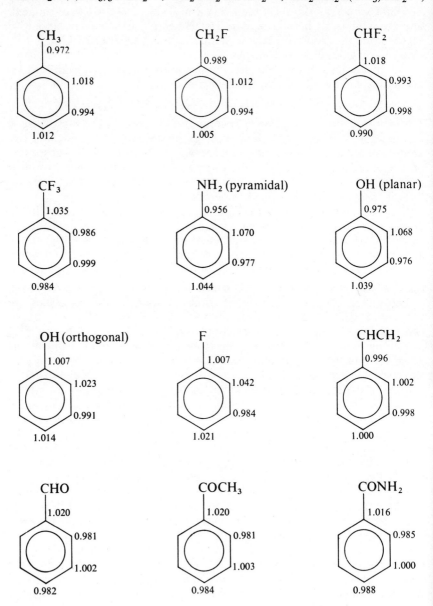

$\overline{CH_2CH_2CH_2CHCH_2^+}$ [82], $\overline{CH_2CH_2CH_2C(CH_3)CH_2^+}$ [82], $C_6H_6^+$ [294] and $C_6H_7^+$ [176]. Other cations that have been studied include $FCH_2CH_2^+$ [339, 342–344], CH_3CHF^+ [342, 343], $CH_2FCH_2CH_2^+$ [82], $HOCH_2CH_2CH_2^+$ [82], $NCCH_2CH_2$ CH_2^+ [82], CO^+ [345], HCO^+ [255, 271, 274, 275], COH^+ [271, 274, 275], H_2COH^+ [123, 255, 346], CH_3O^+ [346], $CH_3OH_2^+$ [223], CH_3CHOH^+ [123], $HOCHOH^+$ [123, 233], H_2OCHO^+ [123, 233], $(CH_3)_2OH^+$ [223], CO_2^+ [347], H_2CN^+ [249], $H_2CNH_2^+$ [249], $CH_3NH_3^+$ [223], $(CH_3)_2NH_2^+$ [223], $(CH_3)_3NH^+$ [223], $\overline{CH_2CH_2NH_2^+}$ [113, 210], $\overline{CH_2CHNH^+}$ [113], $\overline{CHCHNH_2^+}$ [113], $\overline{NHCH_2NH_2^+}$ [261], $\overline{CH_2ONH_2^+}$ [261], pyrrole cation [282], urazole cation [282] and the pyridine cation [292].

Calculations have been reported for the radicals CH [67, 78, 132, 137], CH_3 [67, 78, 117, 118, 137, 307, 315, 348], CH_5 [67, 78], C_2H [78, 137], C_2H_3 [78, 118, 137, 349], C_2H_5 [78, 137], C_7H_7 [176, 350], HCO [351], CH_3CO [90], CN [352], H_2CN [351], CF_3 [117], CH_2NO [349] and CH_2NCO [116].

The anions that have been studied include CH^- [353], CH_3^- [118–120, 136, 145, 307, 312, 318, 354, 355], CH_5^- [320, 324, 355, 356], $C_2H_2^-$ [188], $C_2H_3^-$ [114, 248, 250], $C_2H_5^-$ [145],

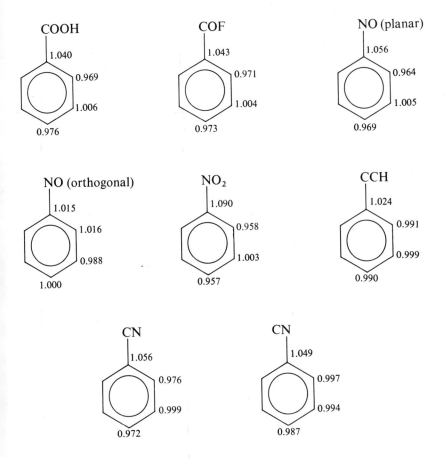

Figure 3.6 Theoretical (STO–3G) π-electron populations for substituted benzenes

$C_3H_3^-$ [113,122], $C_3H_5^-$ [121,357], $C_5H_5^-$ [18,176], $HCOO^-$ [231–233,358], CH_3O^- [145,223], $C_2H_5O^-$ [145], CN^- [224,225,242,243,255,359–361], HCN^- [351], H_2CN^- [249], CH_3N H^- [223], $(CH_3)_2N^-$ [223], NCO^- [186,224,237,242], FCH_3F^- [225,363,364], FCH_3CN^- [225], $\overline{CHNCH}^-$ [113], $\overline{CHCCO}^-$ [279], pyrrole anion[282] and the urazole anion[282].

3.4.4.1 Structures and relative stabilities of hydrocarbon cations

Figure 3.7 summarises some of the more interesting aspects of optimised geometries for hydrocarbon cations obtained with the STO–3G basis set. For those cations for which structural isomers have been examined, the relative energies calculated with the 4–31G basis at the STO–3G optimised geometries are also included.

The CH_4^+ cation is interesting since it illustrates the Jahn–Teller distortion that accompanies the removal of an electron from methane. A D_{2d} structure (50) is found to be most stable[67,78,315–317]. The various structures of CH_5^+ have received a great deal of attention since these are models for the possible intermediates in electrophilic substitution reactions at a saturated carbon atom. Lowest energies are found[67,78,320–324] for a C_s structure (53) but the energies of other forms are not much higher. The C_s form resembles a methyl cation bonded to a hydrogen molecule.

The acetylene (57) and ethylene (60) cations have structures of the same symmetry as the neutral molecules but with some lengthening of the bonds[78]. The vinyl cation is predicted to be planar with a linear CCH linkage[78,314]. Open (classical) forms of both the vinyl (58) and ethyl (61) cations are favoured over the respective bridge forms (59, 62)[78,314,328,329,331]. The relative energies of the latter represent the barriers to 1,2-hydride shifts. $C_2H_7^+$ is another cation which is relevant to recent experimental work on electrophilic substitution in alkanes. The most stable form of this cation is found[334] to be the H-bridged D_{3d} form (64).

A linear structure (66) for C_3H^+ is preferred[335]. The most stable form of $C_3H_3^+$ is the aromatic cyclopropenyl cation (68) which is favoured by 15.3 kcal mol^{-1} over the propargyl cation (69)[335]. The theory predicts[338] that there are three local minima in the $C_3H_5^+$ surface. These are the allyl cation (70), the isopropenyl cation (71) and the 1-propenyl cation (72). The calculations indicate that the cyclopropyl cation (74) is not a local minimum and that it can be converted to the allyl cation via a disrotatory transformation without activation energy[336,338]. Clark and Armstrong[336] have shown using assumed bond lengths and bond angles that the disrotatory mode is preferred to the conrotatory mode in agreement with the Woodward–Hoffman rules. Subsequent studies[338] with complete geometrical optimisation have confirmed this qualitative result but indicate that the geometric parameters do not vary linearly along the reaction coordinate so that geometrical optimisation along this path is important. In particular, the central C—H bond lies in the CCC plane for both the allyl and cyclopropyl cations but is strongly bent out of this plane for the structures along the disrotatory path. The isopropenyl cation (71) is predicted to have an energy only 15.5 kcal mol^{-1} higher than the allyl cation. The activation energy for the conversion

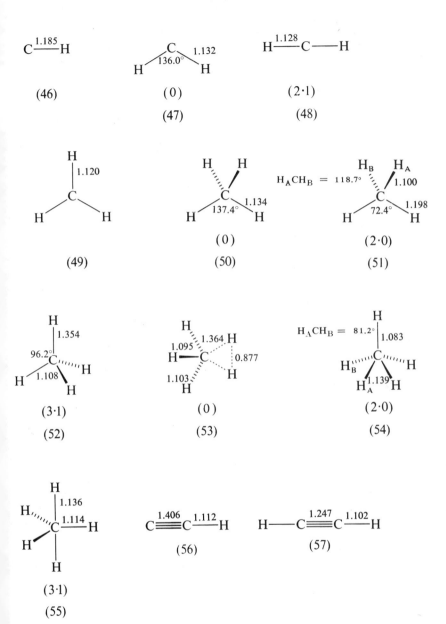

Figure 3.7 STO–3G calculated geometries[67, 78, 334, 335, 338, 341] (bond lengths in Å) and 4–31G calculated relative energies (in kcal mol^{-1}) for hydrocarbon cations (positive charges omitted in diagrams)

100

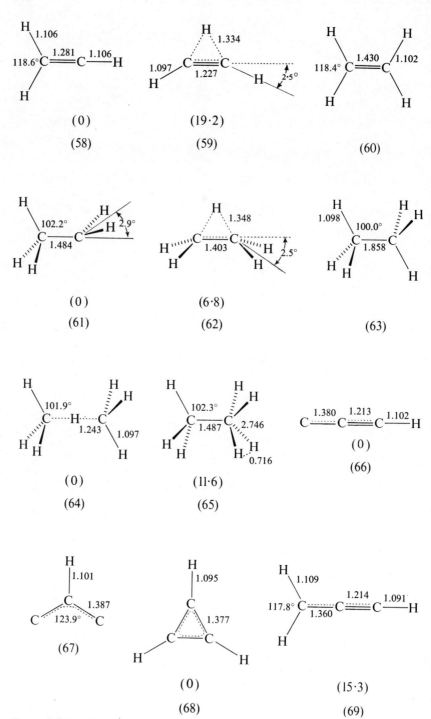

Figure 3.7 (continued)

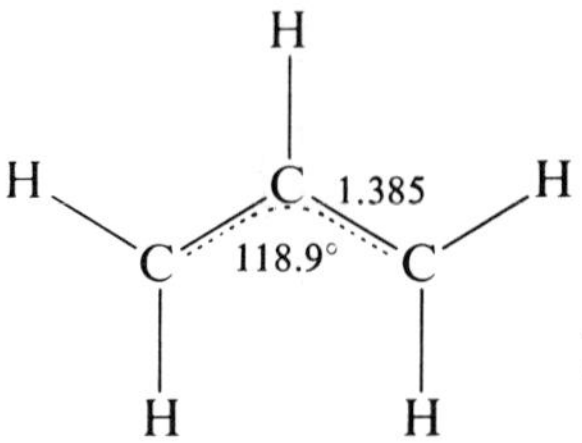

(0)

(70)

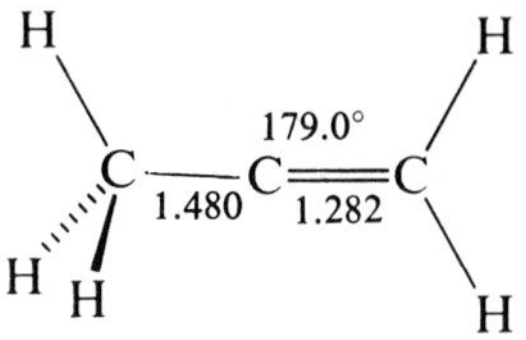

(15·5)

(71)

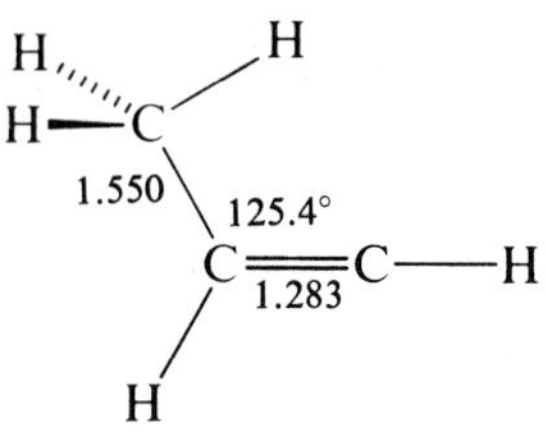

(31·4)

(72)

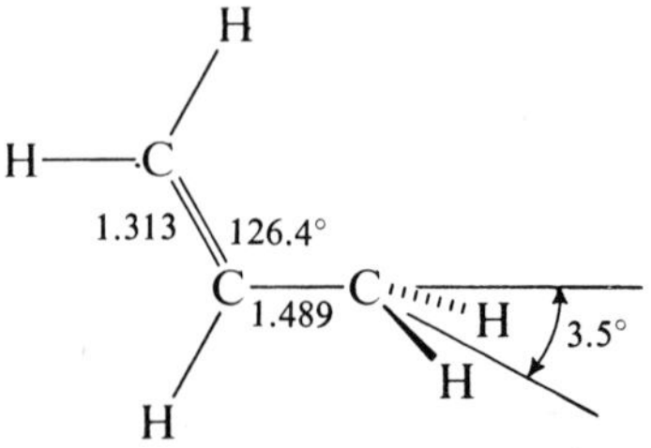

(35·0)

(73)

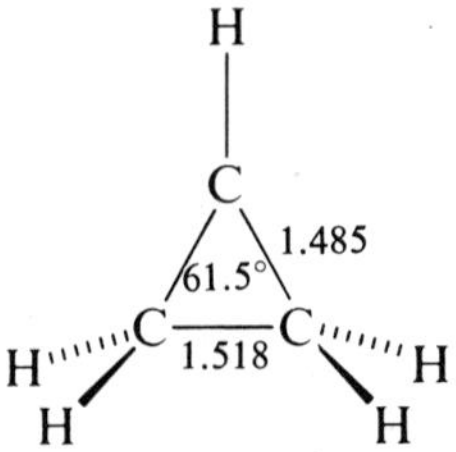

(46·5)

(74)

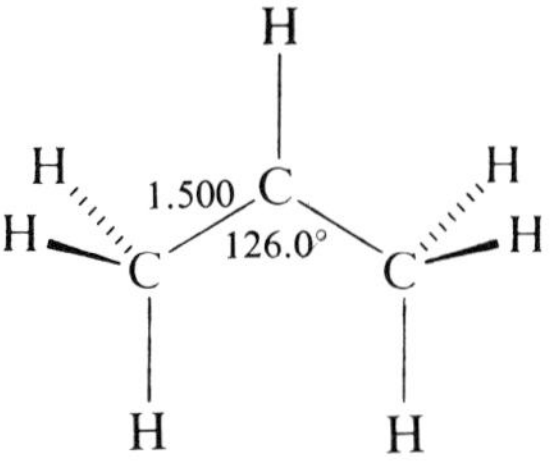

(0)

(75)

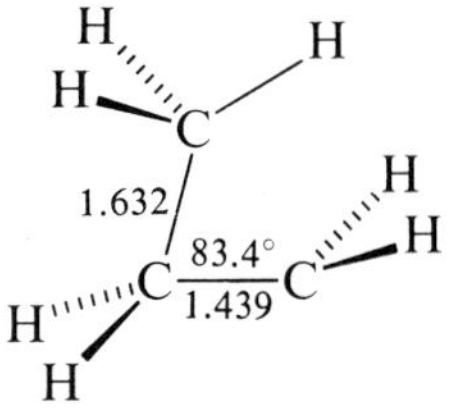

(16·9)

(76)

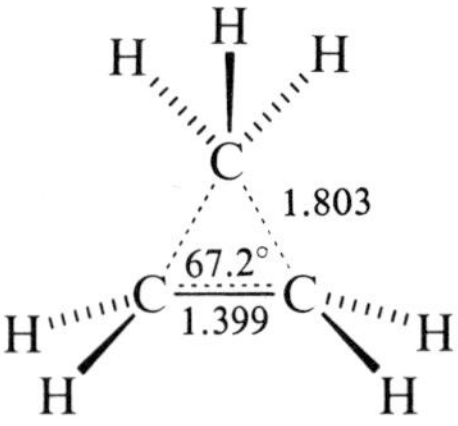

(17·3)

(77)

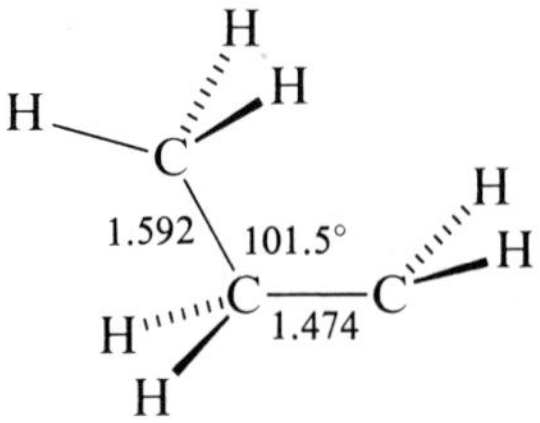

(17·4)

(78)

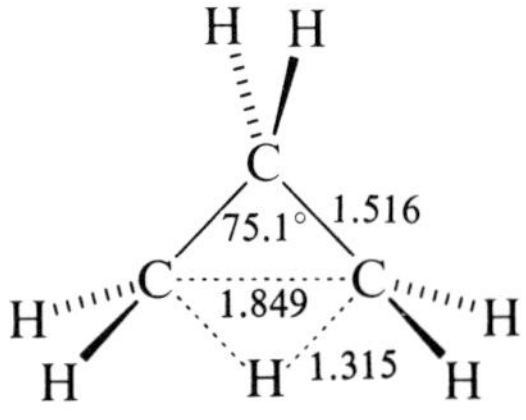

(27·1)

(79)

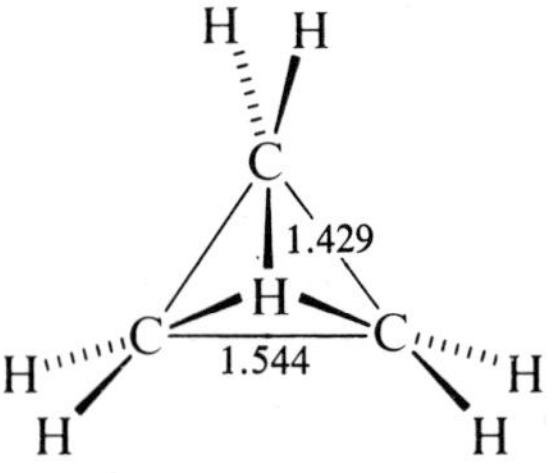

(139·6)

(80)

Figure 3.7 (continued)

of the isopropenyl to the allyl cation is estimated to be somewhat greater than 24 kcal mol^{-1}.

Two local minima in the $C_3H_7^+$ surface are suggested by the theory[130, 341]. The more stable form is the 2-propyl cation (75). The other is a distorted form of a methyl eclipsed 1-propyl cation (76) which looks like a distorted corner protonated cyclopropane. However, the potential surface linking this form to the methyl staggered 1-propyl cation (78) and to corner protonated cyclopropane (77) is very flat and the calculations therefore suggest that these three forms are rapidly interconverting. Corner protonated cyclopropane (77) is found to be more stable than edge protonated cyclopropane (79) while face protonated cyclopropane (80) is quite an unfavourable arrangement of $C_3H_7^+$.

The results in Figure 3.7 present interesting data on bonds containing (approximately) one, three and five electrons. The three electron C≈C bond, of course, occurs in benzene and the appropriate bond lengths in several of the cations are close to the benzene value (1.39 Å). The one- and five-electron bonds are not commonly encountered and their lengths are not well-established experimentally. The results in Figure 3.7 may provide a useful approximate guide to expected values, particularly if taken in conjunction with theoretical (STO–3G) bond lengths for known hydrocarbons (Figure 3.1). The calculated one-electron C---H bond lengths are about 1.3 Å (for example, 1.354 in (52), 1.364 in (53), 1.334 in (59), 1.348 in (62) and 1.315 in (79)), the one-electron C---C bond lengths are c. 1.8 Å and the five-electron C≡C bond lengths are c. 1.22 Å.

3.4.4.2   Other cations

Calculations have been carried out on several substituted ethyl and related cations[82, 339, 340, 342, 344]. It is found[82, 344] that the relative stabilities of the conformations (81) and (82) of these ions are associated with the populations of the formally vacant 2p orbital, 2p(C^+), at the positive carbon. The conformation with the larger 2p(C^+) population is favoured. Electron-withdrawing substituents X generally lead to decreased 2p(C^+) populations in (81) and little change in (82); (82) is then the favoured conformation. Thus, there is a preference for the axis of a formally vacant (or electron deficient) orbital to be *perpendicular* to an electron-withdrawing polar bond. This may be compared with the rule stated in Section 3.4.2.1, namely, there is a preference for the axis of a lone pair (or electron rich) orbital to be *parallel* to an adjacent electron-withdrawing polar bond. For the 1-propyl cation (X = CH_3), (81) is the favoured conformation[82, 340, 344] suggesting that C—C hyperconjugation is more effective than C—H hyperconjugation.

Protonation of CO occurs at either end leading to linear structures HCO^+ and COH^+; the HCO^+ structure is favoured[271, 274, 275]. Interconversion of structures of protonated formaldehyde (83) with H_ACOH *cis* and H_ACOH *trans* is found[123] to take place via a linear transition state (requiring 17.2 kcal mol^{-1}) rather than via rotation (requiring 25 kcal mol^{-1}) due to the strong double-bond character of the C—O bond. The methoxy cation CH_3O^+ is predicted[346] to be considerably less stable than CH_2OH^+. Protonated acetaldehyde is found[123] to be more stable with CCOH *cis* (84) than with CCOH *trans* by 1.4 kcal mol^{-1}; the barrier between these forms is 18.0 kcal mol^{-1}. Calculations on protonated formic acid predict[123, 233] that the *cis–trans* form (85) is most stable but disagree as to the relative stabilities of the

(81)

(82)

(83)

(84)

(85)

(86)

cis–cis and *trans–trans* forms. Both sets of calculations[123, 233] predict that protonation at the carbonyl oxygen (85) is more favourable than protonation at the hydroxy oxygen (86).

3.4.4.3 Radicals

Most calculations predict that the methyl radical is planar[67, 78, 117, 118]. The vinyl radical is also found[78, 118] to be planar but has a non-linear CCH group at the radical centre. The most stable form of the ethyl radical is the open C_s form (see cation structure (61) for the gross shape of this radical); a much higher energy is found[78] for the bridged form of C_2H_5. In contrast to CH_3, the CF_3 radical is predicted[117] to be non-planar (FCF = 112 degrees) with an inversion barrier of 27.4 kcal mol^{-1}.

3.4.4.4 Anions

The validity of single-determinant calculations for anions is somewhat uncertain since this theory often gives higher energies for an anion than for the same neutral species because of the significantly greater correlation energy of the anion. In such cases, one would anticipate that if a sufficiently large basis set were used, the solution obtained would in fact correspond to the neutral species with an electron at infinity. Nevertheless, a number of calculations on anions have been carried out.

There have been several studies on CH_3^-. It has been found[120] that the geometry and inversion barrier of this ion are very sensitive to the basis set chosen. This may be analogous to the situation for ammonia (cf. Section 3.3.5), or it may be caused by the difficulties involved in single-determinant anion calculations mentioned above.

Some of the most important studies of anions deal with investigations of the S_N2 reaction. Van der Lugt and Ros[320], and Mulder and Wright[323] have each carried out calculations on CH_5^- to determine the likely geometry of the transition state or intermediate in an S_N2 reaction. They predict that the D_{3h} structure is considerably more stable (by c. 55 kcal mol^{-1}) than the C_{4v} or C_s forms (see corresponding cation structures (53)–(55) for gross molecular shapes associated with the different symmetries). These results confirm that S_N2 reactions proceed with inversion of configuration. Ritchie and Chappell[356] have followed a reaction path for

$$H—CH_3 + H^- \longrightarrow [H—CH_3—H]^- \longrightarrow H^- + CH_3—H$$

in which the approaching hydride ion attacks at the rear of the C—H bond to be broken and C_{3v} symmetry is maintained throughout. They find an activation energy (corresponding to the relative energy of the D_{3h} transition state) of 48.7 kcal mol^{-1}. Similar calculations on the reaction

$$F—CH_3 + F^- \longrightarrow [F—CH_3—F]^- \longrightarrow F^- + CH_3—F$$

give a lower energy for FCH_3F^- than for $F^- + CH_3F$ when an sp basis is used[363]. However, Dedieu and Veillard[364] and Duke and Bader[225] have each found an activation energy for this reaction when polarisation functions are included in the basis set. Their calculated activation energies are 7.9 and 7.1 kcal mol^{-1} respectively and they suggest that these values are close to the results that would be obtained at the Hartree–Fock limit. Duke and Bader[225] have also studied the reaction

$$F^- + CH_3CN \longrightarrow [FCH_3CN]^- \longrightarrow FCH_3 + CN^-$$

finding an energy change of -5.2 kcal mol^{-1} for the reaction and an activation energy of 17.3 kcal mol^{-1}.

Clark and Armstrong[357] have investigated the electrocyclic transformation of the cyclopropyl to allyl anion with assumed values of bond lengths and angles. They find that a conrotatory mode is favoured over a disrotatory mode in agreement with the Woodward–Hoffmann rules but both require activation energy.

3.5   CONCLUSION

It is clear from this survey of the literature that application of *ab initio* molecular orbital theory to small organic molecules is already extensive and is rapidly becoming more systematic. At present, the greater part of published work is within the framework of the self-consistent single-determinant theory, and in spite of its clear limitations, such a theory is proving valuable in quantitative descriptions of a wide range of molecular properties. As further work develops, it may be hoped that we shall begin to understand in greater detail the way in which results depend on the choice of basis set and hence to appreciate how well various chemical properties are described at the full Hartree–Fock (that is, perfect single-determinant wave function) level. This will permit a clearer separation and systematisation of those aspects of chemistry which are to be specifically associated with correlation and theories beyond the Hartree–Fock level.

Within the single-determinant framework and with the use of some of the simple basis sets already available, extensive studies are now possible for systems with up to about 50 electrons. It seems likely that application of a particular basis will lead to systematic theoretical results for molecular energies, equilibrium geometries and other properties for a wide range of molecules. Since such methods can be applied equally satisfactorily to stable well-characterised molecules and to postulated species of uncertain structure, they are likely to prove of considerable predictive value. The technology is largely available; extensive application will undoubtedly follow.

Note added in proof

Since the completion of this manuscript, a number of additional *ab initio* molecular orbital studies of organic molecules have been reported. These have been added to the list of references[365–379] in order to make it complete to the end of 1971. The molecules for which these additional calculations have been carried out include methane[365], ethane and ethyl fluoride[366], cyclopropane and n-butane[367], carbon trioxide[368], the methanol and methanol–water dimers[369], formic acid dimer[370], furan, pyrrole and 1,2,5-oxadiazole[371], cyclo-octatetraene, the cyclo-octatetraene dianion and bicyclo[4.2.0]octa-2,4,7-triene[372], the methyl and ethyl cations[373], methane and the methyl anion[374], acetylene, vinylidene, the vinyl cation and protonated acetylene[375], the acetylide and diacetylide anions[376], the methyl, fluoromethyl and difluoromethyl cations and related neutral molecules methane, fluoromethane and difluoromethane[377], protonated ethane[378], the methyloxocarbonium and isopropyloxocarbonium ions[379], cyclopropane, methylcyclopropane and the dimethylcyclopropanes[380], and butadiene and its ions[381].

References

1. Streitwieser, A. Jr. (1961). *Molecular Orbital Theory for Organic Chemists*. (New York: Wiley)
2. Hückel, E. (1931). *Z. Physik.*, **70**, 204
3. Pariser, R. and Parr, R. G. (1953). *J. Chem. Phys.*, **21**, 466, 767
4. Pople, J. A. (1953). *Trans. Faraday Soc.*, **49**, 1375
5. Parr, R. G. (1961). *Quantum Theory of Molecular Electronic Structure*. (New York: Benjamin)
6. Hoffmann, R. (1963). *J. Chem. Phys.*, **39**, 1397
7. Pople, J. A., Santry, D. P. and Segal, G. A. (1965). *J. Chem. Phys.*, **43**, S129

8. Dewar, M. J. S. (1969). *The Molecular Orbital Theory of Organic Chemistry*. (New York: McGraw-Hill)
9. Pople, J. A. and Beveridge, D. L. (1970). *Approximate Molecular Orbital Theory*. (New York: McGraw-Hill)
10. Klopman, G. and O'Leary, B. (1970). *Fortschr. Chem. Forsch.*, **15**, 445
11. Jug, K. (1969). *Theoret. Chim. Acta*, **14**, 91
12. Sinanoglu, O. and Wiberg, K. B. (1970). *Sigma Molecular Orbital Theory*. (New Haven and London: Yale University Press)
13. Roothaan, C. C. J. (1951). *Rev. Mod. Phys.*, **23**, 69
14. Roothaan, C. C. J. (1960). *Rev. Mod. Phys.*, **32**, 179
15. Pople, J. A. and Nesbet, R. K. (1954). *J. Chem. Phys.*, **22**, 571
16. Bishop, D. M. (1967). *Advan. Quantum. Chem.*, **3**, 25
17. Hayes, E. F. and Parr, R. G. (1967). *Progr. Theoret. Phys., Suppl.*, **40**, 78
18. Preuss, H. and Diercksen, G. (1967). *Int. J. Quantum Chem.*, **1**, 349
19. Frost, A. A. (1967). *J. Chem. Phys.*, **47**, 3707
20. Christoffersen, R. E. and Maggiora, G. M. (1969). *Chem. Phys. Lett.*, **3**, 419
21. Christoffersen, R. E. (1972). *Advan. Quantum Chem.*, in the press
22. Ransil, B. J. (1960). *Rev. Mod. Phys.*, **32**, 239, 245
23. McLean, A. D. (1960). *J. Chem. Phys.*, **32**, 1595
24. Clementi, E. (1961). *J. Chem. Phys.*, **34**, 1468
25. Pitzer, R. M. and Lipscomb, W. N. (1963). *J. Chem. Phys.*, **39**, 1995
26. Stevens, R. M. (1970). *J. Chem. Phys.*, **52**, 1397
27. Shavitt, I. and Karplus, M. (1965). *J. Chem. Phys.*, **43**, 398
28. Foster, J. M. and Boys, S. F. (1960). *Rev. Mod. Phys.*, **32**, 303
29. Reeves, C. M. and Fletcher, R. (1965). *J. Chem. Phys.*, **42**, 4073
30. O-ohata, K., Taketa, H. and Huzinaga, S. (1966). *J. Phys. Soc. Japan*, **21**, 2306
31. Stewart, R. F. (1969). *J. Chem. Phys.*, **50**, 2485
32. Hehre, W. J., Stewart, R. F. and Pople, J. A. (1969). *J. Chem. Phys.*, **51**, 2657
33. Boys, S. F. (1950). *Proc. Roy. Soc. (London)*, **A200**, 542
34. Clementi, E., Clementi, H. and Davis, D. R. (1967). *J. Chem. Phys.*, **46**, 4725
35. Ditchfield, R., Hehre, W. J. and Pople, J. A. (1970). *J. Chem. Phys.*, **52**, 5001
36. Hehre, W. J., Ditchfield, R. and Pople, J. A. (1970). *J. Chem. Phys.*, **53**, 932
37. Pitzer, R. M. (1967). *J. Chem. Phys.*, **46**, 4871
38. Huzinaga, S. (1965). *J. Chem. Phys.*, **42**, 1293
39. Huzinaga, S. and Sakai, Y. (1969). *J. Chem. Phys.*, **50**, 1371
40. Whitman, D. R. and Hornback, C. J. (1969). *J. Chem. Phys.*, **51**, 398
41. Roos, B. and Siegbahn, P. (1970). *Theoret. Chim. Acta*, **17**, 209
42. Clementi, E. and Davis, D. R. (1966). *J. Comput. Phys.*, **1**, 223
43. Whitten, J. L. (1966). *J. Chem. Phys.*, **44**, 359
44. Schulman, J. M., Moskowitz, J. W. and Hollister, C. (1967). *J. Chem. Phys.*, **46**, 2759
45. Basch, H., Robin, M. B. and Kuebler, N. A. (1967). *J. Chem. Phys.*, **47**, 1201
46. Ritchie, C. D. and King, H. F. (1967). *J. Chem. Phys.*, **47**, 564
47. Salez, C. and Veillard, A. (1968). *Theoret. Chim. Acta*, **11**, 441
48. Huzinaga, S. and Arnau, C. (1970). *J. Chem. Phys.*, **52**, 2224
49. Dunning, T. H. (1970). *J. Chem. Phys.*, **53**, 2823
50. Dunning, T. H. (1971). *J. Chem. Phys.*, **55**, 716
51. Ditchfield, R., Hehre, W. J. and Pople, J. A. (1971). *J. Chem. Phys.*, **54**, 724
52. Moskowitz, J. W. and Harrison, M. C. (1965). *J. Chem. Phys.*, **43**, 3550
53. Neumann, D. and Moskowitz, J. W. (1968). *J. Chem. Phys.*, **49**, 2056
54. Aung, S., Pitzer, R. M. and Chan, S. I. (1968). *J. Chem. Phys.*, **49**, 2071
55. Roos, B. and Siegbahn, P. (1970). *Theoret. Chim. Acta*, **17**, 199
56. Dunning, T. H. (1971). *J. Chem. Phys.*, **55**, 3958
57. Rothenberg, S. and Schaefer, H. F. (1971). *J. Chem. Phys.*, **54**, 2764
58. Mulliken, R. S. (1955). *J. Chem. Phys.*, **23**, 1833
59. Mulliken, R. S. (1962). *J. Chem. Phys.*, **36**, 3428
60. Cusachs, L. Ch. and Politzer, P. (1968). *Chem. Phys. Lett.*, **1**, 529
61. Stout, E. W. and Politzer, P. (1968). *Theoret. Chim. Acta*, **12**, 379
62. Politzer, P. and Harris, R. R. (1970). *J. Amer. Chem. Soc.*, **92**, 6451
63. Nesbet, R. K. (1967). *Advan. Quantum Chem.*, **3**, 1
64. Moccia, R. (1964). *J. Chem. Phys.*, **40**, 2164, 2176, 2186

65. Newton, M. D., Lathan, W. A., Hehre, W. J. and Pople, J. A. (1970). *J. Chem. Phys.*, **52**, 4064
66. Pople, J. A. (1970). *Accounts Chem. Res.*, **3**, 217
67. Lathan, W. A., Hehre, W. J., Curtiss, L. A. and Pople, J. A. (1971). *J. Amer. Chem. Soc.*, **93**, 6377
68. Rauk, A., Allen, L. C. and Clementi, E. (1970). *J. Chem. Phys.*, **52**, 4133
69. Hankins, D., Moskowitz, J. W. and Stillinger, F. H. (1970). *J. Chem. Phys.*, **53**, 4544
70. Stevens, R. M. (1971). *J. Chem. Phys.*, **55**, 1725
71. Clementi, E. and Davis, D. R. (1966). *J. Chem. Phys.*, **45**, 2593
72. Fink, W. H. and Allen, L. C. (1967). *J. Chem. Phys.*, **46**, 2261
73. Pedersen, L. and Morokuma, K. (1967). *J. Chem. Phys.*, **46**, 3941
74. Pitzer, R. M. (1967). *J. Chem. Phys.*, **47**, 965
75. Veillard, A. (1970). *Theoret. Chim. Acta*, **18**, 21
76. Franchini, P. F. and Vergani, C. (1969). *Theoret. Chim. Acta*, **13**, 46
77. Radom, L. and Pople, J. A. (1970). *J. Amer. Chem. Soc.*, **92**, 4786
78. Lathan, W. A., Hehre, W. J. and Pople, J. A. (1971). *J. Amer. Chem. Soc.*, **93**, 808
79. Radom, L., Hehre, W. J. and Pople, J. A. (1972). *J. Amer. Chem. Soc.*, **94**, 2371
80. Fink, W. H. and Allen, L. C. (1967). *J. Chem. Phys.*, **46**, 2276
81. Hoyland, J. R. (1968). *J. Chem. Phys.*, **49**, 1908
82. Radom, L., Pople, J. A., Buss, V. and Schleyer, P.v.R. (1970). *J. Amer. Chem. Soc.*, **92**, 6987
83. Hoyland, J. R. (1968). *J. Chem. Phys.*, **49**, 2563
84. Unland, M. L., Van Wazer, J. R. and Letcher, J. H. (1969). *J. Amer. Chem. Soc.*, **91**, 1045
85. Zeeck, E. (1970). *Theoret. Chim. Acta*, **16**, 155
86. Scarzafava, E. and Allen, L. C. (1971). *J. Amer. Chem. Soc.*, **93**, 311
87. Davidson, R. B. and Allen, L. C. (1971). *J. Chem. Phys.*, **54**, 2828
88. Robb, M. A. and Csizmadia, I. G. (1969). *J. Chem. Phys.*, **50**, 1819
89. Kollmann, P. A. and Allen, L. C. (1970). *Chem. Phys. Lett.*, **5**, 75
90. Veillard, A. and Rees, B. (1971). *Chem. Phys. Lett.*, **8**, 267
91. Christensen, D. H., Kortzeborn, R.N., Bak, B. and Led, J. J. (1970). *J. Chem. Phys.*, **53**, 3912
92. Hehre, W. J., Radom, L. and Pople, J. A. (1972). *J. Amer. Chem. Soc.*, **94**, 1496
93. Hillier, I. H., Saunders, V. R. and Wyatt, J. F. (1970). *Trans. Faraday Soc.*, **66**, 2665
94. Davidson, R. B. and Allen, L. C. (1971). *J. Chem. Phys.*, **55**, 519
95. Dunning, T. H. and Winter, N. W. (1971). *Chem. Phys. Lett.*, **11**, 194
96. Lowe, J. P. (1968). *Progr. Phys. Org. Chem.*, **6**, 1
97. Sovers, O. J., Kern, C. W., Pitzer, R. M. and Karplus, M. (1968). *J. Chem. Phys.*, **49**, 2592
98. Allen, L. C. (1968). *Chem. Phys. Lett.*, **2**, 597
99. Lehn, J. M. (1970). *Fortschr. Chem. Forsch.*, **15**, 311
100. Rauk, A., Allen, L. C. and Mislow, K. (1970). *Angew. Chem. Int. Edn.*, **9**, 400
101. Joshi, B. D. (1965). *J. Chem. Phys.*, **43**, S40
102. Rutledge, R. M. and Saturno, A. F. (1966). *J. Chem. Phys.*, **44**, 977
103. Kaldor, U. and Shavitt, I. (1966). *J. Chem. Phys.*, **45**, 888
104. Bishop, D. M. (1966). *J. Chem. Phys.*, **45**, 1787
105. Clementi, E. (1967). *J. Chem. Phys.*, **46**, 3851
106. Body, R. G., McClure, D. S. and Clementi, E. (1968). *J. Chem. Phys.*, **49**, 4916
107. Pipano, A., Gilman, R. R., Bender, C. F. and Shavitt, I. (1970). *Chem. Phys. Lett*, **4**, 583
108. Kari, R. E. and Csizmadia, I. G., unpublished results
109. Lehn, J. M. and Munsch, B., unpublished results
110. Lehn, J. M. and Munsch, B. (1970). *Chem. Commun.*, 1062
111. Veillard, A., Lehn, J. M. and Munsch, B. (1968). *Theoret. Chim. Acta*, **9**, 275
112. Lehn, J. M., Munsch, B., Millie, P. and Veillard, A. (1969). *Theoret. Chim. Acta*, **13**, 313
113. Clark, D. T. (1970). International symposium on *Quantum Aspects of Heterocyclic Compounds in Chemistry and Biochemistry*, p.238. (Jerusalem: The Israel Academy of Sciences and Humanities)
114. Lehn, J. M., Munsch, B. and Millie, P. (1970). *Theoret. Chim. Acta*, **16**, 351
115. Lehn, J. M. and Munsch, B. (1968). *Theoret. Chim. Acta*, **12**, 91
116. Wood, D. E., Lloyd, R. V. and Lathan, W. A. (1971). *J. Amer. Chem. Soc.*, **93**, 4145
117. Morokuma, K., Pedersen, L. and Karplus, M. (1968). *J. Chem. Phys.*, **48**, 4801

118. Millie, P. and Berthier, G. (1968). *Int. J. Quantum Chem.*, **2S**, 67
119. Kari, R. E. and Csizmadia, I. G. (1967). *J. Chem. Phys.*, **46**, 4585
120. Kari, R. E. and Csizmadia, I. G. (1969). *J. Chem. Phys.*, **50**, 1443
121. Clark, D. T. and Armstrong, D. R. (1969). *Chem. Commun.*, 850
122. Clark, D. T. (1969). *Chem. Commun.*, 637
123. Ros, P. (1968). *J. Chem. Phys.*, **49**, 4902
124. Snyder, L. C. (1967). *J. Chem. Phys.*, **46**, 3602
125. Snyder, L. C. and Basch, H. (1969). *J. Amer. Chem. Soc.*, **91**, 2189
126. Ditchfield, R., Hehre, W. J., Pople, J. A. and Radom, L. (1970). *Chem. Phys. Lett*, **5**, 13
127. Hehre, W. J., Ditchfield, R., Radom, L. and Pople, J. A. (1970). *J. Amer. Chem. Soc.*, **92**, 4796
128. Radom, L., Hehre, W. J. and Pople, J. A. (1971). *J. Amer. Chem. Soc.*, **93**, 289
129. Radom, L., Hehre, W. J. and Pople, J. A. (1971). *J. Chem. Soc. A*, 2299
130. Radom, L., Pople, J. A., Buss, V. and Schleyer, P.v.R. (1971). *J. Amer. Chem. Soc.*, **93**, 1813
131. Radom, L., Lathan, W. A., Hehre, W. J. and Pople, J. A. (1971). *J. Amer. Chem. Soc.*, **93**, 5339
132. For pre-1970 references, see also References 133 and 134
133. Richards, W. G., Walker, T. E. H. and Hinkley, R. K. (1971). *A Bibliography of* ab initio *Molecular Wave Functions*. (London: Oxford University Press)
134. Krauss, M. (1967). *Compendium of* ab initio *Calculations of Molecular Energies and Properties*. NBS Tech. Note 438. (Washington: U.S. Government Printing Office)
135. Hehre, W. J. and Pople, J. A. (1968). *Chem. Phys. Lett.*, **2**, 379
136. Hehre, W. J., Stewart, R. F. and Pople, J. A. (1968). *Symposia Faraday Soc.*, **2**, 15
137. Lathan, W. A., Hehre, W. J. and Pople, J. A. (1969). *Chem. Phys. Lett.*, **3**, 579
138. Rouse, R. A. and Frost, A. A. (1969). *J. Chem. Phys.*, **50**, 1705
139. Newton, M. D., Lathan, W. A., Hehre, W. J. and Pople, J. A. (1969). *J. Chem. Phys.*, **51**, 3927
140. Hehre, W. J. and Pople, J. A. (1970). *J. Amer. Chem. Soc.*, **92**, 2191
141. Schwartz, M. E., Coulson, C. A. and Allen, L. C. (1970). *J. Amer. Chem. Soc.*, **92**, 447
142. Bader, R. F. W. and Preston, H. J. T. (1970). *Theoret. Chim. Acta*, **17**, 384
143. Brundle, C. R., Robin, M. B. and Basch, H. (1970). *J. Chem. Phys.*, **53**, 2196
144. Franchini, P. F., Moccia, R. and Zandomeneghi, M. (1970). *Int. J. Quantum Chem.*, **4**, 487 (1970)
145. Owens, P. H., Wolf, R. A. and Streitwieser, A. (1970). *Tetrahedron Lett.*, 3385
146. Newton, M. D., Switkes, E. and Lipscomb, W. N. (1970). *J. Chem. Phys.*, **53**, 2645
147. Grimmelmann, E. K. and Chesick, J. P. (1971). *J. Chem. Phys.*, **55**, 1690
148. Pulay, P. (1971). *Mol. Phys.*, **21**, 329
149. Christoffersen, R. E., Genson, D. W. and Maggiora, G. M. (1971). *J. Chem. Phys.*, **54**, 239
150. Palke, W. E. and Lipscomb, W. N. (1966). *J. Amer. Chem. Soc.*, **88**, 2384
151. Buenker, R. J., Peyerimhoff, S. D., Allen, L. C. and Whitten, J. L. (1966). *J. Chem. Phys.*, **45**, 2835
152. Kaldor, U. (1967). *J. Chem. Phys.*, **46**, 1981
153. Buenker, R. J., Peyerimhoff, S. D. and Whitten, J. L. (1967). *J. Chem. Phys.*, **46**, 2029
154. Klessinger, M. (1968). *Symposia Faraday Soc.*, **2**, 73
155. Frost, A. A. and Rouse, R. A. (1968). *J. Amer. Chem. Soc.*, **90**, 1965
156. Veillard, A. (1969). *Chem. Phys. Lett.*, **3**, 128
157. Klessinger, M. (1970). *J. Chem. Phys.*, **53**, 225
158. Clementi, E. and Von Niessen, W. (1971). *J. Chem. Phys.*, **54**, 521
159. Levy, B. and Moireau, M. C. (1971). *J. Chem. Phys.*, **54**, 3316
160. Jorgensen, W. L. and Allen, L. C. (1971). *J. Amer. Chem. Soc.*, **93**, 567
161. Hoyland, J. R. (1967). *Chem. Phys. Lett.*, **1**, 247
162. Moskowitz, J. W. and Harrison, M. C. (1965). *J. Chem. Phys.*, **42**, 1726
163. Janiszewski, F. J., Lykos, P. G. and Wahl, A. C. (1968). *U.S. At. Energy Comm.*, ANL-7446
164. Kaldor, U. and Shavitt, I. (1968). *J. Chem. Phys.*, **48**, 191
165. Robin, M. B., Basch, H., Kuebler, N. A., Kaplan, B. E. and Meinwald, J. (1968). *J. Chem. Phys.*, **48**, 5037
166. Petke, J. D. and Whitten, J. L. (1969). *J. Chem. Phys.*, **51**, 3166

167. Baird, N. C. (1970). *Chem. Phys. Lett.*, **6**, 61
168. Brundle, C. R., Robin, M. B., Basch, H., Pinsky, M. and Bond, A. (1970). *J. Amer. Chem. Soc.*, **92**, 3863
169. Klessinger, M. (1970). *Int. J. Quantum Chem.*, **4**, 191
170. Schwartz, M. E. and Rothenberg, S. (1970). *J. Amer. Chem. Soc.*, **92**, 3860
171. Pulay, P. and Meyer, W. (1971). *J. Mol. Spectrosc.*, **40**, 59
172. Meza, S. and Wahlgren, U. (1971). *Theoret. Chim. Acta*, **21**, 323
173. Newton, M. D. and Switkes, E. (1971). *J. Chem. Phys.*, **54**, 3179
174. Rothenberg, S. (1971). *J. Amer. Chem. Soc.*, **93**, 68
175. Janoschek, R. and Preuss, H. (1969). *Int. J. Quantum Chem.*, **3**, 889
176. Preuss, H. and Janoschek, R. (1969). *J. Mol. Structure*, **3**, 423
177. Buenker, R. J. and Whitten, J. L. (1968). *J. Chem. Phys.*, **49**, 5381
178. Buenker, R. J. (1968). *J. Chem. Phys.*, **48**, 1368
179. André, J. M., André, M.Cl., Leroy, G. and Weiler, J. (1969). *Int. J. Quantum Chem.*, **3**, 1013
180. André, J. M., André, M.Cl. and Leroy, G. (1969). *Chem. Phys. Lett.*, **3**, 695
181. Schaad, L. J. (1970). *Tetrahedron*, **26**, 4115
182. McLean, A. D., Ransil, B. J. and Mulliken, R. S. (1960). *J. Chem. Phys.*, **32**, 1873
183. Clementi, E. and Clementi, H. (1962). *J. Chem. Phys.*, **36**, 2824
184. Moskowitz, J. W. (1965). *J. Chem. Phys.*, **43**, 60
185. Moskowitz, J. W. (1966). *J. Chem. Phys.*, **45**, 2338
186. McLean, A. D. and Yoshimine, M. (1967). *IBM J. Res. Develop. Suppl.*, **12**, 206
187. Frost, A. A., Prentice, B. H. and Rouse, R. A. (1967). *J. Amer. Chem. Soc.*, **89**, 3064
188. Griffith, M. G. and Goodman, L. (1967). *J. Chem. Phys.*, **47**, 4494
189. Hoyland, J. R. (1968). *J. Chem. Phys.*, **48**, 5736
190. Switkes, E., Stevens, R. M. and Lipscomb, W. N. (1969). *J. Chem. Phys.*, **51**, 5229
191. Haase, J., Janoschek, R., Preuss, H. and Diercksen, G. (1969). *J. Mol. Structure*, **3**, 165
192. Newton, M. D. and Lipscomb, W. N. (1967). *J. Amer. Chem. Soc.*, **89**, 4261
193. Peyerimhoff, S. D. and Buenker, R. J. (1969). *Theoret. Chim. Acta*, **14**, 305
194. Foster, J. M. and Boys, S. F. (1960). *Rev. Mod. Phys.*, **32**, 305
195. Krauss, M. (1964). *J. Res. Nat. Bur. Stand*, **68A**, 635
196. Harrison, J. F. and Allen, L. C. (1969). *J. Amer. Chem. Soc.*, **91**, 807
197. Bender, C. F. and Schaefer, H. F. (1970). *J. Amer. Chem. Soc.*, **92**, 4984
198. Del Bene, J. E. (1971). *Chem. Phys. Lett.*, **9**, 68
199. Harrison, J. F. (1971). *J. Amer. Chem. Soc.*, **93**, 4112
200. O'Neil, S. V., Schaefer, H. F. and Bender, C. F. (1971). *J. Chem. Phys.*, **55**, 162
201. Buenker, R. J. and Peyerimhoff, S. D. (1969). *J. Phys. Chem.*, **73**, 1299
202. Salem, L. (1970). *Bull. Soc. Chim. France*, 3161
203. Siu, A. K. Q., St. John, W. M. and Hayes, E. F. (1970). *J. Amer. Chem. Soc.*, **92**, 7249
204. Jean, Y. and Salem, L. (1971). *Chem. Commun.*, 382
205. Preuss, H. and Diercksen, G. (1967). *Int. J. Quantum Chem.*, **1**, 361
206. Petke, J. D. and Whitten, J. L. (1968). *J. Amer. Chem. Soc.*, **90**, 3338
207. Kochanski, E. and Lehn, J. M. (1969). *Theoret. Chim. Acta*, **14**, 281
208. Clark, D. T. (1969). *Theoret. Chim. Acta*, **15**, 225
209. Basch, H., Robin, M. B., Kuebler, N. A., Baker, C. and Turner, D. W. (1969). *J. Chem. Phys.*, **51**, 52
210. Bonaccorsi, R., Scrocco, E. and Tomasi, J. (1970). *J. Chem. Phys.*, **52**, 5270
211. Stevens, R. M., Switkes, E., Laws, E. A. and Lipscomb, W. N. (1971). *J. Amer. Chem. Soc.*, **93**, 2603
212. Franchini, P. F. and Zandomeneghi, M. (1971). *Theoret. Chim. Acta*, **21**, 90
213. Robin, M. B., Basch, H., Kuebler, N. A., Wiberg, K. B. and Ellison, G. B. (1969). *J. Chem. Phys.*, **51**, 45
214. André, J. M., André, M. Cl. and Leroy, G. (1969). *Bull. Soc. Chim. Belges*, **78**, 539
215. Wright, J. S. and Salem, L. (1969). *Chem. Commun.*, 1370
216. Schulman, J. M. and Fisanick, G. J. (1970). *J. Amer. Chem. Soc.*, **92**, 6653
217. Hsu, K., Buenker, R. J. and Peyerimhoff, S. D. (1971). *J. Amer. Chem. Soc.*, **93**, 2117
218. Buenker, R. J., Peyerimhoff, S. D. and Hsu, K. (1971). *J. Amer. Chem. Soc.*, **93**, 5005
219. Buenker, R. J. and Peyerimhoff, S. D. (1968). *J. Chem. Phys.*, **48**, 354
220. Buenker, R. J. and Peyerimhoff, S. D. (1969). *J. Amer. Chem. Soc.*, **91**, 4342
221. Hoyland, J. R. (1969). *J. Chem. Phys.*, **50**, 2775

222. Benson, R. C. and Flygare, W. H. (1969). *J. Chem. Phys.*, **51**, 3087
223. Hehre, W. J. and Pople, J. A. (1970). *Tetrahedron Lett.*, 2959
224. Clementi, E. and Klint, D. (1969). *J. Chem. Phys.*, **50**, 4899
225. Duke, A. J. and Bader, R. F. W. (1971). *Chem. Phys. Lett.*, **10**, 631
226. Moffat, J. B. and Vogt, C. (1970). *J. Mol. Spectrosc.*, **33**, 494
227. Unland, M. L., Letcher, J. H., Absar, I. and Van Wazer, J. R. (1971). *J. Chem. Soc. A*, 1328
228. Jorgensen, W. L. and Allen, L. C. (1970). *Chem. Phys. Lett.*, **7**, 483
229. Morokuma, K. (1971). *J. Chem. Phys.*, **55**, 1236
230. Letcher, J. H., Unland, M. L. and Van Wazer, J. R. (1969). *J. Chem. Phys.*, **50**, 2185
231. Peyerimhoff, S. D. and Buenker, R. J. (1969). *J. Chem. Phys.*, **50**, 1846
232. Basch, H., Robin, M. B. and Kuebler, N. A. (1968). *J. Chem. Phys.*, **49**, 5007
233. Hopkinson, A. C., Yates, K. and Csizmadia, I. G. (1970). *J. Chem. Phys.*, **52**, 1784
234. Schwartz, M. E., Hayes, E. F. and Rothenberg, S. (1970). *J. Chem. Phys.*, **52**, 2011
235. McLean, A. D. (1963). *J. Chem. Phys.*, **38**, 1347
236. Peyerimhoff, S. D., Buenker, R. J. and Whitten, J. L. (1967). *J. Chem. Phys.*, **46**, 1707
237. Yoshimine, M. and McLean, A. D. (1967). *Int. J. Quantum Chem.*, **1S**, 313
238. Robb, M. A. and Csizmadia, I. G. (1968). *Theoret. Chim. Acta*, **10**, 269
239. Mely, B. and Pullman, A. (1969). *Theoret. Chim. Acta*, **13**, 278
240. Dreyfus, M., Maigret, B. and Pullman, A. (1970). *Theoret. Chim. Acta*, **17**, 109
241. Moffat, J. B. (1970). *J. Theoret. Biol.*, **26**, 437
242. Bonaccorsi, R., Petrongolo, C., Scrocco, E. and Tomasi, J. (1968). *J. Chem. Phys.*, **48**, 1500
243. Moffat, J. B. and Popkie, H. E. (1968). *Int. J. Quantum Chem.*, **2**, 565
244. Wolfe, S., Rauk, A., Tel, L. M. and Csizmadia, I. G. (1971). *J. Chem. Soc. B*, 136
245. Csizmadia, I. G., Harrison, M. C. and Sutcliffe, B. T. (1966). *Theoret. Chim. Acta*, **6**, 217
246. Rothenberg, S. (1969). *J. Chem. Phys.*, **51**, 3389
247. Sachs, L. M., Geller, M. and Kaufman, J. J. (1969). *J. Chem. Phys.*, **51**, 2771
248. Levy, B., Millie, P., Lehn, J. M. and Munsch, B. (1970). *Theoret. Chim. Acta*, **18**, 143
249. Moffat, J. B. (1970). *Can. J. Chem.*, **48**, 1820
250. Vinh, J., Levy, B. and Millie, P. (1971). *Mol. Phys.*, **21**, 345
251. Newton, M. D. and Palke, W. E. (1966). *J. Chem. Phys.*, **45**, 2329
252. Stamper, J. G. and Trinajstic, N. (1967). *J. Chem. Soc. A*, 782
253. Dunning, T. H. and McKoy, V. (1968). *J. Chem. Phys.*, **48**, 5263
254. Winter, N. W., Dunning, T. H. and Letcher, J. H. (1968). *J. Chem. Phys.*, **49**, 1871
255. Hopkinson, A. C., Holbrook, N. K., Yates, K. and Csizmadia, I. G. (1968). *J. Chem. Phys.*, **49**, 3596
256. Neumann, D. B. and Moskowitz, J. W. (1969). *J. Chem. Phys.*, **50**, 2216
257. Whitten, J. L. and Hackmeyer, M. (1969). *J. Chem. Phys.*, **51**, 5584
258. Dunning, T. H. and Winter, N. W. (1971). *J. Chem. Phys.*, **55**, 3360
259. Pincelli, U., Cadioli, B. and David, D.-J. (1971). *J. Mol. Structure*, **9**, 173
260. Clementi, E., Mehl, J. and Von Niessen, W. (1971). *J. Chem. Phys.*, **54**, 508
261. Bonaccorsi, R., Scrocco, E. and Tomasi, J. (1971). *Theoret. Chim. Acta*, **21**, 17
262. Dreyfus, M. and Pullman, A. (1970). *Theoret. Chim. Acta*, **19**, 20
263. McLean, A. D. (1962). *J. Chem. Phys.*, **37**, 627
264. Burnelle, L. (1964). *Theoret. Chim. Acta*, **2**, 177
265. Pan, D. C. and Allen, L. C. (1967). *J. Chem. Phys.*, **46**, 1797
266. Moffat, J. B. and Collens, R. J. (1967). *Can. J. Chem.*, **45**, 655
267. Moffat, J. B. (1966). *Chem. Commun.*, 789
268. Bruns, R. E. and Person, W. B. (1970). *J. Chem. Phys.*, **53**, 1413
269. Moffat, J. B. and Collens, R. J. (1968). *J. Mol. Spectrosc.*, **27**, 252
270. Clementi, E. and McLean, A. D. (1962). *J. Chem. Phys.*, **36**, 563
271. Jansen, H. B. and Ros, P. (1969). *Chem. Phys. Lett.*, **3**, 140
272. Siu, A. K. Q. and Davidson, E. R. (1970). *Int. J. Quantum Chem.*, **4**, 223
273. Chu, S. Y. and Frost, A. A. (1971). *J. Chem. Phys.*, **54**, 764
274. Forsen, S. and Roos, B. (1970). *Chem. Phys. Lett.*, **6**, 128
275. Jansen, H. B. and Ros, P. (1971). *Theoret. Chim. Acta*, **21**, 199
276. Gelius, U., Allan, C. J., Allison, D. A., Siegbahn, H. and Siegbahn, K. (1971). *Chem. Phys. Lett.*, **11**, 224

277. Hayes, E. F. (1969). *J. Chem. Phys.*, **51**, 4787
278. Lombardi, J. R., Klemperer, W., Robin, M. B., Basch, H. and Kuebler, N. A. (1969). *J. Chem. Phys.*, **51**, 33
279. Clark, D. T. and Lilley, D. M. J. (1970). *Chem. Commun.*, 147
280. Cornille, M. and Horsley, J. (1970). *Chem. Phys. Lett.*, **6**, 373
281. André, J. M., André, M.Cl. and Leroy, G. (1971). *Theoret. Chim. Acta*, **21**, 28
282. Kramling, R. W. and Wagner, E. L. (1969). *Theoret. Chim. Acta*, **15**, 43
283. Siegbahn, P. (1971). *Chem. Phys. Lett.*, **8**, 245
284. Berthier, G., Praud, L. and Serre, J. (1970). International Symposium on *Quantum Aspects of Heterocyclic Compounds in Chemistry and Biochemistry*, p.40. (Jerusalem: The Israel Academy of Sciences and Humanities)
285. Clementi, E. (1967). *J. Chem. Phys.*, **46**, 4731
286. Petke, J. D., Whitten, J. L. and Ryan, J. A. (1968). *J. Chem. Phys.*, **48**, 953
287. Clementi, E. (1967). *J. Chem. Phys.*, **46**, 4737
288. Hackmeyer, M. and Whitten, J. L. (1971). *J. Chem. Phys.*, **54**, 3739
289. Clementi, E., André, J. M., André, M.Cl., Klint, D. and Hahn, D. (1969). *Acta Phys.*, **27**, 493
290. Pullman, A., Dreyfus, M. and Mely, B. (1970). *Theoret. Chim. Acta*, **17**, 85
291. Snyder, L. C., Schulman, R. G. and Neumann, D. B. (1970). *J. Chem. Phys.*, **53**, 256
292. Clementi, E. (1967). *J. Chem. Phys.*, **47**, 4485
293. Schulman, J. M. and Moskowitz, J. W. (1965). *J. Chem. Phys.*, **43**, 3287
294. Schulman, J. M. and Moskowitz, J. W. (1967). *J. Chem. Phys.*, **47**, 3491
295. Diercksen, G. and Preuss, H. (1967). *Int. J. Quantum Chem.*, **1**, 357
296. Janoschek, R., Preuss, H. and Diercksen, G. (1967). *Int. J. Quantum Chem.*, **1S**, 209
297. Buenker, R. J., Whitten, J. L. and Petke, J. D. (1968). *J. Chem. Phys.*, **49**, 2261
298. Praud, L., Millie, P. and Berthier, G. (1968). *Theoret. Chim. Acta*, **11**, 169
299. Peyerimhoff, S. D. and Buenker, R. J. (1970). *Theoret. Chim. Acta*, **19**, 1
300. Christoffersen, R. E. (1971). *J. Amer. Chem. Soc.*, **93**, 4104
301. Preuss, H. (1968). *Int. J. Quantum Chem.*, **2**, 651
302. Buenker, R. J. and Peyerimhoff, S. D. (1969). *Chem. Phys. Lett.*, **3**, 37
303. Millie, P., Praud, L. and Serre, J. (1971). *Int. J. Quantum Chem.*, **4S**, 187
304. Wilhite, D. L. and Whitten, J. L. (1971). *J. Amer. Chem. Soc.*, **93**, 2858
305. Cade, P. E. and Huo, W. M. (1967). *J. Chem. Phys.*, **47**, 614
306. Baird, N. C. and Lemaire, D. (1970). *Theoret. Chim. Acta*, **17**, 158
307. Lykos, P. G., Herman, R. B., Ritter, J. D. S. and Moccia, R. (1964). *Bull. Amer. Phys. Soc.*, **9**, 145
308. Peyerimhoff, S. D., Buenker, R. J. and Allen, L. C. (1966). *J. Chem. Phys.*, **45**, 734
309. Joshi, B. D. (1967). *J. Chem. Phys.*, **46**, 875
310. Kari, R. E. and Csizmadia, I. G. (1967). *J. Chem. Phys.*, **46**, 1817
311. Von Buneau, G., Diercksen, G. and Preuss, H. (1967). *Int. J. Quantum Chem.*, **1**, 645
312. Frost, A. A. (1968). *J. Phys. Chem.*, **72**, 1289
313. Williams, J. E., Sustmann, R., Allen, L. C. and Schleyer, P.v.R. (1969). *J. Amer. Chem. Soc.*, **91**, 1037
314. Sustmann, R., Williams, J. E., Dewar, M. J. S., Allen, L. C. and Schleyer, P.v.R. (1969). *J. Amer. Chem. Soc.*, **91**, 5350
315. Arents, J. and Allen, L. C. (1970). *J. Chem. Phys.*, **53**, 73
316. Handler, G. S. and Joy, H. W. (1969). *Int. J. Quantum Chem.*, **3S**, 529
317. Dixon, R. N. (1971). *Mol. Phys.*, **20**, 113
318. Rutledge, R. M. and Saturno, F. (1965). *J. Chem. Phys.*, **43**, 597
319. Gole, J. L. (1969). *Chem. Phys. Lett.*, **3**, 577; **4**, 808
320. Van der Lugt, W. Th. A. M. and Ros, P. (1969). *Chem. Phys. Lett.*, **4**, 389
321. Dyczmons, V., Staemmler, V. and Kutzelnigg, W. (1970). *Chem. Phys. Lett.*, **5**, 361
322. Lathan, W. A., Hehre, W. J. and Pople, J. A. (1970). *Tetrahedron Lett.*, 2699
323. Mulder, J. J. C. and Wright, J. S. (1970). *Chem. Phys. Lett.*, **5**, 445
324. Guest, M. F., Murrell, J. N. and Pedley, J. B. (1971). *Mol. Phys.*, **20**, 81
325. Michels, H. H., Harris, F. E. and Addison, J. B. (1971). *Int. J. Quantum Chem.*, **4S**, 149
326. Verhaegen, G. (1968). *J. Chem. Phys.*, **49**, 4696
327. Fratev, F., Janoschek, R. and Preuss, H. (1969). *Int. J. Quantum Chem.*, **3**, 873
328. Pfeiffer, G. V. and Jewett, J. G. (1970). *J. Amer. Chem. Soc.*, **92**, 2143
329. Clark, D. T. and Lilley, D. M. J. (1970). *Chem. Commun.*, 549

330. Massa, L. J., Ehrenson, S. and Wolfsberg, M. (1970). *Int. J. Quantum Chem.*, **4**, 625
331. Williams, J. E., Buss, V., Allen, L. C., Schleyer, P.v.R., Lathan, W. A., Hehre, W. J. and Pople, J. A. (1970). *J. Amer. Chem. Soc.*, **92**, 2141
332. Massa, L. J., Ehrenson, S., Wolfsberg, M. and Frishberg, C. A. (1971). *Chem. Phys. Lett.*, **11**, 196
333. Peyerimhoff, S. D. and Buenker, R. J. (1968). *J. Chem. Phys.*, **49**, 312
334. Lathan, W. A., Curtiss, L. A., Hehre, W. J. and Pople, J. A. To be published
335. Radom, L., Pople, J. A. and Schleyer, P.v.R. To be published
336. Clark, D. T. and Armstrong, D. R. (1969). *Theoret. Chim. Acta*, **13**, 365
337. Peyerimhoff, S. D. and Buenker, R. J. (1969). *J. Chem. Phys.*, **51**, 2528
338. Radom, L., Pople, J. A. and Schleyer, P.v.R. To be published
339. Fratev, F., Janoschek, R. and Preuss, H. (1970). *Int. J. Quantum Chem.*, **4**, 529
340. Radom, L., Pople, J. A., Buss, V. and Schleyer, P.v.R. (1970). *J. Amer. Chem. Soc.*, **92**, 6380
341. Radom, L., Pople, J. A., Buss, V. and Schleyer, P.v.R. (1972). *J. Amer. Chem. Soc.*, **94**, 311
342. Clark, D. T. and Lilley, D. M. J. (1970). *Chem. Commun.*, 603
343. Clark, D. T. and Lilley, D. M. J. (1970). *Chem. Commun.*, 1042
344. Radom, L., Pople, J. A. and Schleyer, P.v.R. *J. Amer. Chem. Soc.*, in the press
345. Sahni, R. C. and Sawhney, B. C. (1967). *Trans. Faraday Soc.*, **63**, 1
346. Haney, M. A., Patel, J. C. and Hayes, E. F. (1970). *J. Chem. Phys.*, **53**, 4105
347. Horsley, J. A. and Fink, W. H. (1969). *J. Phys. (B)*, **2**, 1261
348. McDiarmid, R. (1971). *Theoret. Chim. Acta*, **20**, 282
349. Maeder, F., Millie, P. and Berthier, G. (1971). *Int. J. Quantum Chem.*, **4S**, 179
350. Kruglyak, Y. A., Preuss, H. and Janoschek, R. (1970). *Ukrain. Fiz. Zhur.*, **15**, 980
351. Claxton, T. A. (1971). *Trans. Faraday Soc.*, **67**, 897
352. Claxton, T. A. (1969). *Chem. Phys. Lett.*, **4**, 469
353. Cade, P. E. (1967). *Proc. Phys. Soc.*, **91**, 842
354. Joshi, B. D. (1967). *J. Chem. Phys.*, **47**, 2793
355. Ritchie, C. D. and King, H. F. (1968). *J. Amer. Chem. Soc.*, **90**, 838
356. Ritchie, C. D. and Chappell, G. A. (1970). *J. Amer. Chem. Soc.*, **92**, 1819
357. Clark, D. T. and Armstrong, D. R. (1969). *Theoret. Chim. Acta*, **14**, 370
358. Peyerimhoff, S. D. (1967). *J. Chem. Phys.*, **47**, 349
359. Bonaccorsi, R., Petrongolo, C., Scrocco, E. and Tomasi, J. (1969). *Chem. Phys. Lett.*, **3**, 473
360. Doggett, G. and McKendrick, A. (1970). *J. Chem. Soc. A*, 825
361. Moffatt, J. B. and Popkie, H. E. (1970). *J. Mol. Structure*, **6**, 155
362. Arrighini, G. P., Guidotti, C., Maestro, M., Moccia, R. and Salvetti, O. (1969). *J. Chem. Phys.*, **51**, 480
363. Berthier, G., David, D.-J. and Veillard, A. (1969). *Theoret. Chim. Acta*, **14**, 329
364. Dedieu, A. and Veillard, A. (1970). *Chem. Phys. Lett.*, **5**, 328
365. David, D. J. (1971). *Theoret. Chim. Acta*, **23**, 226
366. Allen, L. C. and Basch, H. (1971). *J. Amer. Chem. Soc.*, **93**, 6373
367. Marsmann, H., Robert, J. B. and Van Wazer, J. R. (1971). *Tetrahedron*, **27**, 4377
368. Sabin, J. R. and Kim, H. (1971). *Chem. Phys. Lett.*, **11**, 593
369. Del Bene, J. E. (1971). *J. Chem. Phys.*, **55**, 4633
370. Ady, E. and Brickmann, J. (1971). *Chem. Phys. Lett.*, **11**, 302
371. Palmer, M. H. and Gaskell, A. J. (1971). *Theoret. Chim. Acta*, **23**, 52
372. Wipff, G., Wahlgren, U., Kochanski, E. and Lehn, J. M. (1971). *Chem. Phys. Lett.*, **11**, 350
373. Williams, J. E., Buss, V. and Allen, L. C. (1971). *J. Amer. Chem. Soc.*, **93**, 6867
374. Owens, P. H. and Streitwieser, A., Jr. (1971). *Tetrahedron*, **27**, 4471
375. Hopkinson, A. C., Yates, K. and Csizmadia, I. G. (1971). *J. Chem. Phys.*, **55**, 3835
376. Hopkinson, A. C. and Csizmadia, I. G. (1971). *J. Chem. Soc. (D)*, 1291
377. Baird, N. C. and Datta, R. K. (1971). *Canad. J. Chem.*, **49**, 3708
378. Ray, N. K. (1971). *Theoret. Chim. Acta*, **23**, 111
379. Rees, B., Veillard, A. and Weiss, R. (1971). *Theoret. Chim. Acta*, **23**, 266
380. André, J. M., André, M. C. and Leroy, G. (1971). *Bull. Soc. Chim. Belges*, **80**, 265
381. Hinchliffe, A. (1971). *J. Mol. Structure*, **10**, 379

4
Intermolecular Forces

P. R. CERTAIN and L. W. BRUCH
University of Wisconsin

4.1   INTRODUCTION

A complete consideration of the theoretical aspects of intermolecular forces can include the interaction of hydrogen atoms and the interaction of mica sheets, classical electrostatics and quantum electrodynamics, one-electron systems and the many-body problem, and so on. A limited consideration must therefore be put in perspective. We limit ourselves to *simple* interactions – either to accurate treatments of interactions between simple molecules or to simple treatments of interactions between complicated molecules. The review is divided into two parts: the computation of interactions via quantum mechanical algorithms and the inversion of data obtained by laboratory measurements. Topics with which we shall not deal are: (i) the fundamental concept of intermolecular forces, i.e. the Born–Oppenheimer separation and deviations therefrom; (ii) interactions among excited state molecules; (iii) intermolecular forces in condensed phases; (iv) relativistic intermolecular forces; (v) the calculation of intermolecular forces leading to chemical bonds; (vi) model potentials for molecules with internal structure; and (vii) the critical review of experimental data.

Anyone who has done research in the recent past knows of the problems connected with reviewing an ever-increasing literature. We have written this review without the benefit of computer or staff. This is not a balanced and comprehensive review; we have treated at length the topics with which we are familiar.

The subject of intermolecular forces has been reviewed many times in the last 6 years[1-9]. It is testimony to the breadth of the field that these reviews do not overlap appreciably.

4.2 INTERMOLECULAR INTERACTIONS—QUANTAL CALCULATIONS

In this section we concentrate on various *methods* of computing intermolecular interactions and mention only briefly the *results* of the calculations. It would be very satisfying to be able to reverse the emphasis; however, the calculations are still far from routine and hence most of the research effort has been expended on developing methods rather than studying specific systems. This contrasts with the field of electronic structure calculations, where emphasis on methods is beginning to give way to applications[10].

We are concerned primarily with *ab initio* methods, so that necessarily the discussion is limited to interactions involving atoms and small molecules. The methods we discuss are all fairly standard perturbation and variation procedures. Thus, before beginning the discussion, we should mention the imaginative proposal[11] of Boys and Bernardi for computing intermolecular interaction energies by using Boys' bivariational method. The advantage of the bivariational method is that it allows complex trial functions containing explicit interelectronic coordinates to approximate the exact wave function closely, and yet it avoids many-dimensional integrals. In fact the integrals which do appear are evaluated by rough numerical quadrature. Since the error in the procedure is a product of the errors in the quadrature and the trial wave function, if the latter errors are very small, the total error is small. The extension of the method to intermolecular forces involves ideas of 'counterpoise', i.e. the detailed and delicate cancellation of errors committed in the interacting and non-interacting systems. Model calculations illustrating these ideas have been presented and further applications have been promised.

The development of Boys' ideas lies in the future; the present situation is this. Long-range intermolecular interactions are treated by perturbation theory based on the separated molecules; short-range interactions are dealt with by variational methods based on the Hartree–Fock approximation. The intermediate range, where van der Waals forces are important, requires special consideration. Both variation and perturbation procedures have been developed in recent years. It appears at this time that the variational approach holds the greatest promise for the future. It is very desirable to have a computational method which passes smoothly from the long to intermediate range and this has been accomplished in the calculations of Bertoncini and Wahl[12], and Schaefer and co-workers[13] on the helium–helium interaction. A task for the future is to pass smoothly from intermediate to short range. This is difficult because of the changing nature of electron correlations with nuclear position. A trial wave function which describes the correlations appropriate to the intermediate region generally does not have the flexibility to describe the short-range correlations.

In the sections which follow, the following notation is used without further comment. H denotes the non-relativistic clamped-nuclei electronic

Hamiltonian for interacting molecules, which may be split into an H_0, the Hamiltonian for the non-interacting molecules and a perturbation $V = H - H_0$. If the ground-state eigenvalue of H is E and that of H_0 is E_0, then the interaction energy or potential energy surface is $U(R) = E(R) - E_0$. Here R typically denotes the separation between the molecules, but more generally symbolically denotes a set of distances and angles which specify the positions of the molecules relative to one another.

4.2.1 Long-range perturbation theory

Quantum mechanical perturbation theory has long been used to systemise the study of intermolecular forces. Traditionally, the range of intermolecular separations has been divided into long-, intermediate- and short-range regions, although the precise location of the boundaries separating the regions cannot be specified; different perturbation formalisms are applied to the different regions. The most highly-developed formalism is applicable in the long-range region, where overlap among interacting molecular wave functions is neglected. It is also a formalism which has seen significant computational advances in recent years. We discuss these developments in this section, reserving to later sections a discussion of intermediate- and short-range perturbation theories.

4.2.1.1 Polarisation expansion

The computational advance in the long-range theory has come in the evaluation of dispersion force coefficients from experimental and/or theoretical sum rule data. Before summarising these developments, it is well to recall the bare outlines of the long-range theory in order to indicate the data which are required for the quantitative estimation of other types of long-range forces.

The polarisation or multipole expansion is well-known in principle[9]. It results from a straightforward application of the Rayleigh–Schrödinger perturbation theory to the clamped-nuclei Schrödinger equation for the interacting molecules. The unperturbed problem is the non-interacting molecules and $U(R)$ is expanded in powers of the perturbation, which is all the coulombic interactions among molecules.

$$U(R) = E_1(R) + E_2(R) + E_3(R) + \cdots. \tag{4.1}$$

The perturbation energies are further expanded in inverse powers of the separations between molecules,

$$U(R) = C_1 R^{-1} + C_2 R^{-2} + \cdots - C_6 R^{-6} - C_8 R^{-8} + \cdots, \tag{4.2}$$

and each term is given a classical or quasi-classical interpretation as arising from interactions between either permanent or instantaneous multipole moments of the separated molecules. In this way the electrostatic, induction, dispersion and resonance forces are identified.

The general formalism is applicable to interactions between ground- or excited-state molecules if they are well separated. The theory is of unquestioned utility in understanding a wide variety of experimental results on a qualitative basis and in providing a starting point for semi-empirical theories. The quantitative application of the theory is limited because of complications in the accurate estimation of the various energy terms. For the electrostatic and induction forces, a knowledge of the permanent multipole moments, polarisabilities and hyperpolarisabilities of the isolated molecules is required. In general, these quantities are tensors and each independent component is required to estimate the anisotropy of the long-range forces. In principal, these molecular properties can be obtained from quantum calculation or from experiment, although practice is limited. The resonance forces require a knowledge of a set of molecular wave functions corresponding to the degenerate states which are split by the intermolecular perturbation.

It is not immediately obvious that the dispersion forces can also be expressed in terms of properties of isolated molecules. The reduction of the apparent two-centre problem to one centre (at least for the leading dispersion forces C_6 and C_8) has provided a key step in the evaluation of dispersion force constants. These developments are discussed in detail below.

4.2.1.2 Accurate evaluation of dispersion energies

In his 1967 review article[14], Dalgarno described various methods — accurate variational, Hartree-Fock, double perturbation, and semi-empirical — for estimating dipole–dipole dispersion force coefficients C_6 for atoms and small molecules. Of the various methods the semi-empirical procedure was the most practical and general procedure to apply. Consequently, most of the effort in subsequent years has been aimed at systematising and generalising the method, and, importantly, in estimating error bounds to the results obtained. A recent review[15] of this work has been given by Langhoff and Karplus.

To begin the discussion let us consider the interaction of spherical atoms a and b in their ground states. The expression for C_6 may be written

$$C_6 = \tfrac{3}{2} \sum_{m,n} \frac{f_n^a f_m^b}{\omega_n^a \omega_m^b (\omega_n^a + \omega_m^b)} \tag{4.3}$$

where (f_n, ω_n) are the oscillator strength and frequency for a dipole-allowed transition from the ground state to state n. The original semi-empirical treatment employed photoabsorption data (f_n, ω_n) to approximately sum the series. Since complete data was not, and is not, available, it proved advantageous to adjust the available oscillator strengths to satisfy particular 'sum rules', defined by

$$S(k) = \sum_n \omega_n^k f_n \tag{4.4}$$

where $S(k)$ can be obtained independently of the (f_n, ω_n).

After the initial developments, it became clear[16] that the sum rules, instead of being simply checks on the data, are actually fundamental quantities from which C_6 can be evaluated without specific knowledge of the (f_n, ω_n). This realisation followed the application of the Casimer–Polder identity to uncouple the denominators in equation (4.3) and provide the formula

$$C_6 = \frac{3}{\pi} \int_0^\infty \alpha_a(i\omega)\alpha_b(i\omega)d\omega \qquad (4.5)$$

where $\alpha(i\omega)$ is the dipole polarisability at imaginary frequency,

$$\alpha(i\omega) = \sum_n f_n(\omega_n^2 + \omega^2)^{-1}$$

The imaginary frequencies mean that the strength of the external field changes exponentially with time. Equation (4.5) is a significant computational advance since α involves properties of a single atom instead of the pair. Moreover, α provides a direct link of C_6 to the sum rules $S(k)$. For real frequencies ω, the α has power series expansion in direct powers of ω^2 (the Cauchy expansion);

$$\alpha(\omega) = \sum_{k'=0}^\infty S(-2k'-2)\omega^{2k} \quad \text{(for small values of } \omega) \qquad (4.6)$$

Experimental values of $S(-2k)$ can be obtained by analysis of refractive index data. For high imaginary frequency, only the first two terms in an expansion in inverse powers of ω^2 are finite:

$$\alpha(i\omega) = S(0)\omega^{-2} - S(2)\omega^{-4} + \text{non-analytical terms (for large values of } i\omega) \qquad (4.7)$$

To obtain C_6, the behaviour of α along the real axis must be related to its behaviour along the imaginary axis. The ω cannot simply be changed to $i\omega$, since the Cauchy series converges only for ω less than the first allowed transition frequency. For C_6, a knowledge of α is required along the entire positive imaginary axis; hence analytical continuation is required. This is conveniently performed by a rational approximation

$$\tilde{\alpha}(\omega) = (a_0 + a_1\omega^2 + \cdots + a_{N-1}\omega^{2N-2})/(1 + b_1\omega^2 + \cdots + b_N\omega^{2N}) \qquad (4.8)$$

The $\tilde{\alpha}$ contains $2N$ coefficients which may be chosen so that $2N$ terms in the expansion of $\tilde{\alpha}$ match precisely $2N$ terms of α in the high- and/or low-frequency region. (Mathematically, this is known as the Cauchy interpolation problem and it is not necessarily trivial to solve[17].) There is some choice as to which terms to match and a comparison of various approaches has been given by Langhoff, Gordon and Karplus, who also give references to earlier workers[18]. Having chosen the coefficients, $\tilde{\alpha}$ can be rearranged to

$$\tilde{\alpha}(i\omega) = \sum_{n=1}^N \tilde{f}_n(\tilde{\omega}_n^2 + \omega^2)^{-1}$$

If $\tilde{\alpha}$ is substituted into Equation (4.5), integration yields a formula

$$\tilde{C}_6 = \frac{3}{2} \sum_{n,m}^N \tilde{f}_n^a \tilde{f}_m^b / [\tilde{\omega}_n^a \tilde{\omega}_m^b (\tilde{\omega}_n^a + \tilde{\omega}_m^b)] \qquad (4.10)$$

Although this has the form of Equation (4.3), the $(\tilde{f}_n, \tilde{\omega}_n)$ are obtained directly from the sum rules $S(k)$ and bear no simple relation to the photoabsorption data.

Although the above discussion outlines the basic semi-empirical scheme, it is incomplete since it does not reveal that (i) there are several algorithms for choosing the a_n, b_n in such a way that (assuming exact $S(k)$ are used) $\tilde{\alpha}(i\omega)$ is either an *upper* or a *lower bound* to the exact $\alpha(i\omega)$, and hence $\tilde{C}_6$ is a corresponding bound on the exact C_6; and (ii) it is possible to bound α from a knowledge of almost *any* set of sum rules $S(k)$, not just the even-negative-integer sum rules obtainable from refractive-index data. As regards the first point, it has been shown[18] for a variety of cases that it is possible to obtain tight bounds on C_6 with *five* or *fewer* sum-rule values. It will be a long time before any other method for estimating C_6 can compete with this result. The second point is important for those cases in which experimentally-determined sum rules are unavailable and one must resort to *ab initio* calculations. A computationally revealing expression for $S(k)$ is[19]

$$S(k) = \langle \psi_0 \mid \mu(H_0 - E_0)^{k+1}\mu \mid \psi_0 \rangle \tag{4.11}$$

where $(H_0 - E_0)\psi_0 = 0$ refers to the atomic Schrödinger equation and μ is a component of the atomic dipole moment operator. It is apparent that $2 \geqslant k \geqslant -2$ are the most accessible cases computationally (for $k > 2$, $S(k)$ diverges), in that $S(k)$ is an expectation value with respect to the ground-state wave function (except for $k = -2$, in which case S is simply related to the static polarisability which usually may be obtained experimentally). It is therefore satisfying that C_6 may be bounded, given a knowledge of the ground-state wave-functions of the interacting species.

The discussion so far has assumed that *exact* values of the $S(k)$ are available. If only approximate values $\tilde{S}(k)$ are known, it is still possible to estimate C_6, but the bounds are looser[20, 21]. In employing $\tilde{S}(k)$, it should be realised that the exact $S(k)$, regarded as a function of a continuous variable k, has mathematical properties which must be retained by the $\tilde{S}$ if the bounding properties of the semi-empirical scheme are to be maintained[22].

The methods we have been discussing are not limited to estimating C_6 for ground-state spherical atoms, it is also possible to estimate C_8 coefficients from quadrupole sum rules[23], the anisotropy in C_6 for interactions between asymmetric molecules[18], the Axilrod–Teller three-body dispersion coefficient[15], and long-range relativistic force coefficients[24-28]. The bounding properties apply to these cases also.

Langhoff and Karplus have given tabulations of available estimates of dipole–dipole and triple-dipole dispersion force coefficients and the leading relativistic dispersion force coefficient[15]. The systems for which bounded estimates of C_6 are available are hydrogen, rare gas and alkali atoms and simple diatomic molecules, H_2, N_2 and O_2. Other workers[29-31] have estimated C_6 for interactions involving alkaline-earth atoms.

If one is interested in estimating dispersion force coefficients for molecules in *excited* states, the semi-empirical scheme must be modified somewhat[32]. The bounding properties of the estimates derive ultimately from the fact that, for a ground-state molecule, all of the oscillator strengths are positive. For excited states, the (f_n, ω_n) for transitions to lower states are negative.

In order to estimate C_6 in such cases, one must calculate the contribution from the lower states explicitly, and use the semi-empirical procedures discussed above to estimate and bound the contributions of the higher excited states. Even in the case of ground-state molecules, if some (f_n, ω_n) are known, they may be introduced into α explicitly, and the bounding procedures may be applied to the remainder. Refractive-index data for excited states are not likely to become available, so the sum rules must be computed.

4.2.1.3   Other estimates of long-range interactions

Ab initio calculations are feasible for estimating dispersion force coefficients for small atoms and molecules. Either the coefficients are approximated directly in a two-centre calculation or time-dependent perturbation theory is applied to calculate appropriate polarisabilities. Two-centre calculations have been reported for the interactions of hydrogen atoms[33] and molecules[34], HCl with itself, rare gases or H_2 [35], and for the lowest three states[36] of HeH$^+$ ion. Coupled Hartree–Fock calculations of $\alpha(i\omega)$ have been reported for He–He [37], He–Be [38] and He–Li$^+$ [39] interactions. Exact evaluation of $\alpha(i\omega)$ has been accomplished for hydrogen-atom interactions[40]. Brueckner–Goldstone perturbation theory has been applied to compute $\alpha(i\omega)$ for hydrogen and helium atoms[41].

When accurate estimates of long-range force coefficients are unavailable, rough approximations can be made, employing the procedures discussed by several workers[42, 43]. Dispersion force coefficients may be estimated by the Drude model[44], although the model appears[9] to underestimate C_6 by 10–30% for systems with low polarisability (e.g. rare gases), and by a factor of 2 or 3 for polarisable systems (e.g. alkali atoms). Drüde-model estimates of C_8 also appear to be too small[45]. Kramer and Herschbach[46] have given combination rules for estimating two- and three-body dispersion force coefficients for unlike molecules in terms of C_6 and the dipole polarisabilities of the corresponding like molecules. The rules are accurate to within a few per cent for known cases.

4.2.1.4   Validity of the long-range expansion

Having discussed how one may estimate intermolecular forces in the long-range region by the polarisation expansion, it is now necessary to indicate where the expansion is valid, even though it is difficult to give a precise indication. There are two related expansions in the long-range theory; the interaction energy is expanded in powers of V; and V is expanded in powers of R^{-1}. Qualitatively, both expansions are expected to fail at separations where the isolated molecular wave functions overlap appreciably.

The radius of convergence of the first expansion is not known in the molecular case. A delta function model has been studied and it has been shown that the analogous expansion converges for *all* internuclear separations[47, 48]. For the case of two interacting helium atoms, however, Claverie has given[49] a symmetry argument which proves that the expansion does not converge to the physical ground state, $^1\Sigma_g^+$. It is plausible that Claverie's argument holds for larger systems as well. Malrieu has given further consideration to the convergence problem by explicitly summing high order contributions for a model system[50].

The expansion of $E_2(R)$ in powers of R^{-1} is known to be divergent (more precisely, asymptotic) in the one-electron case, and presumably for larger systems as well. Meath and co-workers[51-54] and Fukui and Yamabe[55] have studied the error involved in truncating the expansion of E_2 after a finite number of terms. Horak and Siskova have evaluated E_2 exactly for H_2^+ by explicitly calculating the contribution of discrete and continuum states to the perturbation sum[56]. These workers find that the truncated multipole expansion is an accurate representation of E_2 for non-bonded systems in the region of the van der Waals minimum, although at shorter separations the error rapidly grows.

Although the polarisation expansion formally involves two divergent expansions, the practical issue involves the ranges of R over which the expansion, truncated in low order, may be used to obtain a useful approximation to $U(R)$. This question is partially answered in an indirect but physical way in the method developed by Le Roy and Bernstein and Stwalley for extracting long-range force coefficients from the distribution of diatomic vibrational levels near the dissociation limit[57-59]. Le Roy and Bernstein have shown that the distribution is governed mainly by the long-range region of the potential and have given a graphical procedure for obtaining the coefficient C_n of the leading R^{-n} term from the slope of a straight line. Stwalley has shown how to obtain the next order non-vanishing coefficient, say, C_m. The fact that the procedure works, i.e. that straight lines are obtained, testifies to the validity of the assumption that $U(R) = C_n R^{-n} + C_m R^{-m}$. Deviations from the straight line indicate the separation at which the assumption fails. Applications so far indicate that the two leading multipole terms give accurate representations of $U(R)$ for $H_2(^3\Sigma_u^+)$, $Mg_2(X^1\Sigma_g^+)$, and Cl_2 $(B^3\Pi(O_u^+))$ for $R \gtrsim 6$ Å. The corresponding potential minima occur for $R = 4.15$ Å, 3.89 Å and 2.44 Å, respectively. Furthermore, the higher multipole terms become important at roughly the same value of R as the exponentially-decreasing overlap terms[59]. Since the two sorts of terms enter with opposite signs, it rarely makes sense to include higher-order multipole terms while neglecting overlap terms.

Therefore, both formal and practical considerations indicate that the long-range expansion is an accurate approximation to $U(R)$ only for R considerably larger than the van der Waals separation. In the next section, we consider various attempts to develop a perturbation formalism to systematically improve upon the long-range results.

4.2.2 Intermediate-range perturbation theory

4.2.2.1 Motivations

The long-range expansion of the intermolecular interaction energy becomes inaccurate as the overlap between the molecular wave functions increases. Then the Pauli Principle demands that the total wave function, Ψ, has permutational symmetry appropriate to the complete Hamiltonian representing the interacting pair. It becomes necessary to consider a new perturbation expansion for the wave function in which the zeroth-order approximation already has the symmetry required by the Pauli Principle. The energy obtained by such an expansion is expected to account for both the multipole interactions and the exponentially-decreasing overlap terms.

A theory which starts with a properly symmetrised wave function is found[60, 61] in the historic work of Heitler and London and of Eisenschitz and London. Here the zeroth-order wave function is an antisymmetrised product of isolated molecule functions. The expectation value of the Hamiltonian with respect to this function defines the Heitler–London energy. Although this is a natural choice, it is now recognised[62, 63] that there is a basic difficulty with the Heitler–London energy, in that the exponential terms are not the correct leading exponential terms in $U(R)$ for those cases for which the correct behaviour is known. Thus a perturbation scheme based on the Heitler–London result is undesirable in the sense that it does not produce in leading order the proper asymptotic behaviour of the quantity of interest.

Nevertheless, almost all workers have chosen the Heitler–London results as a starting point and have attempted to extend the formalism to higher orders. No treatment has demonstrated that the correct exponential terms can be obtained in this way. Moreover, the focus of attention has not been on predicting the correct exponential terms, but rather in attempting to resolve formal difficulties which arise when the Heitler–London result is extended to higher orders.

4.2.2.2   *Mathematical complications*

The most natural choice for the unperturbed Hamiltonian is H_0, the sum of the Hamiltonians for the non-interacting molecules. However, this implies associating particular electrons with definite molecules. Hence the symmetry of H with respect to electron permutations is greater than that of H_0. The difficulties to which this gives rise are known as the *exchange* problem.

A basic difficulty is that the *order* of the perturbation terms is not uniquely defined. If $\mathscr{A}$ is the operator which projects the component with the symmetry of the desired total wave function, the $\mathscr{A}$ commutes with H. However, $\mathscr{A}$ does not commute separately with either H_0 or the perturbation V, but rather

$$[\mathscr{A}, H_0] = [V, \mathscr{A}]. \qquad (4.12)$$

In any conventional perturbation scheme, the left-hand side of equation (4.12) is zeroth-order and the right-hand side is first-order. This equation means that 'order in V' is not a well defined concept and that the apparent order of various terms in a perturbation expansion can be shifted arbitrarily.

It might be thought that 'order' could be defined by powers of overlap integrals, i.e. that terms could be ordered in the sequence 1, e^{-R} e^{-2R}, ... However, the observation that

$$\sum_{n=0}^{\infty} e^{-R} R^n / n! = 1$$

shows this approach to be ambiguous.

Related to the non-uniqueness of order is the difficulty of defining a symmetrised basis set for the expansion of Ψ. A natural choice consists of the set $\mathscr{A}\phi_k$, where the ϕ_k are the complete set of eigenfunctions of H_0,

$$(H_0 - \varepsilon_k)\phi_k = 0, \tag{4.13}$$

where ϕ_k is a simple product of isolated-molecule wave functions. The basis $\mathscr{A}\phi_k$ is non-orthogonal, however, whereas non-degenerate eigenfunctions of a Hermitian Hamiltonian are necessarily orthogonal. Thus it is impossible to define a single, Hermitian unperturbed Hamiltonian, of which every $\mathscr{A}\phi_k$ is an eigenfunction. In particular, since $\mathscr{A}$ and H_0 fail to commute, the $\mathscr{A}\phi_k$ are not eigenfunctions of H_0.

Another lack of uniqueness is due to the fact that the complete set of symmetrised functions $\mathscr{A}\phi_k$ is linearly dependent. To see this, consider a function Ω which has symmetry different from $\mathscr{A}$, i.e. $\mathscr{A}\Omega = 0$. Since the ϕ_k are complete, Ω has the unique expansion

$$\Omega = \sum_k \phi_k C_k \tag{4.14}$$

where the C_k are constants. By hypothesis, $\mathscr{A}\Omega$ vanishes, so that

$$\sum_k \mathscr{A}\phi_k C_k = 0, \tag{4.15}$$

which is a statement of linear dependence of the set $\mathscr{A}\phi_k$. Moreover, the linear dependence of the $\mathscr{A}\phi_k$ is non-trivial in case that it is possible to find an Ω such that $\mathscr{A}\Omega = 0$ and each C_k is non-vanishing. Then it follows that it is impossible to construct a linear independent set by excluding a *finite* number of terms from the original set. For example, $\mathscr{A}\phi_0$ can be eliminated from equation (4.15) by the expansion

$$\mathscr{A}\phi_0 = \sum_k \phi_k a_k = \sum_k \mathscr{A}\phi_k a_k^- \tag{4.16}$$

or

$$\mathscr{A}\phi_0 = \sum_{k>0} \mathscr{A}\phi_k a_k (1 - a_0)^{-1}. \tag{4.17}$$

Substituting into equation (4.15) yields

$$\sum_{k>0} \mathscr{A}\phi_k [C_k + C_0 a_k (1 - a_0)^{-1}] = 0. \tag{4.18}$$

Thus the set $\mathscr{A}\phi_k$, $k > 0$, is also linearly dependent.

This means that no *unique* expansion of the total wave function of the form

$$\Psi = \sum_k \mathscr{A}\phi_k C_k \tag{4.19}$$

is possible.

A related source of ambiguity arises in the observation that *if* one can solve the equation

$$(H - E)\Phi = \chi, \tag{4.20}$$

where $\mathscr{A}\chi = 0$ but χ is otherwise *arbitrary*, then

$$(H - E)\mathscr{A}\Phi = 0 \tag{4.21}$$

This equation can be resolved into perturbation equations in various ways. Although many choices for χ may be made, there is not a clearly defined physical or mathematical significance which can be attached to any particular choice.

The above considerations do not imply that it is impossible to develop a perturbation expansion for intermolecular forces that takes full account of symmetry, but rather that many *different* approaches are possible.

4.2.2.3 Survey of exchange perturbation theories

The oldest perturbation theory which takes the exclusion principle into account was developed by Eisenschitz and London[61]. The formalism has been put into a more modern notation by van der Avoird[64]. These authors assumed the expansion given in equation (4.19). Of course, they recognised the lack of uniqueness of such an expansion, but nevertheless resolved the Schrödinger equation into an infinite set of perturbation equations which can be solved in a well defined, though arbitrary, way. Their expression for the first-order interaction energy agrees with the Heitler–London result. The second-order energy is expressed in a sum over states form which, when evaluated by the Unsöld method, gives the second-order polarisation energies, modified by the effects of exchange. Dalgarno and Lynn[65, 66] have introduced the Unsöld approximation to the second-order energy of a Brillouin–Wigner expansion and obtained the same result as Eisenschitz and London.

Löwdin[67] and Van der Avoird[68, 69] have given elegant wave-operator formalisms which give the same expression for the first- and second-order energy as the Eisenschitz–London expansion. Löwdin's approach also gives a bracketing theorem for the energy. Hirschfelder[70] has derived a Rayleigh–Schrödinger expansion of van der Avoird's equations.

Musher and Salem[71] have also assumed the expansion of Ψ given by equation (4.19). These authors used a Feenberg iteration technique to evaluate the coefficients, however, and obtained a different expression for the second-order energy. This approach does not require H to be separated into $H_0 + V$.

Other examples in which the expansion equation (4.15) is used are the works of Murrell, Randic and Williams[72], of Salem[73], of Pecul[74], and of Dacre and McWeeny[75]. The first authors assumed that equation (4.19) consists of a finite number of terms which includes both covalent and ionic type functions. This allows questions of over-completeness to be avoided, and it is equivalent to solving a finite dimensional secular equation by a perturbation expansion; similar ideas have been advanced by Pecul. Salem makes the assumption that $\langle \mathscr{A}\phi_k | \mathscr{A}\phi_l \rangle = \delta_{kl}$, which does not hold for the functions $\mathscr{A}\phi_k$, but which is useful in assessing the significance of various terms in equation (4.19). Dacre and McWeeny expand the interaction energy in powers of overlap and express their results in terms of a spin Hamiltonian.

A different type of expansion of Ψ was assumed by Murrell and Shaw[76], who used a wave-operator approach, and in an equivalent[77] treatment by Musher and Amos[78, 79], who started from an infinite secular equation. These authors assumed that the zeroth-order component of Ψ has proper symmetry, but that the remainder can be expanded in the unsymmetrised functions ϕ_k. Hence

$$\Psi = \mathscr{A}\phi_0 + \sum_{k>0} \phi_k C_k \qquad (4.22)$$

The expansion in this set of functions is unique, although the higher order terms do not have definite symmetry properties, order by order. Musher and Amos[80] have given a spin-free formulation similar to their original results. The iterative procedure of Carr[81] yields a similar expression for the energy.

A different class of approaches to the exchange problem involves different ways of defining and computing a 'primitive function' whose projection onto the space of desired symmetry is the total wave function Ψ, in the same sense that the zeroth-order function $\mathscr{A}\phi_0$ is the projection of ϕ_0. Hirschfelder and Silbey[70, 82] propose that there is a physically-significant primitive function whose symmetry projections correspond to all the wave functions for the family of states arising from a single electron configuration. Related approaches have been discussed by Herring[62], Musher and Silbey[83], Kirtman[84], Klein[85], Certain and Hirschfelder[86, 87] and Adams[88].

Jansen[89] (see also Byers Brown[90], Ritchie[91, 92], Corinaldesi[93] and Landman and Pauncz[94]) has explicitly constructed an operator Λ which operates on a symmetrised function Ψ to produce a function in which specific electrons are assigned to particular atoms. With Λ he is able to construct an unperturbed non-Hermitian 'label free' Hamiltonian whose eigenfunctions are $\mathscr{A}\phi_k$. This is a basis-set dependent approach since clearly $\mathscr{A}$ has no inverse.

Laughlin and Amos[95] have replaced the original Schrödinger equation by a generalised eigenvalue equation and have devised analogues to several of the perturbation formalisms mentioned above.

Brändas[96] has considered the extension of the perturbation-variation method developed by himself and Goscinski to the exchange perturbation problem. He suggests that the computations could be carried through without recourse to the use of a large basis set.

Boehm and Yaris[97] have given a second quantised formalism which inconsistently omits some exponentially-decreasing Coulomb and exchange terms.

Several workers[98-101] have suggested perturbation formalisms for intermolecular forces within self-consistant field (SCF) theory. Their approaches are not applicable in the intermediate region, however.

Relationships have been explored among some of the formalisms which have been proposed[70, 77, 102, 103].

4.2.2.4 *Applications of exchange perturbation theories*

There have been a number of applications of various exchange perturbation theories to H_2^+ (references 104–107), to H_2 (references 108–113), to He_2 (references 114, 115), to N_2 (reference 75), to a harmonic oscillator model[116, 63], a spin model[117] and a delta-function model[118, 119], to non-bonded hydrogen interactions[120], to hydrogen-bonding[121–124], to charge–transfer interactions[243], to physical adsorption[409], and to the non-additivity of intermolecular forces[125–127]. The largest number of applications have been to H_2 and H_2^+, where comparison with exact energies is possible. The results indicate that no particular formalism is significantly more accurate than all the others. To gain acceptable accuracy (error less than 5%) for H_2, it is necessary to include third- and higher-order energy terms. This requires elaborate and time-consuming computational procedures.

In general, to apply any perturbation theory, further approximations, over and above the truncation of the expansion, must be made. These approximations arise since for molecular systems the perturbation equations cannot be solved exactly. Typically variational principles are used to estimate each order. However, if this step is taken and the same basis set used for each order, the formalism reduces to the expansion of a finite secular equation. The latter problem can be solved exactly in a variety of ways and there is little reason for applying a perturbation formalism. Moreover, since the second-order energy has not been obtained exactly, it cannot be determined whether the incorrect behaviour of the Heitler–London energy has been corrected.

In view of these negative comments, what is the value of an exchange perturbation theory? Consider for a moment the long-range theory, which has great value in the intermediate range in spite of the fact that the multipole expansion itself is invalid, because it suggests an accurate form for a variational wave function. The trial wave function for the helium–helium interaction, to be discussed in Section 4.2.4.2, can be thought of as being obtained by antisymmetrising an orbital approximation to the polarisation wave function. In the same way, the long-range theory is useful in constructing trial functions for approximating properties other than the energy. Exchange perturbation theory can be expected to provide insight into the form of an accurate variational function over and above the terms based on the long-range theory. The information that has been gained thus far is limited by the necessity of using basis sets to approximate the perturbation wave functions, but it seems clear that ionic configurations should be included in the trial wave function. Such terms occur in the variational schemes discussed in Section 4.2.4.2. More useful information could be obtained if progress could be made in the solution of the perturbation equations without recourse to basis sets.

The above remarks are relevant if the theory is used to accurately calculate interaction energies or to obtain analytical forms of perturbation functions to be used as a guide in constructing good variational trial functions. Another motive, however, is to develop a theory which provides a qualitative and semi-quantitative guide to interaction energies of large systems, for which accurate treatments of any type are not feasible. Work along these lines has been reported by Murrell and others[72, 128, 129]. Approximate schemes have been developed for evaluating the Heitler–London energy as an expansion in powers of overlap integrals. The second-order induction and dispersion energies are taken from the long-range theory; i.e. second-order exchange energies are neglected. The formalism has been applied to the hydrogen bond and to charge–transfer interactions. Although it is difficult to judge the general reliability of the method, the scheme gives accurate results[12, 115] for the helium–helium interaction in the region of the van der Waals minimum. At shorter distances, the repulsive branch of $U(R)$ is too soft, because there is no mechanism for turning off the multipole expansion. The neglect of second-order exchange energies may not be accurate in some cases even at long range[130], and Jansen has emphasised their importance in determining solid-state properties[131].

Jansen[131] has formulated a model, based on exchange perturbation theory, for treating two- and three-body exchange forces in solids. A key feature of the model is the inclusion of second-order exchange forces. The model has been successfully applied to problem of rare-gas crystal stability[131], crystal structures and elastic constants of simple ionic solids[132].

4.2.3 Short-range perturbation theory

The ambiguities discussed in the previous section arise in attempts to begin a perturbation expansion for a symmetrical wave function from an unsymmetrical starting point. There are two theories of possible generality, both of which have been applied successfully to the case of H_2, which begin with a fully-symmetrised wave function.

4.2.3.1 Short-range united atom perturbation theory

A perturbation theory for the short-range forces between two atoms was introduced by Morse and Stueckelberg[133] in 1929, one year before London's classic paper on the long-range interaction. At short range $U(R)$ diverges because of the nuclear repulsion, so the meaningful quantity to consider is the electronic interaction energy.

$$W(R) = U(R) - Z_a Z_b / R \qquad (4.23)$$

The original treatment, developed more fully by Bingel[134], expanded $W(R)$ in direct powers of R,

$$W(R) = W_0 + R^2 W_2 + R^3 W_3 + R^4 W_4 + \cdots, \qquad (4.24)$$

where W_0 is the difference between the united-atom and separated-atom limits and W_1 vanishes in all cases by symmetry. The expansion is obtained by a perturbation theory based on a united-atom Hamiltonian, centred at a particular spot on the internuclear axis, followed by an expansion of each energy order in powers of R.

Three difficulties with the expansion (4.24) are that (a) $W(R)$ is not analytical in R because a term $R^5 \ln R$ appears in higher-order energies[135], (b) each energy order is of order R^2, and (c) when truncated, the expansion is not accurate. Byers Brown[136] has analysed the failure of the expansion and has shown that the united-atom theory is poor because a single unperturbed wave function cannot reproduce the correct cusp behaviour at both nuclei of the diatom, no matter where the united atom is placed. However, if ψ^a is a united atom function centred at nucleus A, then

$$\Psi_0 = \Lambda\psi^a = (Z_a\psi^a + Z_b\psi^b)/(Z_a + Z_b) \tag{4.25}$$

has the proper cusp behaviour for $R \to 0$. Here ψ^b is $P\psi^a$ where P inverts ψ^a through the mid-point of the internuclear axis. By employing the operator Λ to carry out a similarity transformation, $\mathscr{H} = \Lambda^{-1}H\Lambda$, Byers Brown has given an elegant derivation of the expansion

$$W = W_0 + \mu W^{\mathrm{I}} + \mu^2 W^{\mathrm{II}} + \cdots, \tag{4.26}$$

where $\mu = Z_aZ_b/(Z_a + Z_b)^2$. The advantages of this expansion are (a) it is a natural expansion in the exactly-solvable one-electron case; (b) in general the nth-order energy is of order R^{2n}; (c) accurate results can be obtained for small atoms at separations up to one Bohr by a first-order treatment[137].

In view of the success of Byers Brown's theory, it is tantalising to consider a similar approach to exchange perturbation theory. In the latter case, however, the obvious choice for Λ in the case of H_2 is $1 + P_{12}$, where P_{12} permutes electron labels. Unlike the short-range case, Λ commutes with H so that Λ leaves H invariant.

Another united-atom perturbation theory has been discussed by Lee, Dutta and Das[138]. These authors treated the HF molecule based on the Ne united atom centred at the position of the F nucleus. The unperturbed Hamiltonian was the Hartree–Fock V^{N-1} Hamiltonian for Ne, and linked cluster perturbation theory was applied to calculate the molecular energy and the electric field at the F nucleus. Although only the equilibrium internuclear separation was considered, the total energy obtained was closer to the 'exact' non-relativistic energy than either Hartree–Fock or configuration interaction (C.I.) results. The electric field due to the electrons failed to cancel the proton field by only 5%.

4.2.3.2   The 1/Z expansion

Another perturbation treatment which preserves the symmetry of the wave function is the $1/Z$ expansion. If the coordinates are scaled properly, the molecular Hamiltonian can be written

$$H = Z^2\left\{\sum_i\left[-\tfrac{1}{2}\nabla_i^2 - \sum_\alpha Z_\alpha/Zr_{\alpha i}\right] + Z^{-1}\sum_{i>j}1/r_{ij}\right\} \tag{4.27}$$

so that Z^{-1} becomes a natural perturbation parameter and the energy and wave function can be computed by standard perturbation theory. Kirtman and Decious[139] and Matcha and Byers Brown[140] have applied this procedure to H_2. The latter authors have computed the energy as far as 5th order at $R = 1.4$ Bohr. The results are comparable with the most accurate variational results. The treatment for H_2 cannot be extended to long range without modification because $(\sigma_g)^2$ does not dissociate properly. A general procedure for modifying the treatment is to apply Van Vleck perturbation theory[141, 142].

For other systems, the Z^{-1} wave function dissociates properly, but the calculations have been done only at short range. For He_2 and LiH, Goodisman[143, 144] finds that the bare-nucleus Z^{-1} expansion converges very slowly at short range. By adopting a screened nucleus zero-order Hamiltonian, however, the perturbation results as far as second order compare favourably with the best variational calculations.

4.2.4 Variational methods

Variational methods are becoming increasingly-powerful tools for computing intermolecular forces. If dispersion terms are neglected, several groups can now calculate potential-energy surfaces which have the expected qualitative features (e.g. correct dissociation limits) and which are being used in meaningful and detailed ways in comparisons with experiment. The surfaces are typically obtained via an *ab initio* valence-bond or molecular-orbital configuration-interaction calculation, in which sufficient configurations are included so that the dissociation products are described consistently with respect to the interacting species. Additional configurations may be included to describe important electron correlations, or to take account of avoided curve crossings. For small systems the orbitals in the configurations can be chosen self-consistently (SCF, MC–SCF[145] or INO[146]).

Several compendiums of systems which have been treated by variational calculations have appeared[147–152]. Recent SCF calculations include the systems $(H_2 + H)$[153, 154] $(Li + HF)$[155], $(Li^+ + H_2)$[156, 157] $(CH_4 + H^+)$[158], $(He_2^+ + H)$[159], $(Be + H_2^+)$[160], $(He + H_2^+)$[161], $(F^- + CH_3F)$[162], $(CN^- + CH_3F)$[162], $(Cl + H_2)$[163], $(Li_2 + Li_2)$[164], $(H_2 + H_2)$[165], $(H^+ + Ar)$[166], (Li or Na + He)[167]. Clementi is continuing his series of SCF studies of interactions between large molecules[168]. It appears to be established that the SCF model is adequate for treating hydrogen bonding[169] and barriers to internal rotation[170, 171]. Lester's treatment of $Li^+ + H_2$ demonstrates that long-range induction forces are described within the SCF approximation if a flexible basis set is employed[156, 157]. Recent C.I. calculations of potential-energy surfaces include $(He + H_2)$[172], $(CH_2 + CH_2)$[173], $(O^+ + N_2)$[174], $(Na + H)$[175], $(H_2 + H^+)$[176], $(Li + F_2)$[177], $(H_2 + H_2)$[178], $(H(2s) + He)$[179]. Space limitations preclude a discussion of these applications in this review. Rather, we are content to discuss recent work on two aspects of intermolecular forces: the non-additivity of forces among closed-shell systems and the calculation of van der Waals interactions between closed-shell molecules.

4.2.4.1 *Non-additive intermolecular forces among closed-shell molecules*

Although elementary forces are two-body forces, the forces between aggregates of particles are many-body in nature. Much of the theory of condensed phases assumes the convergence of a cluster expansion for the total energy $E (abc \ldots n)$ of an aggregate of molecules $a, b, \ldots n$, which is defined by

$$E(abc \ldots n) = \sum_a E_a + \sum_{a>b} E_{ab} + \sum_{a>b>c} E_{abc} + \sum_{a>b>c>d} E_{abcd} + \cdots + E_{abc \ldots n} \tag{4.28}$$

where, for example, $E_{abc} \equiv E(abc) - E_{ab} - E_{ac} - E_{bc} - E_a - E_b - E_c$. Non-additive forces arise from the third and succeeding terms.

The magnitude of E_{abc} has been estimated in both the long- and short-range regions. By using the long-range polarisation expansion as far as third-order Axilrod and Teller[180] obtained the 'triple dipole' term for spherical atoms,

$$E^{(3)}_{abc} = v(abc)(1 + \cos \theta_a \cos \theta_b \cos \theta_c)/(R_{ab}R_{ac}R_{bc})^3 \qquad (4.29)$$

where $v(abc)$ is a constant and θ_a, θ_b, θ_c are the interior angles of the triangle with sides R_{bc}, R_{ac}, R_{ab}. Expressions for higher-order three-body dispersion terms have been given by Bell[181] and by Zucker and Doran[182]. Estimates of $v(abc)$ for the rare gases, alkali metals, and a few simple molecules have recently been given by Langhoff and Karplus[15] using the bounding techniques discussed in Section 4.2.1.2. This work follows previous estimates by Dalgarno[14] and Bell[181]. The sign and magnitude of $E^{(3)}_{abc}$ depends on the geometry, but is typically plus or minus a few per cent of the two-body dispersion forces for the systems considered by Langhoff and Karplus. The triple-dipole term for asymmetric molecules has been considered by Stogryn[183, 184].

An SCF calculation of the energy of He$_3$ has recently been reported by Bader, Navaro and Beltram-Lopez[185]. They find the non-additive energy to be at most a few per cent of the pair-wise additive energies for separations greater than 3 Bohr. This is consistent with the estimate of Williams, Schaad and Murrell[126], who used an exchange perturbation theory. Williams[127] has also studied the non-additivity of the first-order exchange energy for C$_3$. Although none of these calculations claims high accuracy, the results of Bader *et al.* are probably the most reliable.

No work has been reported which treats both the dispersion and exchange contributions to E_{abc} simultaneously although model calculations of Jansen[131, 125] suggest that such a calculation is required to obtain a useful estimate of E_{abc}. In addition, Jansen's results suggest that non-additive exchange effects are particularly small in He$_3$, so that Ne$_3$ or Ar$_3$ are more interesting molecules for study. Thus the simplest interesting case is too large for present-day computing capabilities, so definitive work on non-additivity in rare-gas systems must be postponed.

The recent SCF studies[186] of hydrogen bonding in $(H_2O)_n$ and $(HF)_n$ systems provide an example of very large non-additive effects. It is found that co-operative effects act to stabilise cyclic configurations of these systems.

A model calculation reported by Margenau and Stamper[187] suggests that in some cases four-body forces are larger than three-body forces; this assertion requires further investigation.

4.2.4.2 Variational calculations of van der Waals dispersion forces

The increased power of variational methods is demonstrated by preliminary reports from several groups[12, 13, 188—191] of accurate variational calculations of the van der Waals potential well for He$_2$, HeH, LiHe and Li$_2$. This advance has been made possible by improved technology in integral evaluation and in the manipulation of multiconfigurational wave functions. It appears that an announced goal of the exchange perturbation theory, namely, the accurate evaluation of intermolecular energies by a method which takes full account of the Pauli Principle and which gives accurate van der Waals interactions, will be accomplished instead by these variational techniques.

A common objection to the use of a variational approach in the van der Waals region is that the intermolecular potential obtained in this way is a small difference of two large numbers. The fear is not that the error involved in rounding off will obscure the result, but that, since the absolute error for a fixed nuclear configuration is many times larger than the interaction energy, the fluctuations in the absolute error with nuclear position will be larger than the intermolecular forces. The errors arise primarily because the quality of the trial wave function depends on nuclear position. Recent work has shown that by judicious choice of the variational wave function, the absolute error can be made nearly independent of nuclear position. The calculated potential-energy surface is approximately parallel to the true surface.

The helium–helium interaction is the example which has been studied most thoroughly[12, 13, 188]. It is believed, in general, that the Hartree–Fock configuration $(\sigma_g)^2(\sigma_u)^2$ yields only a repulsive interaction[192], and it surely fails to account for the van der Waals well. It is also well known that an energetically equivalent configuration $(a)^2(b)^2$ involves the localised orbitals

$$
\begin{aligned}
a &= (\sigma_g + \sigma_u)/\sqrt{2}, \\
b &= (\sigma_g - \sigma_u)/\sqrt{2}.
\end{aligned}
\tag{4.29}
$$

At long range a and b evolve into atomic Hartree–Fock orbitals a_{HF} and b_{HF}. Thus, in the separated atom limit the absolute error is twice the correlation energy of a ground-state helium atom, which amounts to some 26 500 K, compared to a van der Waals well depth of 10–12 K. To obtain accurate interaction energies the absolute error must be held constant to 1 p.p.m. as the atoms collide! This has been approached by a C.I. calculation in which the configuration $(a)^2(b)^2$ is fixed and configurations of the form $a\phi_i b\phi_j$ are added, where ϕ_i, ϕ_j are chosen from an orthogonalised basis set of Slater orbitals. In the language of pair-correlation theory[188], only interpair, not intrapair, correlations are included. These are the correlations which are expected to change significantly during the collision. The details of the calculation involve either the MC–SCF method[12] or the INO method[13]. The following simplified analysis does not correspond precisely to either method. However, we believe that it conveys the essential features limitations of either approach.

A more accurate trial function involves, in addition to those considered, the configurations $aab\phi_i, a\phi_i bb, aa\phi_i\phi_j, \phi_i\phi_j bb, a\phi_i\phi_j\phi_k, \phi_i\phi_j\phi_k b, \phi_i\phi_j\phi_k\phi_l$. The primary configuration $(a)^2(b)^2$ has connecting Hamiltonian matrix elements with only the two-electron excitations. Let $V_{\alpha\beta}^{ij}$ denote the matrix element of the molecular Hamiltonian H coupling a^2b^2 with $\alpha\beta\phi_i\phi_j$, $\alpha,\beta = a$ or b; $H_{\alpha\beta}^{ij}$ denote the expectation value of H with respect to $\alpha\beta\phi_i\phi_j$; v_{aa}^{ij} denote the matrix element of the atomic Hamiltonian H_A coupling a_{HF}^2 and $\phi_i\phi_j$; and h_{aa}^{ij} denote the expectation value of H_A with respect to $\phi_i\phi_j$. Then the total molecular energy E obtained by a complete trial function is given by

$$E = E_{HF} + \sum_{ij} \frac{|V_{ab}^{ij}|^2}{E - H_{ab}^{ij}} + 2 \sum_{ij} \frac{|V_{aa}^{ij}|^2}{E - H_{aa}^{ij}} + R(E), \qquad (4.30)$$

where $R(E)$ is a remainder term. Similarly, the corresponding energy for an isolated atom is

$$E^a = E_{HF}^a + \sum_{ij} \frac{|v_{aa}^{ij}|^2}{E^a - h_{aa}^{ij}} + R_a(E^a). \qquad (4.31)$$

The interaction energy is

$$U = E - 2E^a = U_{HF} + \sum_{ij} \frac{|V_{ab}^{ij}|^2}{E - H_{ab}^{ij}}$$

$$+ 2 \sum_{ij} \left\{ \frac{|V_{aa}^{ij}|^2}{E - H_{aa}^{ij}} - \frac{|v_{aa}^{ij}|^2}{E^a - h_{aa}^{ij}} \right\} + R - 2R_a \qquad (4.32)$$

The current method for approximating U is essentially to retain only the first two terms. The first term is the purely repulsive Hartree–Fock potential. The second term gives rise to the dispersion terms $-C_6 R^{-6} - C_8 R^{-8} + \ldots$ in the long range limit. In addition, the second term includes second-order exchange terms which decrease as e^{-2R} at intermediate distances.

The recent calculations[12, 13] have shown that accurate results can be obtained by the method outlined above. The final potentials have not yet been reported, but it may be expected that they will be used to predict beam scattering and thermodynamic data. In order to anticipate the significance of discrepancies which might arise between theory and experiment, it is of interest to examine the residual error in the calculations, assuming the method is carried to the full extent.

The accuracy of the dispersion energies depends on the completeness of the basis orbitals ϕ_i (e.g. to obtain C_6, the basis must include p-functions), but is fundamentally limited by the form of the primary configuration $a^2 b^2$, which describes uncorrelated atoms. Independent calculations[14] of C_6 have shown that the uncoupled Hartree–Fock method overestimates C_6 by 12%, while the coupled Hartree–Fock method underestimates C_6 by 7%. Since $a^2 b^2$ is obtained from a molecular SCF calculation, the present method is expected to agree more closely with the coupled Hartree–Fock value. The third term in equation (4.32) is expected to approximately cancel. An explicit formula in the usual notation for the difference between v_{aa}^{ij} and V_{aa}^{ij} is

$$V_{aa}^{ij} - v_{aa}^{ij} = \sqrt{2} \left[(a\phi_i | a\phi_j) - (a_{HF}\phi_i | a_{HF}\phi_j) \right] \qquad (4.33)$$

At long range, $a \sim a_{HF} - S b_{HF}/\sqrt{2}$ where $S = \langle a_{HF} | b_{HF} \rangle$. Thus the third term in equation (4.32) decreases as $S^2 \sim e^{-2R}$. The term $R - 2R_a$ describes higher multipole and exchange terms.

To summarise, the error in the current approximation involves high powers of R^{-1} and exchange terms which decrease as e^{-2R}. The method includes the lower-order multipole energies, albeit with incorrect coefficients, and part of the exchange terms. The numerical results indicate that the combined error is $c.$ 1 K near the van der Waals minimum. At distances shorter than the van der Waals minimum, the error is expected to gradually increase. For very short separations, the configuration a^2b^2 is inappropriate for describing the united atom beryllium.

Finally, it should be noted that since the potential is obtained as the difference between two approximate energies, it is *not a bound* to the true potential; this obvious fact is sometimes overlooked.

4.2.5 Semi-empirical methods for potential-energy surfaces

In spite of the advances made in recent years in the *ab initio* calculation of potential-energy surfaces, in most cases of interest to the experimentalist the *ab initio* method is inappropriate because either the molecules involved, though small, are too large to be treated by *ab initio* methods or the experimentalist desires only qualitative information about the surface (e.g. the location of a saddle point), which does not justify a large expenditure of computer time. In such cases, it is desirable to have semi-empirical methods available to provide low-resolution information. Many methods have been proposed; we discuss two which have some generality. Another possibility which we do not discuss is the extended Hückel approach, which has been applied in several instances[193-196]. It should be noted that probably the most reliable method is to ask Professor Mulliken[197, 198]!

4.2.5.1 Pseudo-potential methods

The so-called pseudo-potential method arose in a study in solid-state physics, but has now been applied to several atomic and molecular problems[199]. The motivation for the work is the belief that closed-shell atomic cores are insensitive to chemical interactions, Rydberg transitions, collisions, etc., and so may be held frozen for the purposes of calculation. This idea is, of course, an old one and pervades much of chemistry; e.g. the $\sigma-\pi$ separation in organic molecules. The core electrons cannot be forgotten altogether, because in applying a variational method to the resulting Hamiltonian, there is a danger of the valence orbitals collapsing energetically into the core. In a full calculation, this does not happen because the valence orbitals are constrained to be orthogonal to the core. In the pseudo-potential method, the orthogonality constraint is replaced by an additional potential-energy term, the effect of which is to shift the energies of the core orbitals so that they are degenerate with the valence orbital. Then in the transformed Hamiltonian, the ground-state energy is the energy of the valence orbital; a collapse has been avoided. To construct the pseudo-potential requires a knowledge of the core orbitals; in practice it is usual to replace the pseudo-potential by a model potential which the practitioner of the method hopes will mimic satisfactorily the orthogonality constraints. The parameters in the model potential are chosen by fitting to an appropriate set of data, e.g. Rydberg spectra.

Model calculations, based on the pseudo-potential formalism, have been reported for alkali–alkali[200, 201], alkali–noble gas[202], alkali–alkali halide[203] and chlorine ion–methyl bromide[204] interactions, and for van der Waals forces[205, 206]. Although useful results have been obtained, it is difficult to place much confidence in the generality of the results mainly because of the ambiguity in the choice of the model potential. However, if it could be demonstrated that the results are largely insensitive to the precise form of the potential, or that a particular form is to be preferred on general grounds, then the method could be used to generate potentials with the correct qualitative features for systems which are not now accessible.

Two general problems will ultimately limit the method, even with the assumption that the valence–core separation is sufficiently accurate. In the first place, for many interactions involving other than alkali or alkaline-earth atoms the number of valence orbitals is larger than the number of core orbitals and the method becomes intractable for the same reasons as a full calculation. Also, for a large system, even though the core energy is fairly constant, the magnitude of the core energy is enormous, so that a small fluctuation could be as large as the intermolecular forces.

4.2.5.2 Diatomics-in-molecules

Another semi-empirical scheme for estimating energy surfaces is Ellison's diatomics-in-molecules (DIM) method[207], which is an outgrowth of the older atoms-in-molecules (AIM) of Moffitt. In contrast to AIM, however, the only integrals which appear are overlap integrals. This simplification is not achieved through neglect of any of the Coulomb and exchange integrals which appear in AIM, but rather by noticing that the total electronic Hamiltonian for, say, a three atom system can be rearranged to the form

$$H = H_{AB} + H_{AC} + H_{BC} - H_A - H_B - H_C \qquad (4.34)$$

where H_{AB}, etc. are diatomic Hamiltonians and H_A, etc. are atomic Hamiltonians. This rearrangement is possible because only two-body forces occur in H. Now imagine the following variational scheme for approximating the eigenvalues of H. From the atomic wave functions for A, B, C (or orbital approximations thereto) all possible valence-bond wave functions for the diatomic pairs AB, AC, BC are constructed. If more than one function of a given symmetry is obtained for any pair, a configuration-interaction calculation is formed to diagonalise the diatomic Hamiltonian. With the assumption that the resulting eigenfunctions are exact diatomic wave functions, now couple the wave functions of the third atom to each of the three sets of diatomic functions to form VB approximations to the triatomic system. If all possible results are retained, three *equivalent* sets of triatomic wave functions are obtained. They may be designated (AB)C, (AC)B, A(BC) to indicate the order of coupling. Since they are equivalent, a transformation matrix may be defined which relates one set to another. This matrix will involve C.I. coefficients, vector coupling coefficients, and geometrical factors (for non-S-state atoms).

To complete the calculation, the matrix of H in one of the basis sets, say (AB)C, is diagonalised. The action of H_{AB} on any basis state replaces the operator by a diatomic energy. To apply H_{AC}, the (AB)C basis is related to the (AC)B basis via the aforementioned transformation; in this way, the diatomic states of AC enter. Transform back to the (AB)C basis by the inverse transformation. Similar procedures apply to H_{BC}. Finally, to apply the remaining atomic operators, $H_A + H_B + H_C$, simply relate (AB)C to the atomic states. By this procedure, the action of H on each basis state is evaluated. Thus, in each matrix element, the operator H is replaced by diatomic and atomic energies, diatomic C.I. and vector coupling coefficients and geometrical factors. Integration of the matrix element results in overlap integrals between the states of the (AB)C basis.

In the simplest application of the DIM method, no diatomic C.I. is required, the diatomic curves are taken from experiment and the overlap integrals are set to zero. This results in a London-type surface.

For more complicated applications, the number of diatomic curves required multiplies rapidly and usually these curves have to be obtained by calculation rather than from experiment. In any case, the C.I. coefficients require a diatomic energy calculation, even if the curve itself is not used. Also, in the general case, it may be necessary to consider states which arise from coupling excited states of atoms, e.g. (A*B)C, and ionic states, e.g. $(A^+B^-)C$. These features greatly complicate the method and it is not clear how generally it may be applied.

In the past, DIM has been applied mostly to predicting geometries and stabilities of bound molecules, exceptions being treatments of the H_4 and H_3^+ potential-energy surfaces[208, 209]. Recently, three new surfaces have been obtained by the DIM method, namely, new ground- and excited-state H_3^+ surfaces[210, 211] and the ground-state ArH_2^+ surface[212]. In both of these calculations it was necessary to consider surface crossing between different ionic surfaces. For example, in the former calculations, both $H^+ + H_2$ and $H + H_2^+$ surfaces were included. The latter calculation is the most ambitious DIM calculation to date in that full account was taken of the ^2P symmetry of Ar^+. The results obtained by these two groups are encouraging for further applications.

It is difficult to predict when the DIM surfaces will be reliable. For the H_3^+ surfaces, comparison is possible with accurate *ab initio* calculations[210]. The qualitative shapes of the surfaces are well-approximated by the DIM results, although local discrepancies in energy amount to 5–10 kcal mol^{-1} near the minimum of the ground-state surface. Agreement for the excited surface is not as good; for the H_4 DIM surface, the predicted saddle point for the square configuration lies *c.* 50 kcal mol^{-1} below *ab initio* results, although it is closer to the experimental activation energy than is the *ab initio* saddle point[149].

Balint–Kurti and Karplus have revised and modified the AIM method and have recently applied it to calculate the $Li + F_2$ surface[177]. They have also computed the surface by an *ab initio* valence-bond C.I. method, but give reasons why the semi-empirical estimate is expected to be more reliable.

4.2.6   Properties other than the energy

Many molecular properties other than the energy undergo changes as a result of collisions. A number of molecular properties have been studied recently in connection with specific experiments, for example, collision-induced infrared[213] and Raman[214] spectra, and the pressure dependence of dielectric constants[215, 216], Kerr effect constants[217], chemical shifts[218, 219], and hyperfine splittings[220]. The theory of the properties is divided, in general, into long- and short-range regions in a manner analogous to the interaction energy. We discuss how interactions shift atomic hyperfine frequencies, induce dipole moments in rare gas diatoms, and modify atomic polarisabilities.

4.2.6.1   Pressure shifts of atomic hyperfine frequencies

The hyperfine pressure shift in the spectra of paramagnetic atoms in foreign-gas environments is an example of a gas-phase solvent effect. The cases studied have involved paramagnetic atoms in an inert-gas atmosphere. The observable quantity is the shift Δv in the frequency v_0 of a hyperfine transition. The qualitative explanation for the shift was first given by Adrian[221]. The magnitude of the hyperfine frequency is controlled by the Fermi contact term, i.e. Δv depends primarily upon the electron spin density ρ at the nucleus M. The effect of a foreign gas X upon ρ is divided into a short-range and a long-range contribution. At long range the dispersion interactions lead to an expansion of the M s-orbital, and hence a negative shift, at short range exchange effects cause the M s-orbital to contract and leads to a positive shift.

Contact between theory and experiment is made by comparing the fractional pressure shift f_p,

$$f_p = \frac{\partial}{\partial p} \left\langle \frac{\Delta v(R)}{v_0} \right\rangle \propto \int_0^\infty \frac{\Delta v}{v_0} e^{-U(R)/kT} R^2 dR, \qquad (4.35)$$

where

$$\frac{\Delta v}{v_0} = \frac{\rho(R) - \rho(\infty)}{\rho(\infty)}$$

and

$$\rho(R) = \langle \Psi | \sum_i s_{iz} \delta(r_i) | \Psi \rangle.$$

Here $\rho(R)$ is the spin density at the nucleus M, and Ψ is the wave function for the complete system M—X.

Given a theory for Ψ, a theory for ρ is automatically obtained. Das and co-workers[222] have considered the long-range expansion of ρ by substituting the multipole expansion for Ψ. This results in

$$\frac{\Delta v}{v_0} = D_6 R^{-6} + D_8 R^{-8} + D_{10} R^{-10} + \cdots \qquad (4.36)$$

Values of D_6, D_8, D_{10} have been estimated for the hydrogen–rare gas systems. They are negative as expected from Adrian's analysis. The series seems to be very slowly convergent for $R > 7$ Bohr. Das and co-workers have also considered the short-range effect by approximating Ψ as an antisymmetrised product of separated-atom wave function (Heitler–London function). This gives $\Delta v > 0$ at short range, as predicted by Adrian[221].

An MC–SCF calculation of $\Delta\nu$ which includes the long- and short-range effects has been reported by Das and Ray[189] for HeH. The results are claimed to be inaccurate in the long-range region by Ikenberry and Das[223] because of the omission of one-electron excitations which are necessary to describe $\rho(R)$ at long range. Since the MC–SCF method implicitly includes many one-electron excitations, this objection requires clarification.

Currently the most extensive theoretical estimate of f_p for HeH is that of Kunik and Kaldor[224], who computed ρ with a spin-optimised SCF wave function and added the long-range expansion of $\Delta\nu$ for $R > 7$ Bohr. Their results are 20–30% higher than experimental values over a wide temperature range (50–700 K). Most of the error has been attributed to difficulties in defining a cut-off for the dispersion terms.

Calculations of f_p within the SCF approximation have been reported[225] for LiHe and NaHe. Good agreement with experiment was obtained, but Rao and Kestner[226] have shown that this is probably a special result for alkali–helium interactions. In addition, Kunik and Kaldor suggest that the SCF spin densities are actually too small at short range and that this compensates for the neglect of the negative shifts at long range and results in fortuitous agreement with experiment; clearly the definitive calculation is yet to be reported.

4.2.6.2 Interaction dipoles of rare-gas diatoms

During a collision of two dissimilar rare-gas diatoms, an electric dipole moment develops which gives rise to the far-infrared collision-induced spectra of rare-gas mixtures, which were observed first by Kiss and Welch[213], while Bosomworth and Gush[227] have observed the complete absorption bands for He–Ar and Ne–Ar.

The theory of the absorption[228] requires a knowledge of the dipole moment function $\mu(R)$ for the rare-gas diatoms and theories have been developed for both the long-range dispersion region and the short-range overlap region; no satisfactory calculation has yet spanned the two regions smoothly.

The long-range theory was first discussed by Buckingham[228] who showed that $\mu(R)$ has an asymptotic expansion of the form

$$\mu(R) = D_7 R^{-7} + D_8 R^{-8} + \cdots \tag{4.37}$$

Buckingham also made rough estimates of the magnitude of D_7 for the H–He system. More recently, Byers, Brown and Whisnant[229] have presented formulae for D_7 which were developed along the lines discussed in Section 4.2.1.2 for C_6. In this way they have succeeded in expressing D_7 in terms of properties of the isolated atoms and have given an approximation scheme which should yield useful results. They conclude that the dispersion dipole is of the order of 10^{-3} Debye at the collision diameter and of opposite sign to the overlap dipole.

The overlap dipole was also first computed approximately by Buckingham for HeH. Later, Hartree–Fock calculations, which do not describe the dispersion terms and can be fitted to the simple form $\mu = Be^{-R/\rho}$, were performed by Matcha and Nesbet[230] for He–Ar and Ne–Ar. Levine[231] has pointed out that to explain the experimental absorption spectrum it is necessary to include both dispersion and overlap contributions to $\mu(R)$. The obvious approximate procedure is to assume additivity between the overlap and dispersion contributions, although this has no unambiguous theoretical foundations. Clearly more work is needed to establish the behaviour of μ for all values of R.

4.2.6.3 Polarisability tensor of rare-gas diatoms

There is a Raman absorption due to the change in the polarisability $\alpha(R)$ of the rare gas diatom during a collision which corresponds to the collision-induced absorption in the infrared region due to interaction dipoles. Spectra have been reported for the rare gases[214] (Ar, Kr or Xe) and for a number of liquids[232]. The polarisability tensor also governs the virial expansion of the dielectric constant[215] (Clausius–Mossotti function) and the Kerr effect constant[217].

Polarisabilities are second-order properties and are therefore harder to compute than first-order properties such as dipole moments. It is not surprising that the theory of $\alpha(R)$ is even less developed than that of $\mu(R)$. The behaviour of α has been considered both in the dispersion and the overlap regions. The long-range theory was first developed by Jansen and Mazur[233] and Buckingham and Pople[234]. These workers showed that the asymptotic expansion of the components for the homonuclear case is

$$\alpha_{\parallel}(R) = 2\alpha_0 + 4\alpha_0^2 R^{-3} + A_6^{\parallel} R^{-6} + \cdots, \tag{4.38}$$

$$\alpha_{\perp}(R) = 2\alpha_0 - 2\alpha_0^2 R^{-3} + A_6^{\perp} R^{-6} + \cdots. \tag{4.39}$$

Interestingly, the terms of order R^{-3} can be evaluated exactly in terms of α_0, the polarisability of an isolated atom. Approximate[233–235] and model calculations[236] have suggested that $A_6^{\parallel} \sim 8\text{–}15 \, \alpha_0^3$ and $A_6^{\perp} \sim 2\text{–}6 \, \alpha_0^3$. The only systems for which accurate estimates are available are H_2[237], for which $A_6^{\parallel} = 28.08 \, \alpha_0^3$ and $A_6^{\perp} = 13.91 \, \alpha_0^3$, and He_2[238], for which $A_6^{\parallel} = 22.9 \, \alpha_0^3$ and $A_6^{\perp} = 10.8 \, \alpha_0^3$.

The second dielectric virial coefficient depends upon the change in the trace of α, $\bar{\alpha} = \frac{1}{3}(\alpha_{\parallel} + 2\alpha_{\perp}) - 2\alpha_0$, averaged over collisions. The Raman absorption and the Kerr constant virial each depend upon the average of the square of an anisotropy in α, $\beta = \alpha_{\parallel} - \alpha_{\perp}$. Theories relating $\bar{\alpha}$, β to the observable quantities have been developed[233, 234, 239], and it has been shown that, to obtain agreement with experiment, overlap contributions must be considered also. Levine and Birnbaum[240] have suggested by fitting Raman data to an assumed functional form, that the long-range expansion of β overestimates the true function. A similar conclusion is reached in the study[217] of the Kerr effect virial, and it has been clear for some time that the long-range expansion of $\bar{\alpha}$ overestimates the true function, since the second dielectric virial for He_2 and Ne_2 is negative[215], whereas the long-range expansion predicts a positive result.

There have been few calculations of the short-range behaviour of α. For He_2, a coupled Hartree–Fock calculation has been reported by Lim, Linder and Kromhout[241] and an approximate calculation for H_2 ($^3\Sigma_u^+$) has been given by Dupre and McTague[242]. Both results predict that $\bar{\alpha}$ is negative at intermediate values of R, thereby reducing the positive dispersion contribution. Neither calculation included the dispersion terms and it is difficult to judge their accuracy. In the He_2 case the calculated value of $\bar{\alpha}$ still led to a positive value of the second dielectric virial. Clearly more work is needed to establish the behaviour of $\alpha(R)$ and to provide accurate estimates of its magnitude.

4.2.7 Summary

To sum up this section, let us admit that while progress has been made in the *ab initio* and semi-empirical estimation of intermolecular interactions, it remains difficult to give reliable details beyond the broadest outlines of the theory. The fields of beam scattering, i.r. chemiluminescence, chemical lasers, collision-induced spectra and spectral line shapes provide countless examples where detailed knowledge of molecular interactions are required. The variational methods which we have discussed for the energy are now at a level where much is to be learned through specific applications. Further work is required before treatments of electronic properties other than the energy reach a comparable status; the need for accurate theoretical estimates of molecular interactions can be expected to increase in the future.

4.3 INTERMOLECULAR FORCES—DETERMINATIONS

4.3.1 Introduction

Statistical mechanics provides relations between intermolecular forces and the equation of state of fluids and the dissipation coefficients in steady-state transport processes. In principle, this provides a microscopic basis for the calculation of these thermodynamic data. In practice, the relations have been used in the reverse direction to determine the intermolecular potentials. The inversion problem from the data to the potentials has some similarities to the same inversion in nuclear and atomic scattering experiments. The formal treatment of the problem is not as precise as in the scattering case; for thermodynamic data there is usually one more level of averaging in relating potentials to data. At times this has caused some pessimism on questions of sensitivity and uniqueness in the inversion. Still, useful potential functions for a wide variety of molecular systems have been generated from thermodynamic data[1,2]. This portion of our review treats topics in the inversion from experimental data to potentials; the emphasis is on thermodynamic data.

The determination of an intermolecular potential from experimental data usually proceeds by adjusting the parameters in an assumed functional form for the potential. The result is at least a parameterisation of the data and is useful for interpolating and extrapolating the data. The ambition for such work is generally higher: within the Born–Oppenheimer approximation there is a velocity-independent potential for the molecular forces. The hope is that it may be approximated by smooth functions and perhaps even by functions with only a few distinct dependences on separation such as exponentials or inverse power laws. Such determinations adjust many parameters to reproduce data which include uncertainties. There is always the serious question of whether the adjustment achieves an absolute minimum, or only a relative minimum, in the deviation of the result from the unknown unique true potential. This worry would persist as a practical problem even if theoretically sufficient data and very flexible functional forms were used. With some types of data more systematic constructions are available.

Here we focus on the intermolecular potentials for inert gas systems. The two-body potentials are spherically symmetric, but for the solid and liquid phases there are some questions on the contributions of non-additive three-body forces. For helium and argon especially, there is a large body of accurate gas-phase data over a wide range of temperatures. There have been important revisions of some of the data in the last few years. Dymond and Smith[244] have reviewed the virial data and Maitland and Smith[245] have reviewed the viscosity data for many gases, Kestin and Wakeham[246] some of the thermal conductivity data, and Mason and Marrero diffusion[247].

We treat the helium and argon systems at length because of the prospects for quantitative microscopic theories of their condensed phases. Argon may be an idealised version, with neglect of three-body forces, of a classical mechanical fluid of spherically-symmetric molecules interacting via two-body forces. Corrections for quantum effects and approximations to the three-body forces have been made in Monte Carlo and molecular dynamics calculations[248] of the equation of state of the liquid. Helium atoms are less polarisable and the neglect of three-body effects is presumably a better approximation[249]. The treatment of fluid helium requires quantum statistical mechanics and, for instance, quantitative theories of the pair distribution function at finite temperatures have only begun[250].

We have given considerable space in this part of our review to the formalism for the evaluation of thermodynamic properties from assumed potential functions. We have included reference to tabulations of calculations for simple potential models, because those calculations provide useful checks on programmes designed to calculate for more complicated models. We have also included references to calculations which can be used to estimate the validity of semi-classical approximations in statistical mechanics. These are frequently used as asymptotic approximations similar to the WKB approximation in quantum mechanics; some workers have expressed misgivings about their reliability. In many cases there are now fully quantal model calculations available which support their accuracy.

4.3.2 Intermolecular forces in statistical mechanics

We review the theory which relates intermolecular forces to thermodynamic properties. Our treatment is limited to gases with spherically-symmetric potentials (the inert gases) and our emphasis is on the evaluation of the resulting formulae. The topics are in the areas of virial expansions in equilibrium statistical mechanics and Navier–Stokes transport coefficients in steady-state non-equilibrium statistical mechanics. The theoretical foundations of the equilibrium topics are much more firmly established than the transport problems.

In this theory the intermolecular forces appear below some thermal averagings over Boltzmann energy distributions; these averagings blur the precision with which thermodynamic data can be related to microscopic properties of the system. This blurring has been shown in model calculations presented in connection with an experiment by Dondi et al.[251] who observed a quantum statistics effect in the total scattering cross-section of ^{4}He–^{4}He. The effect is an oscillation in the cross-section as a function of the relative velocity of the colliding atoms and the experiment scattered a beam of atoms with narrow velocity distribution from a gas target with a Boltzmann distribution of energies. For scattering energies of 50–150 K (1 K $= 1.16 \times 10^{-4}$ eV) the oscillation would be nearly damped out with a target temperature of 10 K. The experiment used a target at 2 K and the oscillation was quite marked. Such an effect had been predicted by Helbing[252] as a 'higher-order glory' effect, with a semi-classical calculation.

4.3.2.1 Virial equation of state

The equation of state of a uniform gas is a functional relation between the pressure p, the molar volume V, and the absolute temperature T. In classical statistical mechanics, with no forces between the molecules, the relation is Boyle's law for an ideal gas:

$$pV/RT = 1 \qquad (4.40)$$

(R is the gas constant and is related to Boltzmann's constant k in statistical mechanics by Avogadro's number N, $R = Nk$). Experimental data showing the effects of intermolecular forces were first parameterised by intuitively-derived functional relations such as the van der Waals equation. For purposes of statistical mechanics, a more systematic presentation of data on the moderately dense gas is the virial equation of state[253]:

$$PV/RT = 1 + B(T)/V + C(T)/V^2 + \cdots \qquad (4.41)$$

The temperature-dependent coefficients $B(T)$, $C(T)$, ... are the second, third, ... virial coefficients. (The first virial coefficient is the absolute temperature). A major achievement of classical statistical mechanics has been the derivation of concise expressions for these coefficients in terms of inter-molecular potentials; results are presented by Mayer and Mayer[254]. More recently, the theory has been treated again as an example of linked cluster expansions in statistical mechanics. From estimates for the general coefficient in the series for a variety of potential models, proofs have been given that the series converges for some finite values of the density. Some of the convergence theorems are also proved for quantum statistical mechanics. These topics are reviewed by Ruelle[255] and by Lebowitz[256].

Our interest is in the information on intermolecular forces contained in the leading virial coefficients. The expressions for these in classical statistical mechanics, in terms of the Mayer f-function for a potential $\phi(r)$, defined by $f(r) = \exp\{-\beta\phi(r)\} - 1$ where $\beta = (kT)^{-1}$ are

$$B_{Cl}(T) = -\tfrac{1}{2}\int d^3 r\, f(r) \tag{4.42}$$

$$C_{Cl}(T) = -\tfrac{1}{3}\int d^3 r\, d^3 s\, f(r) f(s) f(|\mathbf{r}-\mathbf{s}|) \tag{4.43}$$

There is a corresponding result for the pair distribution function[253]:

$$g_{Cl}(r_{12}) = \exp\{-\beta\phi(r_{12})\}\left[1 + \frac{N}{V}\int d^3 r_3\, f(r_{13}) f(r_{23}) + O(N/V)^2\right] \tag{4.44}$$

The expressions in equations (4.42)–(4.44) are valid in classical mechanics for two-body spherically-symmetric interactions, such that the integrals converge. For quantum statistical mechanics[257, 258], the expressions become more difficult to evaluate. Recent work, which we treat in the following subsections, has shown that in many cases semi-classical expansions provide quantitatively-accurate approximations to the quantum results. Early work with these expansions, in even powers of Planck's constant h for smooth potentials, has been reviewed by deBoer[257]; a recent treatment has been given by Larsen et al.[259] and by Boyd et al.[260].

For argon, calculations are usually made with the classical expressions, correcting for quantum effects by a semi-classical approximation. For helium at 1–4 K, the full quantum treatment is necessary; much data is available for this temperature range.

(a) *Second virial coefficient* – The expression for $B_{Cl}(T)$ in equation (4.42) is simple enough that it has been evaluated for many potential models; tabulations have been published[261–263]. Some qualitative relations between $B_{Cl}(T)$ and $\phi(r)$ are known[264–266].

Integration of equation (4.42) by parts brings it almost to the form of a Laplace transform:

$$B_{Cl}(T) = -\tfrac{2}{3}\pi\beta \int_0^\infty \exp\{-\beta\phi(r)\} r^3 \frac{d\phi}{dr}\, dr \tag{4.45}$$

If the potential $\phi(r)$ is a monotonic function of r, this is a Laplace transform of the function $r^3(\phi)$. However, the potential functions for the inert gases have branches corresponding to repulsive and attractive forces; for some atomic excited-state interactions there are even more branches to the potential functions. In the absence of further assumptions about the analyticity of the potential function the classical second virial coefficient does not invert to a unique potential function.

Jonah and Rushbrooke[267] have presented an example of an inversion by use of equation (4.43). They fitted ^{4}He $B(T)$ data for the range 273–1473 K with a smooth function for which the inverse Laplace transform was tabulated. They did the analysis twice: first, they simply took the inverse Laplace transform of the fit, ignoring the fact that for a general potential equation (4.45) is not a Laplace transform. Secondly, they adjusted the data by shifting it according to their estimates, based on potential models, of contributions in the data from quantum effects and from the region of negative potential. They regarded the result of the inverse transform of the adjusted data as being a more realistic potential; it was in reasonable agreement with other model potentials. In both cases they found that the extrapolations of their data fits to high temperatures gave unreasonable results for the potential at small separations.

A frequent method of analysis of $B(T)$ data has assumed a functional form for $\phi(r)$ and determines the parameters in it by a fit to $B(T)$. The remarkably long-lived potential of deBoer and Michels[268] for helium is an example of such a construction. More recent constructions of helium potentials which include information on the attractive branch of the potential are also in this class[269, 270]. While the information is restricted to the long-range portion of the attraction and is not enough to assure a unique inversion of $B_{Cl}(T)$ to the remaining portion of the potentials, non-rigorous ideas of smoothness of the unknown potential gave some confidence in the practicality of the procedure. The resulting potentials have successfully described other data not included in their construction[271, 272].

This discussion of the inversion is based on the expression for $B_{Cl}(T)$. Another property[265] of $B_{Cl}(T)$ is that it does not show a maximum, as a function of temperature, for a potential model which consists of a rigid-sphere repulsion plus an arbitrary potential well. This result, when applied to data for helium showing a maximum at 200–300 K, leads to an interpretation of the data to show the finite penetrability of the repulsion in the helium intermolecular force. The result does not survive in a quantum treatment[273]; there a maximum is obtained for a model of rigid sphere plus square well[274].

The quantal evaluation of the second virial coefficient $B(T)$ has been based on the expression in terms of scattering phase shifts given by Uhlenbeck and Beth[275]. These enter because in the quantum formulation $B(T)$ is given in terms of a change in density of states of the two-particle system caused by the interactions. Extensive calculations with this formulation have been made[260, 276, 277] and cover the temperatures where the transition from the quantum region to the classical region occurs. For smooth potentials (Lennard–Jones, Morse, and other combinations)[260, 276–278] the result is that a semi-classical (Wigner–Kirkwood) expansion is accurate to 1% for helium at 50 K and that quantum effects contribute only 4% at 300 K. These results do not hold for rigid-sphere models; one estimate[279] with helium parameters gave quantum contributions of 10% at 1000 K.

For realistic potential models the use of semi-classical approximations down to c. 50 K for helium has been established by the modern phase-shift evaluations. The misgivings of Murrell and Shaw[114] on this point have not been borne out. For some models, bounds on the magnitude of quantum effects have been presented by Bruch[273].

The estimate for the rigid-sphere model reflects the increased quantum correction for potentials with much steeper repulsions than are used in realistic potential models. Hill and Luchinsky[280] have shown that the high-temperature expansion for rigid spheres can be obtained as the mathematical limit, $n \to \infty$ of a high-temperature expansion for the r^{-n} potential. The high-temperature limit at fixed n reproduces the Wigner–Kirkwood expansion for the r^{-n} potential. Hill and Luchinsky[280] suggest their expansion may be an improvement on the Wigner–Kirkwood expansion for moderate values of n, but there have been no numerical comparisons of the methods yet.

The calculations[260, 281] also show the suppression of the quantum statistics term in $B(T)$. This term appears as the second virial coefficient in the equation of state of ideal quantum gases and has frequently been included in prescriptions[1] for semi-classical approximations to $B(T)$. It is now well-established, especially through the work of Boyd, Larsen and Kilpatrick[260], that this term should be omitted in such approximations for realistic intermolecular potentials. This is an effect which is not properly treated by the usual semi-classical theory, so that in the prescription one of a pair of cancelling terms was retained. For helium the statistics term is a negligible part of $B(T)$ at temperatures above 5 K. The analytical studies of Larsen et al.[282] and of Hill[283] support these remarks.

(b) *Third virial coefficient* – The third virial coefficient is a property in which we might hope to identify contributions from non-additive three-body forces. It is difficult to extract this property from experimental data and also difficult to compute it.

In classical statistical mechanics $C_{Cl}(T)$ in equation (4.43) can be reduced to a triple integral and it has been tabulated for some models[1]. The h^2 term in a semi-classical series has also been tabulated[284]. However, in quantum statistical mechanics there is still no useful formalism for the evaluation of $C(T)$ from a model two-body potential. Such a calculation will be necessary for the interpretation of the measurements of Sherman and Kerr[285] for the $C(T)$ values of ^{4}He and ^{3}He at 3–4 K.

The status of the quantum theory of $C(T)$ is that the formal expressions for $C(T)$ in terms of quantum cluster integrals have long been available, but a formulation similar to the phase-shift expression for $B(T)$ is still incomplete. Larsen and Mascheroni[286] have developed expressions in terms of three-body phase shifts and have given limited results for an idealised square-well model. Servadio[287] has noted that there are subtle mathematical limits in this formulation which need more discussion. It appears that a discussion of experimental data with realistic potentials using this formulation is still distant.

A contrasting approach to problems of liquid helium temperatures by working down from temperatures in the classical regime has been used by Jordan and Fosdick[250]. Their method is a folding together of semi-classical approximations in a factorisation of the three-particle density operator. In their work they evaluated the resulting high-dimensional multiple integrals by Monte Carlo methods. The accuracy of these evaluations becomes less for lower temperatures and the labour required increases rapidly. The lowest temperature for which they quote a calculated third virial coefficient is 10 K. Their calculations were for the deBoer–Michels–Lennard-Jones 12–6 potential for ^{4}He.

The results of Jordan and Fosdick[250] can be compared with the values obtained by a semi-classical expansion carried to the term in h^2. The two approaches give results in agreement to 10% down to 20 K and deviate widely below this temperature. While there have been speculations that quantum effects may be larger in $C(T)$ than in $B(T)$, there is no evidence for this in these calculations.

(c) *Pair distribution function* – The Fourier transform of the pair distribution function is measurable in x-ray diffraction experiments on fluids.

A Fourier inversion of the data would give the pair distribution function and if the density-independent term in classical mechanics (equation (4.44)), could be extracted, inversion to the potential would be easy.

For argon, such experiments have been performed with x-rays[288] and neutrons[289, 290] on the gas near critical conditions. With the present level of accuracy, these results have been used more for discussion of the theory[291] of the dense fluid than for determination of details of the pair potential. The experiments of Achter and Meyer[292] and of Hallock[293] on helium include moderate densities where a virial expansion may be useful but are at temperatures for which quantum statistical mechanics is necessary.

In classical mechanics, tabulations of the first density-dependent terms in equation (4.44) were presented for a Lennard–Jones 12–6 potential by Henderson[294] and by Henderson and Oden[295]. These calculations do not present difficulties for other models.

In quantum mechanics even the evaluation of the density-independent term corresponding to the Boltzmann factor in equation (4.44) is lengthy. Precise evaluations for Lennard-Jones 12–6 potential models for helium and hydrogen have been presented by Larsen, Witte and Kilpatrick[259] and by Poll and Miller[296]. These use evaluations of the continuum wave functions of the Hamiltonian of the interacting pair of particles. For the helium model, Fosdick and Jordan[250, 297, 298] have used their method of factorisation of the density operator for this term and for the first density-dependent term. Again, their method is most suitable for the semi-classical regime, but they were able to get a modest agreement with the evaluation of Larsen et al.[259] at 2 K.

Larsen et al.[259] and Poll and Miller[296] showed comparisons of their results for the density-independent term with semi-classical approximations. For ^{4}He at 2 K, the semi-classical approximation becomes accurate at large separations. For hydrogen at 40 and 80 K, the semi-classical approximation is accurate over much of the range of separations. It fails for separations where the variation of the potential is appreciable over a thermal wavelength; a modification which improves the accuracy there and is an upper bound on the term at all separations has been suggested by Bruch[299].

Achter and Meyer[292] measured the scattering of x-rays from ^{4}He gas at 4.2 K and 0.98 atm. They comment on the uncertainties in the Fourier inversion to the pair correlation function $g(r)$, but do show the result of the inversion. There are some differences between this result and what might be expected from the calculations of Jordan and Fosdick[250] for the deBoer–Michels potential. In the experiment, the first maximum in $g(r)$ has height 1.70 and is located at 4.0 Å; the calculation gives 1.60 at 3.80 Å. In the experiment there is a minimum at 6.2 Å followed by a second maximum at 7.5 Å; the $g(r)$ value obtained from the calculations shows no minimum for separations up to 7.2 Å, which is the maximum separation in the triplet calculations. Some reservations should be attached to interpretations[292] of the experimental data as showing the presence of a second maximum caused by triplet clusters. Both the measurements and the calculations discussed in this paragraph contain large uncertainties.

Hallock[293] has measured the x-ray scattering from ^{4}He at 5 K and two densities. From such data, the density-independent term might be extracted, but the present uncertainties in the data are sufficiently large that this analysis has not been performed.

4.3.2.2 Intermolecular forces in transport theory

We base our discussion on the binary collision approximation and the Boltzmann equation[1, 300]. In the steady-state this leads to density-independent transport coefficients; we do not discuss the density-dependent corrections derived from generalisations of the Boltzmann equation[301-303].

There are three levels of approximation in this theory. The first, the acceptance of the Boltzmann equation and the irreversibility in the Boltzmann H-theorem, is fundamental. The second, a linearisation of the Boltzmann equation based on small amplitude disturbances or small departures from local equilibrium, is presumably an approximation which can be mimicked by experiment. The third is the series solution of the inhomogeneous integral equation thus obtained and is an approximation which can be extended systematically and compared against exact solutions for some idealised models.

(a) *Steady-state transport properties* — The coefficient of thermal conductivity in the Fourier law of heat conduction and the coefficients of viscosity and diffusion are defined in a linear response approximation. Chapman and Cowling[300] present series solutions of the Boltzmann equation for these quantities. We do not discuss the correlation function expressions[304].

For classical rigid spheres the Boltzmann equation for cases leading to the thermal conductivity or viscosity can be transformed to third- and fourth-order differential equations[305-307]. These have been numerically integrated to yield precise values for the transport coefficients and a tabulation of the perturbation to the local Maxwellian velocity distribution. The relative contribution of the increment in the distribution function becomes increasingly large for increasing velocities, although it gives no net contribution when an average over orientations of the velocity is made. A similar feature is observed in the solution for the classical Maxwell potential[308]. It appears that the linearisation procedure applied to the Boltzmann equation is to be viewed as a linearisation in some average sense and not at all velocities.

The series solution of the linearised Boltzmann equation is given in terms of the Sonine polynomials (which, according to recent normalisations[309], are now the same as generalised Laguerre polynomials). The procedure of Chapman and Cowling takes terms in the series in a way which leads to variational lower bounds on the viscosity and thermal conductivity. Mason[310] has developed an approximation suggested by Kihara; this requires fewer calculations than the variational calculation and in the applications so far has yielded comparable accuracies.

Tables of classical collision integrals are available for many models[261, 311]. The formalism has been extended to the regime where quantum effects are important in particle scatterings by using a quantal evaluation of the transport cross-sections[312, 313]. The derivation of this prescription is not appreciably less rigorous than the derivation of the classical mechanical collision term.

Many of the quantum effects can be included by the use of the Wigner distribution function[314] in place of the classical phase-space distribution function. Both the classical and quantal treatments follow the Bogoliubov procedure[315] of achieving a closure in the hierarchy of distribution function equations by assuming initially uncorrelated particles in a scattering event. The treatment of bound-state effects, which could be described as high-order multiple scatterings, or as a pair of interacting particles which do not trace back to an initially uncorrelated pair of particles at much earlier times, is incomplete. This becomes part of the topic of density-dependent corrections to the Boltzmann equation. There is no conspicuous experimental evidence of failures of the present theory for the density-independent term.

For the inert gases other than helium, calculations have usually been made with fully classical formulations. For argon one quantal calculation[316] of the viscosity at 100 K has been reported; the results differed from the classical calculations by 0.2%. In contrast to the equilibrium properties no quantitatively-useful semi-classical approximation is available to reduce the labour of these calculations. This was shown in quantal calculations of the collision integrals of the Lennard-Jones 12–6 potential[317, 318] and also appears in a calculation of the high-energy transport cross-section for rigid spheres[319].

There is a qualitative relation between low-temperature viscosity data and the coefficient C_6 of the long-range van der Waals attraction in the intermolecular force. This relation is derived in classical mechanics[266, 320–323]; the way in which the low-temperature limit is approached depends on the model. Available viscosity data for argon, krypton and xenon do not extend to sufficiently low temperatures to permit a model-independent extrapolation to yield C_6. The construction to extract C_6 from the data has been performed in a variety of ways; a summary is given by Clarke and Smith[323]. The results are C_6 values in agreement with theoretical calculations; Clarke and Smith were able to achieve agreement to 10% with a model-dependent analysis. A similar construction fails for helium, where quantum effects are important in the low-temperature data and s-wave scattering makes important contributions.

Dymond[324] has presented an analysis which uses high-temperature viscosity data for inert gases to determine the intermolecular repulsion. He presented his results as a band of separations where the potentials should lie; much of the uncertainty is due to the dependence of the analysis on the potential models assumed. Maitland and Smith[325] have presented similar results. Dymond[326] has also used a corresponding states analysis to argue that the repulsion for neon is steeper than for the heavier inert gases.

Although there are differences in the scatterings of ^{3}He and ^{4}He due to quantum statistics[327], the effect in the transport properties is exceedingly small at temperatures above 5 K[328]. There is a suppression of statistics effect here, after the thermal averaging, similar to that discussed for the second virial coefficient. However, symmetry oscillations in the differential cross-section of ^{4}He–^{4}He have been observed at relative kinetic energies of 300 K by Siska et al.[329]. This and the symmetry oscillation in the total cross-section observed by Dondi et al.[251] are examples of effects which are nearly averaged out in the thermodynamic data.

(b) *Dispersion of sound* — Intermolecular forces also enter in the dispersion of sound; Foch and Ford[330] have reviewed the theory. The status is that after the scattering processes are included through the coefficients of viscosity and thermal conductivity, the remaining dependence of the theory on intermolecular forces is small. At present, comparisons of the theory with experimental results for dispersion and absorption are made as tests of the theory of the Boltzmann equation, rather than for information on forces.

4.3.2.3 Rules-of-thumb

There have been frequent speculations[2] on how many thermodynamic properties are necessary in principle to achieve a unique inversion to the potential. In practice, we also have the questions of how sensitive the types of data are to parameters in the potential models and the functional forms used in the models. A general body of experience has grown up on the practical questions and has been made more precise by model calculations with few parameter potentials[331-336]. The most extensive such work has been presented by Hanley and Klein[333-336]. The scope of their work is within classical statistical mechanics and the three-parameter models which in practice have not been found to be sufficiently flexible to fit simultaneously extensive gas-phase data for an inert gas such as argon[337]. Some of these models were reviewed by Fitts[338].

In their calculations, Hanley and Klein take the thermodynamic properties calculated for a Lennard-Jones 12–6 model as a reference set and study the ability of the commonly used three-parameter potentials (Lennard-Jones m–6, Morse, exponential-6, and Kihara potentials) to reproduce this set. They find that the parameters can be adjusted in each of the families to achieve comparable fits, but with comparable shortcomings. The discussion is in terms of the well-depth parameter ε of the Lennard-Jones 12–6 and a reduced temperature $T^* = T/\varepsilon$. They find for all their models: (i) there is a range of $T^* = 2$–10 for the second virial coefficient, 2–5 for the viscosity, and 2–4 for thermal diffusion, for which properties are insensitive to the potential function[334]. (ii) None of their models can reproduce the Lennard-Jones 12–6 results for temperature ranges both above and below the intermediate insensitive range[334]. (iii) A very low level of smear in their reference set is sufficient to remove the ability to distinguish between models[334]. They recover[336] the result[339] that the thermal diffusion coefficient, even with appreciable uncertainties, is extremely sensitive to the model.

They consider[335] whether additional information is to be obtained by fitting pairs of properties rather than a single property and reach the remarkable conclusion that it is not. This refers specifically to the persistence of insensitive intermediate temperature ranges even with pairs of properties. As a general result, it is not in agreement with prior experience nor with the lack of a unique inversion for $B_{Cl}(T)$. They suggest that their models are misleading here.

Gas data for krypton and xenon cover T^* ranges of 0.6–11 and 0.5–7 so that rule (i) imposes serious restrictions on the extent of potential information available from these data.

Rule (ii) corresponds to the experience that attempts to fit experimental data for high and low temperatures simultaneously, with few parameters to adjust, are difficult. These are also temperature ranges where the scatter in experimental data is largest.

These model calculations emphasise that for many purposes the various three-parameter potential functions are equivalent and that, to distinguish between models, information at high and low temperatures is needed. There remains the problem of finding a sufficiently flexible few-parameter function to describe the available data. Hanley and Klein[340], who treat argon as an experimental realisation of their model, increase their flexibility by adding a fourth parameter. They have applied their potential model, the m–6–8, to heavier inert gases[341]. Another approach, in the analysis of scattering data[329], is to join functional forms suitable in various separation ranges by spline functions. This can be done to yield few-parameter potentials.

4.3.3 Non-thermodynamic data

We discuss some molecular scattering and spectroscopy experiments which are contributing important information on intermolecular potentials. These techniques have only recently been developed to the state of giving detailed information on intermolecular potentials for inert gases at thermal energies. With both types of data, there are semi-classical constructions[342–344] of the potential from sufficiently extensive data. The potential for Na–Hg has been constructed this way from scattering data[345] and the potential for Mg_2, a van der Waals molecule, has been constructed from spectroscopic data[346] by a Rydberg–Klein–Rees (RKR) method.

4.3.3.1 Spectroscopy

(a) *Ultraviolet spectroscopy* – For helium, Tanaka and Yoshino[347, 348] and Smith[349, 350] have studied the ultraviolet spectrum in the wavelength range 500 to 800 Å. The bulk of the derived information has been on the potential energy of interaction of a ground state and an electronically excited helium atom. There is a slight dependence of the observed continuum–continuum emission spectrum on the ground-state interaction at short separations and Smith and Chow[351] derived a ground-state repulsion energy from the data. The result was far from the empirical models derived from scattering and thermodynamic data. Sando[352] has since shown the data to be consistent with more usual estimates of the ground-state potential; his analysis differs from Smith's in the numbering of states of the excited-state potential. The results for the ground-state potential thus are sensitive to details of the description of the excited-state potential.

Tanaka and Yoshino[353] have reported high-resolution absorption studies of argon gas in the wavelength range 780 to 1080 Å. In this data band structure attributable to bound vibrational states in the ground- and excited-states intermolecular potentials was observed. Some rotational structure was also resolved, but the analysis of it is incomplete.

The vibrational structure for the ground-state potential gives a linear Birge–Sponer plot, with six levels observed. By assuming the validity of a Taylor series expansion of the potential about its minimum, we can extract the curvature of the potential at its minimum from the slope and intercept of the plot. This is as direct an inversion from data to a property of the potential as we are able to cite in this review.

The ground-state data has been analysed for potential function information by calculations which reproduce the Birge–Sponer plot[354–356]. Maitland and Smith[357] treated the data differently: they obtained the width of the potential well as a function of its depth from an RKR-analysis of the vibrational spectrum.

Tanaka and Yoshino report[353] that they have data showing the bound-state structure of the ground-state potentials of other inert gases; but this has not yet been published.

(b) *Optical spectroscopy* — Balfour and Douglas[346] have observed the absorption spectrum of magnesium vapour in the range 3140–3880 Å. They observed vibrational and rotational structure attributed to bound states of Mg_2, a molecule bound in the ground state through van der Waals forces. Thirteen vibrational levels and rotational structure of the ground state were identified. There were sufficient data for the application of an RKR construction of the ground state; the resulting potential had a shape very similar to a Morse potential. Balfour and Whitlock[358] have presented a preliminary report of similar work on Ca_2.

(c) *Infrared spectroscopy* — McKellar and Welsh have observed[359] effects attributed to bound states of molecules[360] of isotopes of hydrogen $(H_2)_2$, $(D_2)_2$, and $H_2–D_2$. The data show effects of anisotropies in the intermolecular potentials. Welsh and co-workers[361, 362] also have presented data on the interactions of hydrogen and inert gases.

4.3.3.2 Molecular scatterings at thermal energies

There is an extensive general theory for the inversion of scattering data to a potential function, either from data for one phase shift at all energies or for all phase shifts at one energy. In both cases there is additional information needed, as well as constraints on the moments of the potential, to be able to generate a unique potential from the data. In the first case, bound-state information is needed; in the second case, a restriction on the asymptotic form of the potential is needed. Recent work on the second case is summarised by Sabatier[363]; Newton[364] has reviewed the theory up to 1966.

For the treatment of molecular scattering data, this general theory has not been used; the extraction of a complete set of phase shifts from such data has not yet been feasible. Firsov[365] presented a construction based on the WKB approximation which becomes complicated for non-monotonic potentials. Buck[344] has further developed this method to achieve a practical method for the inversion of heavy-atom scattering data to a potential function. His formulation is based heavily on the use of rainbow scattering data in the construction of the scattering deflection function. There are at least nine parameters in the construction, but some of these are in the spline-function interpolations which join the small impact-parameter, rainbow region and large impact-parameter regions of the deflection function. Buck and Pauly applied the method[345] to the determination of the potential for Na–Hg, for which an impressively large body of high-quality experimental data, including both differential and total scattering cross-section measurements, was available. They used more information than the minimum necessary for the application of Buck's method, to have some redundancy in the analysis. Buck, Kick and Pauly have also applied Buck's method to K–Hg and Cs–Hg[366].

Bernstein and Muckerman[367] have reviewed scattering experiments up to 1967. Here we review scattering data for like pairs of inert gas atoms at thermal energies, a field in which the quality of work has increased rapidly since then.

(a) *Total cross-section measurements* – A long-standing discrepency between high energy (eV) helium scattering experiments and *a priori* potential calculations was markedly reduced by a re-analysis of the experiments by Jordan and Amdur[368]. There are new experiments extending to this energy range by Gengenbach et al.[369].

We have already described the first Genoa experiment[251] which observed the statistics oscillation in the total cross-section of ^{4}He–^{4}He. That experiment was extended to a wider range of velocities by Cantini et al.[370] and by Bennewitz et al.[371–373]. Dohmann[374] has given a rough argument relating the velocity period of the oscillation to a rigid sphere diameter in the potential. The length derived from the experiments is 2.0 Å, in agreement with the range of values 1.8–2.2 Å used in rigid-sphere theories of superfluid helium. The data are more refined than this crude analysis indicates and have given important information on the low-energy repulsion.

Bennewitz et al.[371–373] have measured the cross-section for ^{4}He–^{4}He down to 0.3 meV (3.5 K). Oates and King[375] reported a measurement at lower energies, but the resulting cross-section is much larger than anticipated and may be suspect.

Bennewitz et al.[372] have measured the cross-section of ^{3}He–^{3}He at energies from 5 to 20 K. The results are still of low accuracy, but do not show any evidence of differences for the interaction of ^{3}He–^{3}He and ^{4}He–^{4}He.

(b) *Differential cross-section measurements* – There has been an abrupt increase in measurements of the differential cross-sections for inert gas scatterings. We limit our presentation to the results of two of the experimental groups, Lee at Chicago and Scoles at Genoa, with the purpose of indicating the types of effects observed and the analysis given.

Siska et al. at Chicago have measured the differential cross-section for ^{4}He–^{4}He and Ne–Ne out to wide angles for two collision energies in both cases[329, 376]. The results show symmetry oscillations reflecting the indistinguishability of the scattering atoms as well as diffraction effects. For helium the relative energies in the collisions are 230 and 730 K and for neon, 280 and 710 K. They fit their data with Morse–spline–van der Waals potentials. For helium, they concluded by model calculations that the data were not very sensitive to the depth of the potential minimum and reflected more the low-energy repulsion, for neon the data were of higher quality and were more sensitive to the attractive part of the potential. The results for neon are of particular interest because there have been no multiparameter neon-potential constructions from thermodynamic data.

Parson and Lee[377, 378] measured the differential cross-section of Ar–Ar at an average centre-of-mass collision energy of 730 K and observed symmetry oscillations at wide angles as well as rainbow structure. They argued that their data indicated that the long-range attractive part of the Barker–Bobetic potential[379] is somewhat too strong. Their argument refers to the separations from 4–5 Å where the Barker–Bobetic potential is more attractive than estimates based on multipole coefficients would give. If this can be made precise, it may be informative for other applications of multipole coefficients.

Lee and his co-workers also have data for other inert gas pairs and mixed pairs of inert gases which they are analysing in terms of combination rules for potential parameters of like pairs[380].

Cavallini *et al.* at Genoa[381–383] have measured differential cross-sections for He–He, Ar–Ar and Kr–Kr to centre-of-mass scattering angles of 30 degrees. For the latter two pairs, they observe rainbow oscillations and parametrise the results by a rule of Düren and Schlier[384] which gives the rainbow angle in terms of the slope of the potential at its inflection point. This rule, obtained by approximation to results for the Lennard-Jones 12–6 potential, appears to be a more crude parameterisation than the results warrant.

4.3.4 Intermolecular potentials for inert gases

In this section we review the status of proposed intermolecular potentials for helium, argon and krypton. In terms of approximate values for the potential minima of 10, 130 and 190 K respectively, the reduced temperature (T^*) ranges of available gas-phase data for these elements are 0.1–200, 0.5–18 and 0.6–10.

4.3.4.1 Argon

For argon we limit our review to intermolecular potentials proposed since 1966. Our reason is the important revisions since then in the virial and viscosity data[385–387] to which such potentials are fitted.

Barker and co-workers[248, 379, 388], Dymond and Alder[389], and Klein and Hanley[340] have proposed multiparameter potentials fitted primarily to thermodynamic data. Dymond and Alder[389] and Klein and Hanley[340] determined their potentials using only gas-phase data. The Barker potentials include solid-phase data in the construction and include corrections for three-body forces in the assumed[390] form of the Axilrod–Teller triple-dipole interaction. Of these groups, Hanley and Klein with four parameters have the fewest adjustable parameters in their fits.

Dymond and Alder[389] attempted to overcome the restrictions on potential shapes inherent in the previously-used functional forms by working with a numerically-tabulated potential. They were aware of the difficulties in judging sensitivities of their fits to their adjustments of the potential. Later data[353, 391], which probe the region of the potential minimum, show strong evidence that the shape of the potential well in their model is faulty. In particular, the data of Tanaka and Yoshino do not reflect the anharmonicities in their model.

Klein and Hanley[340] adjusted their four parameters in a potential with inverse power law dependences on separation. McGee[392] has calculated the Birge–Sponer plot for this potential and found fair agreement with the Tanaka–Yoshino[353] data, with some disagreement on the spacings of the lowest levels. Maitland and Smith[357] with an RKR analysis of the same data concluded that the potential well in this model is too narrow for much of its depth.

The Barker–Pompe[388] and Barker–Bobetic[379] potentials had five and seven adjustable parameters to fit thermodynamic data and incorporated the theoretical value of the dipole–dipole coefficient C_6 and theoretical estimates for C_8 and C_{10}. They had two additional parameters to match the high-energy molecular-beam results. The Barker–Bobetic potential was a refinement of the Barker–Pompe potential which was constrained to yield the experimental cohesive energy of solid argon. The Barker–Fisher–Watts potential[248] is an interpolation between the first two Barker potentials[379, 388] which was designed to improve the fit of the equation of state of the liquid at high pressures. It also brought the scattering predictions into closer agreement with the data of Parson and Lee[377, 378]. The changes from the Barker–Bobetic potential are similar to those suggested by Maitland and Smith[357] on the basis of their analysis of the Tanaka–Yoshino data.

The result of Barker's efforts is a description of large quantities of experimental data. The question occurs whether all the data should be fitted starting from the theory he uses, either because of possible experimental errors or because of possible errors in the description of such effects as three-body forces. Barker did not fit the old high-temperature viscosity data[393], which have since been revised. The resolution of the discrepancies in predictions of properties of the solid such as the bulk modulus[379] is likely to include refinements in the theory as well as the experiments.

The data of Parson and Lee[377, 378] on the differential scattering cross-section of Ar–Ar at one energy have been analysed twice for potential functions. The initial analysis with five parameters yielded a potential with a poor fit to the low-temperature virial coefficient data[394]. A later analysis by Parson, Siska and Lee[378] included adjustments to fit virial data and the Tanaka–Yoshino data[353]. While different aspects of the scattering can be qualitatively ascribed to different features of the potential, the precision with which the assignment holds is not enough to justify an independent parameter for each such feature of the data.

When the parameters of the potential minima of the potentials discussed here are compared, it becomes apparent that the different groups are beginning to arrive at comparable answers. It is quite noticeable that the three Barker potentials are very similar and represent slight adjustments about the original Barker–Pompe potential to achieve better fits of additional data. The construction of the argon potential from thermodynamic data has apparently reached an advanced state and further progress in the specification of the potential will include the scattering data at thermal energies. To determine a potential of comparable accuracy from scattering data alone without thermodynamic data will probably require considerable advances in the experiments. An application of Buck's construction[344] to argon scattering data is desirable and will probably occur in the near future.

4.3.4.2 Krypton

Bobetic, Barker and Klein[395] have reported an eight-parameter pair potential for krypton constructed in a way similar to the Barker constructions of argon potential models. Here also there have been revisions in the virial and viscosity data for the gas. They found difficulties in trying to incorporate high-energy molecular scattering data of Amdur and Mason and the bulk modulus of the solid.

There have been two measurements of the differential cross-section for Kr–Kr at energies high in the thermal range. Cavallini et al[383] reported a measurement at 760 K relative energy and compared the data with results calculated for the Bobetic–Barker–Klein potential: the data were imperfectly fitted; Bobetic et al. interpreted this as confirming some misgivings they had in their results for the thermal expansion of solid krypton[395].

Schafer et al.[396] reported the differential cross-section of Kr–Kr at 720 K relative energy and fit their data with a Morse–spline–van der Waals potential. These potentials are not in as close agreement yet as the argon results; they differ considerably from the Lennard-Jones 12–6 potential presented by Sherwood and Prausnitz[397]. Krypton also is a candidate for Buck's construction[344].

4.3.4.3 Helium

There have been fewer potentials proposed for helium than for argon, perhaps because much of the gas data requires consideration of quantum effects in statistical mechanics. Also, one of the early models, the Lennard-Jones 12–6 potential of deBoer and Michels[268], has given a good description of much gas-phase data; some effort is needed to produce a model which gives a better overall description.

Haberlandt[398] fitted a Lennard-Jones 8–6 potential to $B(T)$ data in the range 45–300 K. Schiff and Verlet[399] used this model in a variational calculation of the ground-state energy of liquid ^{4}He and obtained a value below the experimental value. They therefore excluded it as a reasonable model; such a result has not yet been obtained for the helium models based on more extensive data.

The Yntema–Schneider potential[400], based on virial data from the range 300–1500 K, has been used in calculations[401] of the properties of liquid ^{3}He with the thought that the exponential dependence in the repulsion is more realistic than the $1/r^{12}$ in the deBoer–Michels potential. A misprint in the parameters of the repulsion in London's book[402] has caused confusion in this work. deBoer[403] estimated that the second virial coefficient at low temperatures for this potential deviates widely from the experimental data; we know of no other gas-phase calculations for this potential which give strong support to it as a realistic model.

Massey[404] adjusted the parameters in a Lennard–Jones 12–6 potential to fit, with an approximate theory, the binding energy of liquid ^{4}He. The calculated second virial coefficient in the range 2–4 K deviates by 10 % from the experimental data; the deviation from the viscosity at 2–4 K is 5–10 %. The second virial coefficient can be accurately estimated for Lennard–Jones 12–6 parameters near the deBoer–Michels values by use of the information in Table II of Kilpatrick et al.[276].

There are two exponential-six models constructed to fit the gas-phase data, the MR-1 and MR-5 in a series constructed by Mason and Rice[277, 405]. For the temperature range 2–300 K, when both viscosity data and virial data are considered[406], neither is a great improvement on the deBoer–Michels potential.

Beck[270] proposed a four-parameter potential with an exponential repulsion and C_6 and C_8 coefficients taken from theoretical calculations. He included $B(T)$ data from 50 to 1500 K and high-energy potential information in his construction. His potential well has about the same depth as in the deBoer–Michels potential (10 K). His potential has the same successful data fits as the deBoer–Michels potential and an improvement in the fitting of data above 500 K and for the thermal diffusion coefficient[272].

Bruch and McGee[269] started from a consideration of the piecewise potential construction of Bernstein and Morse[407] and proposed a series of two-parameter potentials with wells 10 % to 20 % deeper than the deBoer–Michels value. These potentials joined together functional forms containing an exponential repulsion with a van der Waals potential containing the theoretical calculations of C_6 and C_8. Their first constructions were based on virial data, but after some successes and failures noted by Keller and Taylor[271], the constructions were repeated to achieve a good overall fit to the gas data from 1–2000 K. Keller and Taylor asserted that the deeper wells of the earlier Bruch–McGee work could be excluded; this is not clear in the later Bruch–McGee work[278].

There are two Bruch–McGee potentials, MDD-1 and MDD-2 in their notation, which together seem to indicate the amount of uncertainty in helium potentials, with a few parameters, derived from thermodynamic data. These have been used in liquid helium calculations[408] to generate a measure of the uncertainty in the results to be ascribed to uncertainties in the potential models. The differences in the fits of the Beck and MDD-2 potentials are small. The MDD-1, MDD-2 and Beck potentials yield viscosities somewhat higher than the data of Guevara et al.[385] near 2000 K, which are higher than the extrapolation of the Dawe–Smith data[386]. The high energy repulsion in these models may be suspect.

At the time when the last of these potentials was being constructed, the first scattering data which were at all sensitive to the potential at thermal energies were reported. Since then, Cantini et al.[370] and Siska et al.[329] have reported determinations of the low-energy repulsion from scattering data. Cantini et al. presented their results for the potential in the form of a band of separations of width of about 0.04 Å for energies of from 20–300 K. Siska et al. analysed their differential cross-section data with the exponential–

spline–Morse–spline–van der Waals function. They appear to have been most sensitive to the potential at an energy of about 600 K. Bennewitz *et al.*[371, 373] have presented a potential for separations greater than 2.0 Å which is based on total cross-section data.

Most of the scattering data for helium are at energies where many partial waves contribute and the very low-energy scattering data are still imprecise. There is no rainbow scattering in helium and Buck's construction does not apply directly. The thermodynamic data of good accuracy at temperatures of 2–4 K will be difficult to supplant by scattering data in potential determinations. Progress in the specification of the helium potential will probably come from the additional information made available by scattering experiments and from additional theoretical understanding of the origin of the van der Waals potential minimum.

4.3.4.4 Summary

In Table 4.1 we have listed the position r_{min}, and depth ε of the potential minimum and the finite position σ of zero potential for four inert gas pairs. We list values obtained from scattering and from macroscopic data; the agreement is quite striking.

Table 4.1   Parameters of the potential minimum

System	σ Å	r_{min} Å	ε K	ε meV	Source
He–He	2.64–2.68	2.97–3.02	10.3–11.5	0.888–0.99	Macroscopic data[269, 270, 278]
	2.68	2.98	10.4	0.90	Total cross-sections[371]
Ne–Ne	2.68–2.73	3.1	34–43	2.9–3.7	Macroscopic data[1, 326]
	2.73	3.03	45.8	3.95	Differential cross-section[329]
Ar–Ar	3.36	3.76	142	12.2	Macroscopic data[248]
	3.345	3.76	140.7	12.1	Differential cross-section[378]
Kr–Kr	3.59	4.01	197	17.0	Macroscopic data[395]
	3.70	4.10	191	16.5	Differential cross-section[396]

We make these qualifications on the material presented in Table 4.1: (a) the values cited for the krypton scattering determination are preliminary; (b) the argon scattering determination included non-scattering data; (c) other macroscopic data argon potentials, might have been cited with differences of 0.05 Å in positions and 10 K in well depth; we prefer the Barker–Fisher–Watts values[248]; (d) it is only a recent[326] macroscopic data neon potential that has a well depth within 10% of the value from scattering; (e) with the possible exception of the neon entries, we consider the potentials from the two sources are in agreement, given the known uncertainties in the determinations; (f) a scattering determination for xenon is in progress.

Consideration will soon have to be given to the question of how accurately the inert-gas potentials need be known for applications. The scattering determinations will continue to be refined; the values cited in Table 4.1 are mostly from the first burst of results. The question is the point of diminishing returns: when is the point reached? One time-dependent answer is that the potential when used as input to a theory should be an order-of-magnitude better than the theory.

4.4 CONCLUSION

We are impressed that, in spite of the length of this review, we have treated only exceedingly simple atoms and molecules. For such systems, the determinations and calculations of interaction energies are coming into agreement. The quantitative and even qualitative knowledge of electronic properties other than the energy is still only crude.

The computational development has been of general methods designed to describe the qualitative features of the interactions. Future developments will be in the direction of improved treatment of correlation effects in specific systems. The practicality of the methods will become clearer as applications are made.

There are still some open questions in the general theory which should not depend on quantitative studies. The smoothness of the potential defined in the Born–Oppenheimer approximation has not been settled. The potential calculated by Hartree–Fock theory for inert gases has not been proved to be strictly repulsive. Basis set independent information on the analytical behaviour of molecular wave functions is limited.

Note added in proof.

The reader's attention is called to the following supplementary references: Section 4.1, Reference 410; 4.2.1.3, 411; 4.2.1.4, 412; 4.2.3.1, 413; 4.2.4, 414–420; 4.2.4.1, 421–424; 4.2.4.2, 425; 4.2.5, 426; 4.2.5.1, 427; 4.3.1, 428; 4.3.2.1, 429; 4.3.4.3, 430–431.

References

1. Hirschfelder, J. O., Curtiss, C. F. and Bird, R. B. (1954). *Molecular Theory of Gases and Liquids.* (New York: John Wiley)
2. *Intermolecular Forces* (1965), *Discuss. Faraday Soc.,* **40,**
3. Chu, B. (1967). *Molecular Forces,* Based on the Baker Lectures of Peter J. W. Debye. (New York: Interscience Publishers)
4. *Molecular Forces* (1967). Study week arranged by the Pontifical Academy of Sciences
5. Hirschfelder, J. O., ed. (1967). *Advan. in Chem. Phys.,* **12,** (New York: John Wiley)
6. Schlier, C. (1969). *Ann. Rev. Phys. Chem.,* **20,** 191
7. Buckingham, A. D. and Utting, B. D. (1970). *Ann. Rev. Phys. Chem.,* **21,** 287
8. Lester, W. A., ed. (1970). *Potential Energy Surfaces in Chemistry.* (San Jose: IBM Research Laboratory)
9. Margenau, H. and Kestner, N. R. (1971). *Theory of Intermolecular Forces, 2nd Edition.* (New York: Pergamon Press)
10. Schaefer, H. F. (1972). *The Electronic Structure of Atoms and Molecules.* (Reading: Addison-Wesley)
11. Boys, S. F. and Bernardi, F. (1970). *Mol. Phys.,* **19,** 553
12. Bertoncini, P. and Wahl, A. C. (1970). *Phys. Rev. Lett.,* **25,** 991
13. Schaefer, H. F., McLaughlin, D. R., Harris, F. E. and Alder, B. J. (1970). *Phys. Rev. Lett.,* **25,** 988

14. Dalgarno, A. (1967). *Advan. Chem. Phys.*, **12**, 143
15. Langhoff, P. W. and Karplus, M. (1970). *The Pade Approximant in Theoretical Physics*, ed. by G. A. Baker and J. L. Gammel, 41. (New York: Academic Press)
16. Bell, R. J. and Kingston, A. E. (1967). *Proc. Phys. Soc. (London)*, **90**, 901
17. Meinguet, J. (1970). *Approximation Theory*, Ed. by A. Talbot, 137. (New York: Academic Press)
18. Langhoff, P. W., Gordon, R. G. and Karplus, M. (1971). *J. Chem. Phys.*, **55**, 2126
19. Langhoff, P. W. (1971). *Chem. Phys. Lett.*, **12**, 217
20. Weinhold, F. (1970). *J. Chem. Phys.*, **54**, 1874
21. Yates, A. C. and Langhoff, P. W. (1970). *Phys. Rev. Lett.*, **25**, 1317
22. Barnsley, M. (1971). *J. Math. Phys.*, **12**, 957
23. McQuarrie, D. A., Terebey, J. and Shire, S. J. (1969). *J. Chem. Phys.*, **51**, 4683
24. Chang, T. Y. and Karplus, M. (1970). *J. Chem. Phys.*, **52**, 4698
25. Getzin, P. M. and Karplus, M. (1970). *J. Chem. Phys.*, **53**, 2100
26. Alexander, M. H. (1970). *Phys. Rev. A*, **1**, 1397
27. Pack, R. T. (1970). *Chem. Phys. Lett.*, **6**, 555
28. Langhoff, P. W. (1971). *Chem. Phys. Lett.*, **12**, 223
29. Stwalley, W. C. (1971). *J. Chem. Phys.*, **54**, 4517
30. Bezpal'ko, R. M. and Gutman, I. I. (1971). *Izu. Vuz. Fiz.*, **3**, 151
31. Shabanova, Z. N. (1969). *Optika I. Spectrosk.*, **27**, 383; *Optics and Spectrosc.*, **27**, 205
32. Abdulnur, S. F. (1971). *Int. J. Quantum Chem.*, **5**, 525
33. Adamov, M. N., Balmakov, M. D. and Rebane, T. K. (1969). *Int. J. Quantum Chem.*, **3**, 13
34. Kochanski, E. (1971). *Chem. Phys. Lett.*, **10**, 543
35. Girardet, C. and Robert, D. (1971). *J. Mol. Struct.*, **7**, 31
36. Piela, L. (1969). *Int. J. Quantum Chem.*, **3**, 945
37. Kaneko, S. (1971). *J. Chem. Phys.*, **54**, 819
38. Arrighini, G. P., Biondi, F. and Guidotti, C. (1971). *J. Chem. Phys.*, **55**, 4090
39. Broussard, J. T. and Kestner, N. R. (1970). *J. Chem. Phys.*, **53**, 1507
40. Deal, W. J. and Young, R. H. (1970). *Mol. Phys.*, **19**, 427
41. Dutta, N. C., Ishihara, T., Matsubara, C., and Das, T. P. (1970). *Int. J. Quantum Chem.*, **3S**, 367
42. Claverie, P. and Rein, R. (1969). *Int. J. Quantum Chem.*, **3**, 537
43. Schweig, A. (1969). *Int. J. Quantum Chem.*, **3**, 823
44. Van der Merwe, J. H. and van der Merwe, A. J. (1969). *J. Math. Phys.*, **10**, 539
45. Davison, W. D. (1968). *J. Phys. B.*, **1**, 139
46. Kramer, H. L. and Hershbach, D. R. (1970). *J. Chem. Phys.*, **53**, 2792
47. Claverie, P. (1969). *Int. J. Quantum Chem.*, **3**, 349
48. Certain, P. R. and Byers Brown, W. (1972). *Int. J. Quantum Chem.*, **6**, 131
49. Claverie, P. (1971). *Int. J. Quantum Chem.*, **5**, 273
50. Malrieu, J. P. (1971). *Int. J. Quantum Chem.*, **5**, 435, 455
51. Kreek, H. and Meath, W. J. (1969). *J. Chem. Phys.*, **50**, 2289
52. Singh, T. R., Kreek, H. and Meath, W. J. (1970). *J. Chem. Phys.*, **52**, 5565; erratum (1970), *J. Chem. Phys.*, **53**, 4121
53. Kreek, H., Pan, Y. H. and Meath, W. J. (1970). *Mol. Phys.*, **19**, 513
54. Pan, Y. H. and Meath, W. J. (1971). *Mol. Phys.*, **20**, 873
55. Fukui, K. and Yamabe, T. (1968). *Int. J. Quantum Chem.*, **2**, 359; erratum (1971), *Int. J. Quantum. Chem.*, **5**, 478
56. Horak, Z. J. and Siskova, J. (1970). *Chem. Phys. Lett.*, **6**, 375
57. Le Roy, R. J. and Bernstein, R. B. (1971). *J. Mol. Spectrosc.*, **37**, 109
58. Le Roy, R. J. (1971). *J. Mol. Spectrosc.*, **39**, 175
59. Stwalley, W. C. (1970). *Chem. Phys. Lett.*, **7**, 600
60. Heitler, W. and London, F. (1927). *Z. Physik*, **44**, 455
61. Eisenschitz, R. and London, F. (1930). *Z. Physik*, **60**, 491
62. Herring, C. (1966). *Magnetism, Vol. 2B*, Ed. by A. Rado and H. Suhl. (New York: Academic Press)
63. Damburg, R. J. and Propin, R. Kh. (1971). *J. Chem. Phys.*, **55**, 612
64. Van der Avoird, A. (1967). *Chem. Phys. Lett.*, **1**, 24
65. Dalgarno, A. and Lynn, N. (1956). *Proc. Phys. Soc. (London)*, **69**, 821
66. Lynn, N. (1958). *Proc. Phys. Soc. (London)*, **72**, 201

67. Löwdin, P. O. (1969). *Advan. Chem. Phys.,* **14,** 283
68. Van der Avoird, A. (1967). *J. Chem. Phys.,* **47,** 3649
69. Van der Avoird, A. (1967). *Chem. Phys. Lett.,* **1,** 411
70. Hirschfelder, J. O. (1967). *Chem. Phys. Lett.,* **1,** 326, 363
71. Musher, J. I. and Salem, L. (1966). *J. Chem. Phys.,* **44,** 2943
72. Murrell, J. N., Randič, M. and Williams, D. R. (1965). *Proc. Roy. Soc. (London),* **A284,** 566
73. Salem, L. (1965). *Discuss. Faraday Soc.,* **40,** 150
74. Pecul, K. (1971). *Chem. Phys. Lett.,* **9,** 316
75. Dacre, P. D. and McWeeny, R. (1970). *Proc. Roy. Soc. (London),* **A317,** 435
76. Murrell, J. N. and Shaw, G. (1967). *J. Chem. Phys.,* **46,** 1768
77. Epstein, S. T. and Johnson, R. E. (1968). *Chem. Phys. Lett.,* **1,** 599
78. Musher, J. I. and Amos, A. T. (1967). *Phys. Rev.,* **164,** 31
79. Amos, A. T. and Musher, J. I. (1967). *Chem. Phys. Lett.,* **1,** 149
80. Amos, A. T. and Musher, J. I. (1969). *Chem. Phys. Lett.,* **3,** 721
81. Carr, W. J. (1963). *Phys. Rev.,* **131,** 1947
82. Hirschfelder, J. O. and Silbey, R. (1966). *J. Chem. Phys.,* **45,** 2188
83. Musher, J. I. and Silbey, R. (1968). *Phys. Rev.,* **174,** 94
84. Kirtman, B. (1968). *Chem. Phys. Lett.,* **1,** 631
85. Klein, D. J. (1971). *Int. J. Quantum. Chem.,* **4S,** 271
86. Certain, P. R. and Hirschfelder, J. O. (1970). *J. Chem. Phys.,* **52,** 5992
87. Hirschfelder, J. O. (1971). *Int. J. Quantum Chem.,* **4S,** 257
88. Adams, W. H. (1971). *Chem. Phys. Lett.,* **11,** 441
89. Jansen, L. (1967). *Phys. Rev.,* **162,** 63
90. Byers Brown, W. (1967). *Chem. Phys. Lett.,* **2,** 105
91. Ritchie, A. B. (1968). *J. Chem. Phys.,* **49,** 2167
92. Ritchie, A. B. (1968). *Phys. Rev.,* **171,** 125
93. Corinaldesi, E. (1962). *Nuovo Cimento,* **30,** 105
94. Landman, U. and Pauncz, R. (1971). *Chem. Phys. Lett.,* **9,** 489
95. Laughlin, C. and Amos, A. T. (1971). *J. Chem. Phys.,* **55,** 4837
96. Brändas, E. (1971). *Int. J. Quantum Chem.,* **4S,** 285
97. Boehm, R. and Yaris, R. (1971). *J. Chem. Phys.,* **55,** 2620
98. Devaquet, A. (1970). *Mol. Phys.,* **18,** 233
99. Sustman, R. and Binsch, G. (1970). *Mol. Phys.,* **20,** 1
100. Von Niessen, W. (1971). *J. Chem. Phys.,* **55,** 1948
101. Bacon, J. and Santry, F. P. (1971). *J. Chem. Phys.,* **55,** 3743
102. Lekkerkerker, H. N. W. and Laidlaw, W. G. (1970). *J. Chem. Phys.,* **52,** 2953
103. Amos, A. T. (1970). *Chem. Phys. Lett.,* **5,** 587
104. Van der Avoird, A. (1967). *Chem. Phys. Lett.,* **1,** 429
105. Brändas, E. and Goscinski, O. (1969). *J. Chem. Phys.,* **51,** 975
106. McQuarrie, D. A. and Hirschfelder, J. O. (1967). *J. Chem. Phys.,* **47,** 1775
107. Sanders, W. A. (1969). *J. Chem. Phys.,* **51,** 491
108. Certain, P. R., Hirschfelder, J. O., Kolos, W. and Wolniewicz, L. (1968). *J. Chem. Phys.,* **49,** 24
109. Hirschfelder, J. O. and Certain, P. R. (1968). *Int. J. Quantum Chem.,* **2S,** 125
110. Goscinski, O. and Brändas, E. (1968). *Chem. Phys. Lett.,* **2,** 299
111. Hirschfelder, J. O. and Certain, P. R. (1968). *Chem. Phys. Lett.,* **2,** 539
112. Piela, L. (1971). *Int. J. Quantum Chem.,* **5,** 85
113. Alexander, M. H. and Salem, L. (1967). *J. Chem. Phys.,* **46,** 430
114. Murrell, J. N. and Shaw, G. (1968). *Mol. Phys.,* **15,** 325
115. Murrell, J. N. and Shaw, G. (1967). *Mol. Phys.,* **12,** 475
116. Certain, P. R. (1968). *J. Chem. Phys.,* **49,** 35
117. Epstein, S. T. and Johnson, R. E. (1968). *Chem. Phys. Lett.,* **1,** 602
118. Certain, P. R., Hirschfelder, J. O. and Epstein, S. T. (1969). *Chem. Phys. Lett.,* **4,** 401
119. Epstein, S. T. and Rosenthal, C. M. (1970). *Chem. Phys. Lett.,* **6,** 551
120. Lekkerkerker, H. N. W. and Laidlaw, W. G. (1970). *Trans. Faraday Soc.,* **66,** 1830
121. Van Duijneveldt, F. B. and Murrell, J. N. (1967). *J. Chem. Phys.,* **46,** 1759
122. Van Duijneveldt, F. B. (1968). *J. Chem. Phys.,* **49,** 1424
123. Van Duijneveldt-van de Rijdt, J. G. C. M. and Van Duijneveldt, F. B. (1968). *Chem. Phys. Lett.,* **2,** 565

124. Shaw, G. (1969). *Int. J. Quantum Chem.*, **3**, 219
125. Jansen, L. and Lombardi, E. (1967). *Chem. Phys. Lett.*, **1**, 417
126. Williams, D. R., Schaad, L. J. and Murrell, J. N. (1967). *J. Chem. Phys.*, **47**, 4916
127. Williams, D. R. (1968). *J. Chem. Phys.*, **49**, 4478
128. Murrell, J. N. and Shaw, G. (1969). *J. Chem. Phys.*, **49**, 4731
129. Murrell, J. N. and Teixeria-Dias, J. J. C. (1970). *Mol. Phys.*, **19**, 521
130. Certain, P. R. (1971). *J. Chem. Phys.*, **55**, 3045
131. Jansen, L. (1965). *Advan. in Quantum Chem. Vol. 2*, Ed. by P. O. Lowdin. (New York: Academic Press)
132. Lombardi, E. and Ritter, R. (1970). *Chem. Phys. Lett.*, **7**, 143
133. Morse, P. M. and Stueckelberg, E. C. G. (1929). *Phys. Rev.*, **33**, 932
134. Bingel, W. A. (1959). *J. Chem. Phys.* **30**, 1250
135. Byers Brown, W. and Steiner, E. (1966). *J. Chem. Phys.*, **44**, 3934
136. Byers Brown, W. (1968). *Chem. Phys. Lett.*, **1**, 655
137. Byers Brown, W. and Power, J. D. (1970). *Proc. Roy. Soc. (London)*, **A317**, 545
138. Lee, T., Dutta, N. C. and Das, T. P. (1970). *Phys. Rev. Lett.*, **25**, 204
139. Kirtman, B. and Decious, D. R. (1966). *J. Chem. Phys.*, **44**, 830
140. Matcha, R. L. and Byers Brown, W. (1968). *J. Chem. Phys.*, **48**, 74
141. Kirtman, B. (1968). *J. Chem. Phys.*, **49**, 3890
142. Certain, P. R. and Hirschfelder, J. O. (1970). *J. Chem. Phys.*, **52**, 5977
143. Goodisman, J. (1969). *J. Chem. Phys.*, **50**, 903
144. Goodisman, J. (1969). *J. Chem. Phys.*, **51**, 3540
145. Wahl, A. C. and Das, G. (1970). *Advan. Quantum Chem.*, **5**, 261
146. Bender, C. F. and Davidson, E. R. (1966). *J. Phys. Chem.*, **70**, 2675
147. Krauss, M. (1967). *Compendium of* ab initio *Calculations of Molecular Energies and Properties*, NBS Technical Note 438, (Washington: U.S. Government Printing Office)
148. Allen, L. C. (1969). *Ann. Rev. Phys. Chem.*, **20**, 315
149. Krauss, M. (1970). *Ann. Rev. Phys. Chem.*, **21**, 39
150. Clark, R. G. and Stewart, E. T. (1970). *Quart. Rev. Chem. Soc.*, **24**, 95
151. Krauss, M. (1971). *Proceedings, Conference on Potential Energy Surfaces in Chemistry*, 6. (San Jose: IBM Research Laboratory)
152. Richards, W. G., Walker, T. E. H. and Hinkley, R. K. (1971). *Bibliography of* ab initio *Calculations*. (Oxford: Oxford University Press)
153. Trivedi, P. C. (1970). *Physica*, **48**, 486
154. Liu, B. (1971). *Int. J. Quantum Chem.*, **5S**, 123
155. Lester, W. A. (1970). *J. Chem. Phys.*, **53**, 1611
156. Lester, W. A. (1970). *J. Chem. Phys.*, **53**, 1511
157. Lester, W. A. (1971). *J. Chem. Phys.*, **54**, 3171
158. Guest, M. F., Murrell, J. N. and Pedley, J. B. (1971). *Mol. Phys.*, **20**, 81
159. Poshusta, R. D. and Siems, W. F. (1971). *J. Chem. Phys.*, **55**, 1995
160. Poshusta, P. D., Klint, D. W. and Liberles, A. (1971). *J. Chem. Phys.*, **53**, 252
161. Brown, P. J. and Hayes, E. F. (1971). *J. Chem. Phys.*, **55**, 922
162. Duke, A. J. and Bader, R. F. W. (1971). *Chem. Phys. Lett.*, **10**, 631
163. Rothenberg, S. and Schaefer, H. F. (1971). *Chem. Phys. Lett.*, **10**, 565
164. Janoschek, R. (1969). *Acta Phys. Hungar.*, **27**, 373
165. Tapia, O., Bessis, G., Bratoz, S. (1971). *Int. J. Quantum Chem.*, **4S**, 289
166. Roach, A. C. and Kuntz, P. J. (1970). *Chem. Commun.*, 1336
167. Krauss, M., Maldonado, P. and Wahl, A. C. (1971). *J. Chem. Phys.*, **54**, 4944
168. Clementi, E., Mehl, J. and von Niessen, W. (1971). *J. Chem. Phys.*, **54**, 508
169. Kollman, P. A. and Allen, L. C. (1971). *J. Amer. Chem. Soc.*, **93**, 4991
170. Davidson, R. B. and Allen, L. C. (1971). *J. Chem. Phys.*, **55**, 519
171. Dunning, T. H. and Winter, N. W. (1971). *Chem. Phys. Lett.*, **11**, 194
172. Gordon, M. D. and Secrest, D. (1970). *J. Chem. Phys.*, **52**, 120; erratum (1970). *J. Chem. Phys.*, **53**, 4408
173. Basch, H. (1971). *J. Chem. Phys.*, **55**, 1700
174. Pipano, A. and Kaufman, J. J. (1971). *Abstracts of papers of the VIIth International Conference on the Physics of Electronic and Atomic Collisions*, 966
175. Lewis, E. L., McNamara, L. F. and Michels, H. H. (1971). *Phys. Rev. A*, **3**, 1939
176. Csizmadia, I. G., Kari, R. E., Polanyi, J. C., Roach, A. C. and Robb, M. A. (1970). *J. Chem. Phys.*, **52**, 6205

177. Balint-Kurti, G. G. and Karplus, M. (1971). *Chem. Phys. Lett.*, **11**, 203
178. Tapia, O. (1971). *Chem. Phys. Lett.*, **10**, 613
179. Slocomb, C. A., Miller, W. H., Schaefer, H. F. (1971). *J. Chem. Phys.*, **55**, 926
180. Axilrod, B. M. and Teller, E. (1943). *J. Chem. Phys.*, **11**, 299
181. Bell, R. J. (1970). *J. Phys. B*, 751
182. Doran, M. B. and Zucker, I. J. (1971). *J. Phys. C*, **4**, 307
183. Stogryn, D. E. (1970). *Phys. Rev. Lett.*, **24**, 971
184. Stogryn, D. E. (1971). *Mol. Phys.*, **22**, 81
185. Bader, R. F. W., Novaro, O. A. and Beltran-Lopez, V. (1971). *Chem. Phys. Lett.*, **8**, 568
186. Del Bene, J. E. and Pople, J. A. (1971). *J. Chem. Phys.*, **55**, 2296
187. Margenau, H. and Stamper, J. (1967). *Advan. Quantum Chem.*, **3**, 129
188. Kestner, N. R. and Sinanoglu, O. (1966). *J. Chem. Phys.*, **45**, 194
189. Das, G. and Ray, S. (1970). *Phys. Rev. Lett.*, **24**, 1391
190. Das, G. and Wahl, A. C. (1971). *Phys. Rev. A*, **4**, 825
191. Kutzelnigg, W. and Gelus, M. (1970). *Chem. Phys. Lett.*, **7**, 296
192. Butler, D. and Kestner, N. R. (1970). *J. Chem. Phys.*, **53**, 1704
193. Kaufman, J. J., Harkins, J. J. and Koski, W. S. (1969). *J. Chem. Phys.*, **50**, 771
194. Hoffman, R., Wan, C. C. and Neagu, V. (1970). *Mol. Phys.*, **19**, 113
195. Zhogolev, D. A. and Matyash, I. V. (1971). *Chem. Phys. Lett.*, **10**, 444
196. Baetzold, R. C. (1971). *J. Chem. Phys.*, **55**, 4355
197. Mulliken, R. S. (1970). *J. Chem. Phys.*, **52**, 5170
198. Mulliken, R. S. (1971). *J. Chem. Phys.*, **55**, 288
199. Weeks, J. D., Hazi, A. and Rice, S. A. (1969). *Advan. Chem. Phys.*, **16**, 283
200. Bardsley, J. N. (1970). *Chem. Phys. Lett.*, **7**, 517
201. Roach, A. C. and Baybutt, P. (1970). *Chem. Phys. Lett.*, **7**, 7
202. Baylis, W. E. (1969). *J. Chem. Phys.*, **51**, 2665
203. Roach, A. C. and Child, M. S. (1968). *Mol. Phys.*, **14**, 1
204. Woolley, A. N. and Child, M. S. (1970). *Mol. Phys.*, **19**, 625
205. Kutzelnigg, W. (1969). *Chem. Phys. Lett.*, **4**, 435
206. Dalgarno, A., Bottcher, C. and Victor, G. A. (1970). *Chem. Phys. Lett.*, **7**, 265
207. Ellison, F. O. (1963). *J. Amer. Chem. Soc.*, **85**, 3540
208. Wu, A. A. and Ellison, F. O. (1968). *J. Chem. Phys.*, **48**, 1491
209. Abrams, R. B., Patel, J. C., and Ellison, F. O. (1968). *J. Chem. Phys.*, **49**, 450
210. Preston, R. K. and Tully, J. C. (1971). *J. Chem. Phys.*, **54**, 4297
211. Krenos, J., Preston, R., Wolfgang, R., and Tully, J. (1971). *Chem. Phys. Lett.*, **10**, 17
212. Kuntz, P. J. and Roach, A. C. (1972). *Trans. Faraday Soc.*, **68**, 259
213. Kiss, Z. J. and Welsh, H. L. (1959). *Phys. Rev. Lett.*, **2**, 166
214. McTague, J. P. and Birnbaum, G. (1971). *Phys. Rev. A*, **3**, 1376
215. Orcutt, R. H. and Cole, R. H. (1967). *J. Chem. Phys.*, **46**, 697
216. Kerr, E. C. and Sherman, R. H. (1970). *J. Low Temp. Phys.*, **3**, 451
217. Buckingham, A. D. and Dunmur, D. A. (1968). *Trans. Faraday Soc.*, **64**, 1776
218. Raynes, W. T. (1969). *Mol. Phys.*, **17**, 169
219. Jameson, C. J. and Jameson, A. K. (1971). *Mol. Phys.*, **20**, 957
220. Wright, J. J., Balling, L. C. and Lambert, R. H. (1970). *Phys. Rev. A*, **1**, 1018
221. Adrian, F. J. (1960). *J. Chem. Phys.*, **32**, 972
222. Rao, B. K., Ikenberry, D. and Das, T. P. (1970). *Phys. Rev. A*, **2**, 1411
223. Ikenberry, D. and Das, T. P. (1971). *Phys. Rev. Lett.*, **27**, 79
224. Kunik, D. and Kaldor, U. (1971). *J. Chem. Phys.*, **55**, 4127
225. Ray, S., Das, G., Maldonado, P. and Wahl, A. C. (1970). *Phys. Rev. A*, **2**, 2196
226. Rao, B. K. and Kestner, N. R. (1971). *Phys. Rev. A*, **4**, 1322
227. Bosomworth, D. R. and Gush, H. P. (1965). *Can. J. Phys.*, **43**, 751
228. Buckingham, A. D. (1959). *Propriétiés optiques et acoustiques des fluides comprimés et actions intermoléculaires*, 57. (Paris: Centre national de la Recherche Scientifique)
229. Byers Brown, W. and Whisnant, D. M. (1970). *Chem. Phys. Lett.*, **7**, 329
230. Matcha, R. L. and Nesbet, R. K. (1967). *Phys. Rev.* **160**, 72
231. Levine, H. B. (1968). *Phys. Rev. Lett.*, **21**, 1512
232. Bucaro, J. A. and Litovitz, T. A. (1971). *J. Chem. Phys.*, **54**, 3846
233. Jansen, L. and Mazur, P. (1955). *Physica*, **21**, 193 and 208
234. Buckingham, A. D. and Pople, J. A. (1955). *Trans. Faraday Soc.*, **51**, 1029
235. Buckingham, A. D. (1956). *Trans. Faraday Soc.*, **52**, 1035

236. Levine, H. B. and McQuarrie, D. A. (1968). *J. Chem. Phys.*, **49**, 4181
237. Tulub, A. V., Balmakov, M. D., and Khallaf, S. A. (1971). *Sov. Phys. Doklady*, **16**, 18; *Dokl. Akad. Nauk SSSR*, **196**, 75
238. Certain, P. R. and Fortune, P. J. (1971). *J. Chem. Phys.*, **55**, 5818
239. Levine, H. B. (1972). *J. Chem. Phys.*, **56**,
240. Levine, H. B. and Birnbaum, G. (1971). *J. Chem. Phys.*, **55**, 2914
241. Lim, T. K., Lindner, B. and Kromhout, R. A. (1970). *J. Chem. Phys.*, **52**, 3831
242. Dupre, D. B. and McTague, J. P. (1969). *J. Chem. Phys.*, **50**, 2024
243. Cook, E. G. and Schug, J. C. (1970). *J. Chem. Phys.*, **53**, 723
244. Dymond, J. H. and Smith, E. B. (1969). *The Virial Coefficients of Gases.* (Oxford: Clarendon Press)
245. Maitland, G. C. and Smith, E. B. (1971). *The Viscosities of Eleven Common Gases: A Critical Compilation.* (Oxford: Physical Chemistry Laboratory)
246. Kestin, J. and Wakeham, W. (1970). *Proceedings of the Fifth Symposium on Thermophysical Properties.* (American Society of Mechanical Engineers, New York), p. 55
247. Mason, E. A. and Marrero, T. R. (1970). *Advan. Atom. Mol. Phys.*, **6**, 155
248. Barker, J. A., Fisher, R. A., and Watts, R. O. (1971). *Mol. Phys.*, **21**, 657
249. Murphy, R. D. and Barker, J. A. (1971). *Phys. Rev. A*, **3**, 1037
250. Jordan, H. F. and Fosdick, L. D. (1968). *Phys. Rev.*, **171**, 128
251. Dondi, M. G., Scoles, G., Torello, F. and Pauly, H. (1969). *J. Chem. Phys.*, **51**, 392
252. Helbing, R. K. B. (1969). *J. Chem. Phys.*, **50**, 493
253. Hill, T. L. (1956). *Statistical Mechanics.* (McGraw-Hill, New York)
254. Mayer, J. E. and Mayer, M. G. (1940). *Statistical Mechanics.* (New York: John Wiley and Sons)
255. Ruelle, D. (1969). *Statistical Mechanics.* (New York: W. A. Benjamin Inc)
256. Lebowitz, J. L. (1968). *Ann. Rev. Phys. Chem.*, **19**, 389
257. deBoer, J. (1949). *Rep. Progr. in Phys.*, **12**, 305
258. Kihara, T. (1958). *Advan. in Chem. Phys.*, **1**, 267
259. Larsen, S. Y., Witte, K. and Kilpatrick, J. E. (1966). *J. Chem. Phys.*, **44**, 213
260. Boyd, M. E., Larsen, S. Y., and Kilpatrick, J. E. (1969). *J. Chem. Phys.*, **50**, 4034
261. Leonas, V. B. and Samiulov, E. V. (1966). *High Temp. (USSR)*, **4**, 664
262. Boyd, M. E. (1971). *J. Res. Nat. Bur. Stand. (U.S.)*, **75A**, 57
263. Gallagher, J. S. and Klein, M. (1971). *J. Res. Nat. Bur. Stand. (U.S.)*, **75A**, 337
264. Keller, J. B. and Zumino, B. (1959). *J. Chem. Phys.*, **30**, 1351
265. Frisch, H. L. and Helfand, E. (1960). *J. Chem. Phys.*, **32**, 269
266. Rowlinson, J. S. (1965). *Discuss. Faraday Soc.*, **40**, 19
267. Jonah, D. A. and Rowlinson, J. S. (1966). *Trans. Faraday Soc.*, **62**, 1067
268. deBoer, J. and Michels, A. (1939). *Physica*, **5**, 945
269. Bruch, L. W. and McGee, I. J. (1967). *J. Chem. Phys.*, **46**, 2959
270. Beck, D. E. (1968). *Mol. Phys.*, **14**, 311; (1968). **15**, 332
271. Keller, J. M. and Taylor, W. L. (1969). *J. Chem. Phys.*, **51**, 4829
272. Taylor, W. L. and Keller, J. M. (1971). *J. Chem. Phys.*, **54**, 647
273. Bruch, L. W. (1971). *J. Chem. Phys.*, **54**, 4281
274. Theumann, A. (1970). *J. Math. Phys.*, **11**, 1772
275. Beth, E. and Uhlenbeck, G. E. (1937). *Physica*, **4**, 915
276. Kilpatrick, J. E., Keller, W. E., Hammel, E. F. and Metropolis, N. (1954). *Phys. Rev.*, **94**, 1103
277. Kilpatrick, J. E., Keller, W. E. and Hammel, E. F. (1955). *Phys. Rev.*, **97**, 9
278. Bruch, L. W. and McGee, I. J. (1970). *J. Chem. Phys.*, **52**, 5884
279. Lieb, E. (1967). *J. Math. Phys.*, **8**, 43
280. Hill, R. N. and Luchinsky, H., to be published
281. Boyd, M. E., Larsen, S. Y. and Kilpatrick, J. E. (1966). *J. Chem. Phys.*, **45**, 499
282. Larsen, S. Y., Kilpatrick, J. E., Lieb, E. H. and Jordan, H. F. (1965). *Phys. Rev. A*, **140**, 229
283. Hill, R. N. (1968). *J. Math. Phys.*, **9**, 1534
284. Kim, S. and Henderson, D. (1966). *Proc. Nat. Acad. Sci. (U.S.)*, **55**, 705
285. Sherman, R. H. and Kerr, E. C., to be published
286. Larsen, S. Y. and Mascheroni, P. L. (1970). *Phys. Rev. A*, **2**, 1018
287. Servadio, S. (1971). *Phys. Rev. A*, **4**, 1256
288. Mikolaj, P. G. and Pings, C. J. (1967). *J. Chem. Phys.*, **46**, 1412

289. Bruin, C. and Hasman, A. (1971). *Phys. Lett.*, **33A**, 338
290. Hasman, A. and Zandfeld, P. (1971). *Phys. Lett.*, **34A**, 112
291. Levesque, D. and Verlet, L. (1968). *Phys. Rev. Lett.*, **20**, 905
292. Achter, E. K. and Meyer, L. (1969). *Phys. Rev.*, **188**, 291
293. Hallock, R. B. (1970). *Bull. Amer. Phys. Soc.*, **15**, 59 (and private communication)
294. Henderson, D. (1965). *Mol. Phys.*, **10**, 73
295. Henderson, D. and Oden, L. (1966). *Mol. Phys.*, **10**, 405
296. Poll, J. D. and Miller, M. S. (1971). *J. Chem. Phys.*, **54**, 2673
297. Fosdick, L. D. and Jordan, H. F. (1966). *Phys. Rev.*, **143**, 58
298. Fosdick, L. D. and Jacobson, R. C. (1971). *J. Comp. Phys.*, **7**, 157
299. Bruch, L. W. (1971). *J. Chem. Phys.*, **55**, 5101
300. Chapman, S. and Cowling, T. G. (1970). *The Mathematical Theory of Non-uniform Gases*, (3rd Edn.) (Cambridge: University Press)
301. Hanley, H. J. M., McCarty, R. D. and Sengers, J. V. (1969). *J. Chem. Phys.*, **50**, 857
302. Hanley, H. J. M., Ed. (1969). *Transport Phenomena in Fluids*. (New York: Marcel Dekker)
303. Curtiss, C. F. (1967). *Ann. Rev. Phys. Chem.*, **18**, 125
304. Zwanzig, R. W. (1965). *Ann. Rev. Phys. Chem.*, **16**, 67
305. Cotter, J. R. (1952). *Proc. Roy. Irish Acad.*, *A***55**, 1
306. Pekeris, C. L. and Altermann, Z. (1957). *Proc. Nat. Acad. Sci. (U.S.)*, **43**, 998
307. Brooker, P. I. and Green, H. S. (1968). *Aust. J. Phys.*, **21**, 543
308. Wang Chang, C. S. and Uhlenbeck, G. E. (1952), reprinted in *Studies in Statistical Mechanics*, **5**, 43 (1970)
309. Abramowitz, M. and Stegun, I. A. (1964). *Handbook of Mathematical Functions*, Nat. Bur. Stand. Appl. Math. Ser. 55 (Washington, D.C.: U.S. Gov. Printing Office)
310. Mason, E. A. (1954). *J. Chem. Phys.*, **22**, 169
311. Klein, M. and Smith, F. J. (1968). *J. Res. Nat. Bur. Stand. (U.S.)*, **72A**, 359
312. Uehling, E. A. and Uhlenbeck, G. E. (1933). *Phys. Rev.*, **43**, 552
313. Imam-Rahajoe, S. and Curtiss, C. F. (1967). *J. Chem. Phys.*, **47**, 5269
314. Mori, H., Oppenheim, I. and Ross, J. (1962). *Studies in Statistical Mechanics*, Ed. by deBoer, J. and Uhlenbeck, G. E. Vol. 1. (Amsterdam: North Holland)
315. Bogoliubov, N. N. (1946). *Problems of a Dynamical Theory in Statistical Physics*. (Moscow)
316. Barker, J. A., Bobetic, M. V. and Pompe, A. (1971). *Mol. Phys.*, **20**, 347
317. Imam-Rahajoe, S., Curtiss, C. F. and Bernstein, R. B. (1965). *J. Chem. Phys.*, **42**, 530
318. Munn, R. J., Smith, F. J., Mason, E. A. and Monchick, L. (1965). *J. Chem. Phys.*, **42**, 537
319. Boyd, M. E. and Larsen, S. Y. (1971). *Phys. Rev. A*, **4**, 1155
320. Munn, R. J. (1965). *J. Chem. Phys.*, **42**, 3032
321. Mason, E. A., Munn, R. J. and Smith, F. J. (1965). *Discuss. Faraday Soc.*, **40**, 27
322. Hanley, H. J. M. and Childs, G. E. (1969). *J. Chem. Phys.*, **50**, 4600
323. Clarke, A. G. and Smith, E. B. (1969). *J. Chem. Phys.*, **51**, 4156
324. Dymond, J. H. (1968). *J. Chem. Phys.*, **49**, 3673
325. Maitland, G. C. and Smith, E. B. (1970). *J. Chem. Phys.*, **52**, 3848
326. Dymond, J. H. (1971). *J. Chem. Phys.*, **54**, 3675
327. Halpern, O. and Buckingham, R. A. (1955). *Phys. Rev.*, **98**, 1626
328. Buckingham, R. A. and Gal, E. (1968). *Advan. Atom. Mol. Phys.*, **4**, 37
329. Siska, P. E., Parson, J. M., Schafer, T. P. and Lee, Y. T. (1971). *J. Chem. Phys.*, **55**, 5762
330. Foch, J. D. and Ford, G. W. (1970). *Studies in Statistical Mechanics*, **5**, 101. (Amsterdam: North Holland)
331. Munn, R. J. (1965). *Discuss. Faraday. Soc.*, **40**, 130
332. Smith, F. J., Mason, E. A. and Munn, R. J. (1965). *J. Chem. Phys.*, **42**, 1334
333. Klein, M. (1966). *J. Res. Nat. Bur. Stand. (U.S.)*, **70A**, 259
334. Hanley, H. J. M. and Klein, M. (1967). *Nat. Bur. Stand. Tech. Note*, 360
335. Klein, M. and Hanley, H. J. M. (1968). *Trans. Faraday Soc.*, **64**, 2927
336. Hanley, H. J. M. and Klein, M. (1969). *J. Chem. Phys.*, **50**, 4765
337. Sevast'yanov, R. M. and Zykov, N. A. (1971). *High Temp.*, **9**, 40
338. Fitts, D. D. (1966). *Ann. Rev. Phys. Chem.*, **17**, 59
339. Mason, E. A., Munn, R. J. and Smith, F. J. (1966). *Advan. Atom. Mol. Phys.*, **2**, 33
340. Klein, M. and Hanley, H. J. M. (1970). *J. Chem. Phys.*, **53**, 4722
341. Hanley, H. J. M. and Klein, M., to be published
342. Steele, D. Lippincott, E. R., and Vanderslice, J. T. (1962). *Rev. Mod. Phys.*, **34**, 239

343. Mason, E. A. and Monchick, L. (1967). *Advan. Chem. Phys.,* **12**, 329
344. Buck, U. (1971). *J. Chem. Phys.,* **54**, 1923
345. Buck, U. and Pauly, H. (1971). *J. Chem. Phys.,* **54**, 1929
346. Balfour, W. J. and Douglas, A. E. (1970). *Can. J. Phys.,* **48**, 901
347. Tanaka, Y. and Yoshino, K. (1963). *J. Chem. Phys.,* **39**, 3081
348. Tanaka, Y. and Yoshino, K. (1969). *J. Chem. Phys.,* **50**, 3087
349. Smith, A. L. (1968). *J. Chem. Phys.,* **49**, 4817
350. Chow, K. W., Smith, A. L. and Waggoner, M. G. (1971). *J. Chem. Phys.,* **55**, 4208
351. Chow, K. W. and Smith, A. L. (1971). *J. Chem. Phys.,* **54**, 1556
352. Sando, K. M. (1971). *Mol. Phys.,* **21**, 439
353. Tanaka, Y. and Yoshino, K. (1970. *J. Chem. Phys.,* **53**, 2012
354. Bruch, L. W. and McGee, I. J. (1970). *J. Chem. Phys.,* **53**, 4711
355. Chen, C. T. and Present, R. D. (1971). *J. Chem. Phys.,* **54**, 3645
356. Docken, K. (1970), private communication, to be published.
357. Maitland, G. C. and Smith, E. B. (1972), to be published
358. Balfour, W. J. and Whitlock, R. F. (1971). *Chem. Commun.,* 1231
359. McKellar, A. R. W. and Welsh, H. L. (1971). *Physics in Canada,* **27**, 54
360. Watanabe, A., Hunt, J. L. and Welsh, H. L. (1971). *Can. J. Phys.,* **49**, 860
361. McKellar, A. R. W. and Welsh, H. L. (1971). *J. Chem. Phys.,* **55**, 595
362. Kudian, A. K. and Welsh, H. L. (1971). *Can. J. Phys.,* **49**, 230
363. Sabatier, P. C., *J. Math. Phys.,* to be published
364. Newton, R. G. (1966). *Scattering Theory of Waves and Particles,* (McGraw-Hill, New York)
365. Firsov, O. B. (1953). *JETP,* **24**, 279
366. Buck, U., Kick, M. and Pauly, H. (1971). *Abstracts of VII ICPEAC,* p. 543. (Amsterdam: North Holland)
367. Bernstein, R. B. and Muckerman, J. T. (1967). *Advan. in Chem. Phys.,* **12**, 389
368. Jordan, J. E. and Amdur, I. (1967). *J. Chem. Phys.,* **46**, 165
369. Gengenbach, R., Hahn, C., Toennies, J. P. and Welz, W. (1971). *Abstracts of VII ICPEAC,* p. 653. (Amsterdam: North Holland)
370. Cantini, P., Dondi, M. G., Scoles, G. and Torello, F., to be published
371. Bennewitz, H. G., Busse, H. and Dohmann, H. D. (1971). *Chem. Phys. Lett.,* **8**, 235
372. Bennewitz, H. G., Busse, H., Dohmann, H. D. and Schrader, W. (1971). *Abstracts of VII, ICPEAC,* p.651. (Amsterdam: North Holland)
373. Bennewitz, H. G., Busse, H., Dohmann, H. D. and Schrader, W., to be published
374. Dohmann, H. D. (1970). *Thesis,* University of Bonn
375. Oates, D. E. and King, J. G. (1971). *Phys. Rev. Lett.,* **26**, 735
376. Siska, P. E., Parson, J. M., Schafer, T. P., Tully, F. P., Wong, Y. C. and Lee, Y. T. (1970). *Phys. Rev. Lett.,* **25**, 27
377. Parson, J. M. and Lee, Y. T. (1971). *Third International Symposium on Molecular Beams,* (Cannes)
378. Parson, J. M., Siska, P. E. and Lee, Y. T. (1972). *J. Chem. Phys.,* **56**, 1511
379. Bobetic, M. V. and Barker, J. A. (1970). *Phys. Rev. B,* **2**, 4169
380. Schafer, T. P., private communication
381. Cavallini, M., Meneghetti, L., Scoles, G. and Yealland, M. (1970). *Phys. Rev. Lett.,* **24**, 1469
382. Cavallini, M., Gallinaro, G., Meneghetti, L., Scoles, G. and Valbusa, U. (1970). *Chem. Phys. Lett.,* **7**, 303
383. Cavallini, M., Dondi, M. G., Scoles, G. and Valbusa, U. (1971). *Symposium on Molecular Beams.* (Cannes)
384. Düren, R. and Schlier, Ch. (1965). *Discuss. Faraday Soc.,* **40**, 56
385. Guevara, F. A., McInteer, B. B. and Wageman, W. E. (1969). *Phys. Fluids,* **12**, 2493
386. Dawe, R. A. and Smith, E. B. (1970). *J. Chem. Phys.,* **52**, 693
387. Kalelkar, A. S. and Kestin, J. (1970). *J. Chem. Phys.,* **52**, 4248
388. Barker, J. A. and Pompe, A. (1968). *Aust. J. Chem.,* **21**, 1683
389. Dymond, J. H. and Alder, B. J. (1969). *J. Chem. Phys.,* **51**, 309
390. Dymond, J. H. and Alder, B. J. (1971). *J. Chem. Phys.,* **54**, 3472
391. Baratz, B. and Andres, R. P. (1970). *J. Chem. Phys.,* **53**, 6145
392. McGee, I. J. (1971), unpublished results
393. Mason, E. A. and Rice, W. E. (1954). *J. Chem. Phys.,* **22**, 843

394. Barker, J. A. and Klein, M. L. (1971). *Chem. Phys. Lett.*, **11**, 501
395. Bobetic, M. V., Barker, J. A. and Klein, M. L. (1972). *Phys. Rev. B*, **5**, (in the press)
396. Schafer, T. P., Siska, P. E. and Lee, Y. T. (1971). *Abstracts of VII ICPEAC*, p. 546. (Amsterdam: North Holland)
397. Sherwood, A. E. and Prausnitz, J. M. (1964). *J. Chem. Phys.*, **41**, 429
398. Haberlandt, R. (1965). *Phys. Lett.*, **14**, 197
399. Schiff, D. and Verlet, L. (1967). *Phys. Rev.*, **160**, 208
400. Yntema, J. L. and Schneider, W. G. (1950). *J. Chem. Phys.*, **18**, 646
401. Østgaard, E. (1968). *Phys. Rev.*, **176**, 351
402. London, F. (1964). *Superfluids*, Vol. II. (New York: Dover reprint)
403. deBoer, J. (1958). *Physica*, **24**, S90
404. Massey, W. (1966). *Phys. Rev.*, **151**, 153
405. Mason, E. A. and Rice, W. E. (1954). *J. Chem. Phys.*, **22**, 522
406. Keller, W. E. (1957). *Phys. Rev.*, **105**, 41
407. Bernstein, R. B. and Morse, F. A. (1964). *J. Chem. Phys.*, **40**, 917
408. Murphy, R. D. and Watts, R. O. (1970). *J. Low Temp. Phys.*, **2**, 507
409. Van der Avoird, A. (1968). *Thesis*, Eindhoven
410. Kihara, T. (1970). *Physical Chemistry, An Advanced Treatise, Vol. V*, Ed. by H. Eyring. (New York: Academic Press)
411. Victor, G. A. and Sando, K. (1971). *J. Chem. Phys.*, **55**, 5421
412. Chatosinski, G., Kolos, W., Petelenz, B. and Piela, L. (1971). *Chem. Phys. Lett.*, **12**, 233
413. Singh, T. R. (1971). *Chem. Phys. Lett.*, **11**, 598
414. Bertoncini, P. J., Das, G. and Wahl, A. C. (1970). *J. Chem. Phys.*, **52**, 5112
415. Julienne, P. S. and Krauss, M. (1971). *Chem. Phys. Lett.*, **11**, 16
416. Gilbert, T. L. and Wahl, A. C. (1971). *J. Chem. Phys.*, **55**, 5247
417. Gangi, R. A. and Bader, R. F. W. (1971). *J. Chem. Phys.*, **55**, 5369
418. Schaefer, H. F., Wallach, D. and Bender, C. F. (1972). *J. Chem. Phys.*, **55**, 5247
419. Benson, M. J. and McLaughlin, D. R. (1972). *J. Chem. Phys.*, **56**, 1322
420. Kunik, D. and Kaldor, U. (1972). *J. Chem. Phys.*, **56**, 1741
421. Navaro, O. A. and Beltran-Lopez, V. (1972). *J. Chem. Phys.*, **56**, 815
422. Brailsford, D. F. and Ford, B. (1971). *Chem. Phys. Lett.*, **11**, 60
423. Block, R., Roël, R. and Ter Maten, G. (1971). *Chem. Phys. Lett.*, **11**, 425
424. Present, R. D. (1971). *Contemp. Phys.*, **12**, 595
425. McLaughlin, D. R. and Schaefer, H. F. (1971). *Chem. Phys. Lett.*, **12**, 244
426. Kaufman, J. J. and Predny, R. (1971). *Int. J. Quantum Chem.*, **5S**, 235
427. Bottcher, C., Allison, A. C. and Dalgarno, A. (1971). *Chem. Phys. Lett.*, **11**, 307
428. Bhaduri, R. K., Nogami, Y. and Ross, C. K. (1971). *J. Phys. C.*, **4**, 2734
429. Gibson, W. G. (1972). *Phys. Rev. A*, **5**, 862; Hallock, R. B. *Proc. 12th International Conference on Low Temperature Physics*, p.159. Ed. by E. Kanda. (Academic Press of Japan, Tokyo, 1971)
430. Murphy, R. D. (1972). *Phys. Rev. A*, **5**, 331
431. Barker, J. A. (1972). *Chem. Phys. Lett.*, **14**, 242

5
Molecular Properties

W. N. LIPSCOMB
Harvard University

5.1 INTRODUCTION

Molecular polarisation, which occurs when an external homogeneous electric field acts on a molecule, is described by the polarisability α. The Langevin–Debye formula[1] for the electric polarisation

$$P = \frac{4\pi}{3} N(\alpha + \frac{\mu^2}{3kT})$$

allows separation of the induced polarisation, as measured by α, and the molecular dipole moment μ by measurements at different temperatures. Corrections to α for molecular vibrations are small for diatomic molecules, *c.* 5% for most polyatomic molecules, and occasionally larger for especially 'floppy' polyatomic molecules containing polar groups. Our attention will be directed towards the electronic polarisability. Molecular beam methods permit measurements of diagonal components of the polarisability tensor α, from which the average polarisability α is determined by

$$\alpha = \tfrac{1}{3}(\alpha_{xx} + \alpha_{yy} + \alpha_{zz})$$

For diatomics only two components are unique: $\alpha_{xx} = \alpha_{yy} = \alpha_\perp$ and $\alpha_{zz} = \alpha_{||}$, where the internuclear axis is in the z-direction.

When a molecule in which not all electrons are paired is placed in a magnetic field, there is a large temperature-dependent change in the energy due to the interaction of the external magnetic field with both the unpaired electron spins and any uncompensated orbital angular momentum. In addition there is a temperature-independent term which is present even for closed-shell molecules, i.e. for molecules in which the orbital angular momentum is quenched and all electrons are paired. We shall consider below only closed-shell molecules, and hence only the temperature-independent susceptibility

$$\chi = \tfrac{1}{3}(\chi_{zz} + \chi_{yy} + \chi_{zz})$$

which, for diatomic molecules, has only two independent diagonal components, $\chi_{||} = \chi_{zz}$ and $\chi_\perp = \chi_{xx} = \chi_{yy}$. The polarisability and magnetic susceptibility constants define the energy $E^{(2)}$ of a molecule in an electric field F_e or a magnetic field H as

$$E^{(2)} = -\tfrac{1}{2}F_e \cdot \alpha \cdot F_e$$

and

$$E^{(2)} = -\tfrac{1}{2}H \cdot \chi \cdot H$$

respectively. If the field is regarded as the perturbation, the energy change is to be found in second-order perturbation theory, as indicated by the superscript.

If a nucleus which has a magnetic moment is present, its magnetic field may also influence the energy change produced by an external field. The magnetic shielding constants σ are defined by

$$E^{(2)} = \mu \cdot \sigma \cdot H$$

Since the magnetic field produced by the nucleus is proportional to its magnetic moment μ, the shielding constants σ also are second-order molecular properties, even though the external field is present only to the first power in this equation for $E^{(2)}$. Physically one expects that the susceptibility will depend largely on the wave function in the outer parts of the molecule, whereas the magnetic shielding will depend mostly on the wave function near the shielded nucleus.

Even for closed-shell spin-paired molecules there are distinct diamagnetic and paramagnetic contributions, both temperature independent, to the magnetic susceptibility and shielding. While the sum of these contributions is independent of the choice of origin of the vector potential (a kind of gauge invariance, in the language of the trade), the individual contributions are not. Nevertheless, there are two kinds of additional molecular constants to be associated with some of these individual contributions for particular choices of origin. First, the rotational magnetic moment μ_J induced by rotation of the electronic cloud and the nuclei has as a natural origin the centre of rotation and, of course, will be proportional to the rotational quantum number J. The associated molecular constant $\mu_J J$ is non-zero because the magnetic effects associated with molecular rotation of the nuclei and of the electron clouds do not cancel, even for a neutral molecule. This constant is, to a very good approximation, independent of the diamagnetic currents induced by an external magnetic field and, hence, is directly related to the paramagnetic susceptibility evaluated with the origin at the centre of molecular rotation. Secondly, the nuclear spin-rotation constant C is a measure of the interaction of the nuclear magnetic moment with the magnetic field created by molecular rotation and, hence, is related to the part of the magnetic shielding associated with excitations, that is, to the paramagnetic shielding evaluated with the nucleus as the appropriate origin. These relationships are[2], in c.g.s. units for a diatomic molecule AB,

$$\chi^{\mathrm{p}} = \frac{e^2 R^2 N}{6mc^2}\left[\frac{Z_A Z_B}{Z_A + Z_B} + \frac{(Z_A + Z_B)(D^2 - d^2)}{R^2} - \frac{\mu' \mu_J}{J \mu_N}\right]$$

$$\sigma_A^{\mathrm{p}} = -\frac{e^2}{3mc^2}\left[\frac{Z_B}{R} + \frac{2\pi C_A M S_A \mu' R^2}{e \mu_N \mu_A}\right]$$

where R is the internuclear distance, Z_A and Z_B are nuclear charges, D is the distance from the centre of mass to the centroid of nuclear charge, d is the distance from the centre of mass to the centroid of electronic charge, μ' is the reduced mass of the molecule in proton mass units, μ_N is the nuclear magneton, S_A is the spin of nucleus A, M is the proton mass, and μ_A is the nuclear magnetic dipole moment of nucleus A in nuclear magnetons.

A physical basis of these effects for paramagnetic terms is the unquenching of electronic orbital angular momentum in the molecule[1, 3, 4]. Consider first the formation of a closed-shell diatomic molecule in the absence of an external field. Let each isolated atom have a net spin and orbital angular momentum. As these atoms approach one another to form a molecule in a $^1\Sigma$ ground state, the electron spins become oppositely paired and an orbital angular momentum is created from exchange effects between these two atoms. The electronic angular momenta quantise according to the molecular symmetry (here, $\Lambda = 0$), and the net orbital angular momentum perpendicular to the internuclear axis is quenched due to torques arising from the non-spherical nature of the molecular potential. This quenching is such that the average value of the square of the total angular momentum is non-zero, even though (in the absence of a field) the expectation value of any component of the angular momentum is zero. The total angular momentum of the free molecule is necessarily a constant, but there are fluctuations between the electronic orbital angular momentum perpendicular to the internuclear axis and the angular momentum due to the molecular rotation. Now turn on a uniform magnetic field. The tendency of this field is to align this net orbital angular momentum, which is thus partially unquenched. The degree of unquenching depends primarily on the energy separations between the ground state and excited states of appropriate symmetries in the molecule, and is thus a temperature-independent paramagnetic contribution to the susceptibility. As a result of this partial unquenching, the average value of the component of the angular momentum in the direction of the field is no longer zero. Thus the rotational magnetic moment and spin-rotation constants are related to the temperature-independent paramagnetic parts of the susceptibility and shielding, respectively.

Superimposed on these effects we have the diamagnetic parts of the susceptibility and shielding. This is the usual Larmor precession, arising from Lenz's Law, in which electric currents are induced in a direction which produces a force opposing the introduction of a molecule into the magnetic field. Excitations are not involved in these diamagnetic terms, only averages over the ground state wave function of r^2 or $1/r$ for the averaged (over all orientations) susceptibility or shielding, respectively. Since most closed-shell molecules are diamagnetic this term tends to dominate molecular susceptibilities, but a notable exception is the BH molecule, as discussed later.

5.2 THEORY

This section is composed for those who like to see the Hamiltonian; results are given in the following section. The Hamiltonian is:

$$H = \frac{1}{2m}\sum_k\left[\frac{\hbar}{i}\nabla_k + \frac{e}{c}A_k\right]^2 + V + \sum_k eF_e\cdot r_k$$

where

$$V = -\sum_{v,k} Z_v e^2/r_{vk} + (\tfrac{1}{2})\sum_{k \neq l} e^2/r_{kl}$$

is the usual potential energy of interaction of electrons k with nuclei v, and electrons k with electrons l. Either the perturbation due to the uniform electric field F_e is present, or that due to the uniform magnetic field H is introduced through the vector potential

$$A_k = \tfrac{1}{2}H \times r_k + \mu \times r_k/r_k^3$$

This Hamiltonian defines precisely the level of approximation. Assuming, for the moment, $F_e = 0$ and $\mu = 0$, so that the only perturbation* arises from $A_k = \tfrac{1}{2}H \times r_k$, we find that the term involving $A_k\cdot A_k$ yields

$$H^{(2)} = \frac{e^2}{2mc^2}A_k\cdot A_k = \frac{e^2}{8mc^2}\sum_k H\cdot(r_k^2 I - r_k r_k)\cdot H$$

where I is the unit dyadic, and that the term containing $A_k\cdot\nabla_k$ yields

$$H^{(1)} = \frac{e}{2mc}\sum_k H\cdot M_k$$

---

*If the electric field is the only perturbation the second-order energy is

$$E^{(2)} = -e^2\sum_{n \neq 0}\frac{\langle\Psi_0|\sum_k F_e\cdot r_k|\Psi_n\rangle^2}{E_n - E_0}$$

where M_k is the angular momentum operator. The usual second order perturbation theory then led Van Vleck[1] to the expression

$$E^{(2)} = \frac{e^2}{8mc^2} \langle \Psi_0 | \sum_k H \cdot (r_k^2 I - r_k r_k) \cdot H | \Psi_0 \rangle$$

$$- \frac{e^2}{4m^2c^2} \sum_{n \neq 0} \frac{\langle \Psi_0 | \sum_k H \cdot M_k | \Psi_n \rangle \langle \Psi_n | \sum_k M_k \cdot H | \Psi_0 \rangle}{E_n - E_0}$$

where the first term is the diamagnetic (Larmor precession) contribution, and the second term is the temperature-independent paramagnetic contribution to the susceptibility. This expression has the form

$$E^{(2)} = -\tfrac{1}{2} H \cdot \chi \cdot H$$

Magnetic shielding was developed similarly by Ramsey[2], as the cross term between μ and H from the full form of the vector potential A_k. The result is

$$E^{(2)} = \frac{e^2}{2mc^2} \langle \Psi_0 | \sum_k \frac{\mu}{r_k^3} \cdot (r^2 I - r_k r_k) \cdot H | \Psi_0 \rangle$$

$$- \frac{e^2}{m^2c^2} \sum_{n \neq 0} \frac{\langle \Psi_0 | \sum_k (\mu/r_k^3) \cdot M_k | \Psi_n \rangle \langle \Psi_n | \sum_k M_k \cdot H | \Psi_0 \rangle}{E_n - E_0}$$

which is of the proper form

$$E^{(2)} = \mu \cdot \sigma \cdot H$$

The first terms in each of the quantum mechanical expressions are of the form $\langle \Psi_0 | r_k^2 | \Psi_0 \rangle$ for the diamagnetic part* of the susceptibility, and $\langle \Psi_0 | r_k^{-1} | \Psi_0 \rangle$ for the diamagnetic part of the shielding. These integrals are easily evaluated. But the additional terms in the expression for $E^{(2)}$ add integrals of the form $\langle \Psi_0 | M_x | \Psi_n \rangle$, $\langle \Psi_0 | M_x/r_k^3 | \Psi_n \rangle$ and $\langle \Psi_0 | x_k | \Psi_n \rangle$, and are quite intractable because they require accurate wave functions for a large number of excited states of the unperturbed system including the continuum. These excited state and continuum wave functions are quite diffuse, compared with the ground state Ψ_0 of the unperturbed system, and hence local perturbations of the orbitals by an external field are not well represented by these expansions, which converge very slowly.

The perturbed or fully coupled Hartree–Fock theory[5] of these second-order magnetic and electric properties has led to reasonable agreement between theory and experiment. The most important physical basis for this method is that the perturbing operators depend only on the coordinates of each electron separately, so that the changes in the molecular wave function caused by the perturbing field affect the e^2/r_{ij} terms indirectly through the self-consistency condition. In addition, the atomic orbitals (AOs) are

---

*The Larmor precession has angular momentum which arises not from mixing of the zero-angular momentum ground state with excited states having angular momentum, but rather by modification of the angular momentum operator from $r \times (\hbar/i)\nabla$ in the absence of a field to $r \times [(\hbar/i)\nabla + (e/c)A]$ for an electron in the presence of a field. The unmodified angular momentum operator has zero expectation value for a non-degenerate ground state, but the modified operator $(A \neq 0)$ has a non-zero value which gives rise to diamagnetism.

optimised by exponent variation, and molecular orbitals (MOs) are varied by self-consistent adjustment of coefficients of the component atomic orbitals in the linear combinations (LCAO); hence, the perturbations are made in the localised regions where they are expected to occur. Finally, the fully coupled theory has been shown[6] to reduce, when further approximations are made, to related, but less complete, methods especially for studies of the chemical shift[6-11]. We now outline the fully coupled Hartree–Fock theory[5, 12].

A single-determinantal antisymmetrised wave function Ψ_0 of the LCAO type is used, in the form

$$\Psi_0 = | \Phi_1\bar{\Phi}_1\Phi_2\bar{\Phi}_2 \ldots \Phi_n\bar{\Phi}_n |$$

where $\Phi_1 = \phi_1\alpha(1)$, $\bar{\Phi}_1 = \phi_1\beta(2)$, etc., are the spin orbitals which form the diagonal elements of the determinant, and a normalising factor is implied. There are n orthonormal, usually real, linear combinations of atomic orbitals χ_p of the type

$$\phi_i = \sum_{p=1}^{m} C_{pi}\chi_p$$

where the χ_p are Slater-type orbitals of the form

$$\chi = Nr^{n-1}P_l^m(\cos\theta) \begin{Bmatrix} \sin m\phi \\ \text{or} \\ \cos m\phi \end{Bmatrix} e^{-\zeta r}$$

The electronic energy is

$$E_0 = 2 \sum_{i=1}^{n} H_{ii} + \sum_{i,j=1}^{n}(2J_{ij} - K_{ij})$$

where

$$H_{ii} = \langle \phi_i | H | \phi_i \rangle$$

in which

$$H_i = \frac{1}{2m}\left(\frac{\hbar}{i}\nabla_i\right)^2 - \Sigma\frac{Z_v e^2}{r_{vi}}$$

is the one-electron part of the unperturbed Hamiltonian, and where

$$J_{ij} = \langle \phi_i(1)\phi_j(2)\frac{e^2}{r_{12}}\phi_i(1)\phi_j(2) \rangle = \langle \phi_i(1) | J_j | \phi_i(1) \rangle$$

$$K_{ij} = \langle \phi_i(1)\phi_j(2)\frac{e^2}{r_{12}}\phi_i(2)\phi_j(1) \rangle = \langle \phi_i(1) | K_j | \phi_i(1) \rangle$$

is the two-electron part, which also defines the operators J_j and K_j. Minimisation of the unperturbed ground state energy E_0 with respect to variation of the coefficients C_{pi} yields the Fock equations

$$F\phi_i = \varepsilon_i\phi_i$$

where $F = H + G$, $H = \Sigma H_i$ is the one-electron part, and $G = \Sigma(2J_j - K_j)$ is the two-electron part of the Fock operator. A relation between the ε_i and the total energy is

$$E = \sum_i H_{ii} + \sum_i \varepsilon_i$$

Thus, the sum over the Hartree–Fock eigenvalues is not the total energy, but ε_j is by Koopmans' theorem, a fairly good approximation to the ionisation potential from orbital j. This zero-order theory is due to Roothaan[13] and Hall[14].

The quantities F, ε_j, ϕ_j are expanded to second order in the presence of a field, and the first-order and second-order wave functions are expanded in terms of the zero-order wave functions,

$$F = F^{(0)} + KF^{(1)} + K^2 F^{(2)}$$

$$\varepsilon_j = \varepsilon_j^{(0)} + K\varepsilon_j^{(1)} \times K^2 \varepsilon_j^{(2)}$$

$$\phi_j = \phi_j^{(0)} + K\phi_j^{(1)} + K^2 \phi_j^{(2)}$$

where $\phi_j^{(1)} = \sum_k C_{kj}^{(1)} \phi_k^{(0)}$ and $\phi_j^{(2)} = \sum_k C_{kj}^{(2)} \phi_k^{(0)}$

The second-order energy is then[5, 12]

$$E^{(2)} = \sum_{j=1}^{n} \left[2H_{jj}^{(2)} + \sum_{p=1}^{m} C_{pj}^{(1)*} H_{pj}^{(1)} + H_{jp}^{(1)} C_{pj}^{(1)} \right]$$

where there are $N = 2n$ electrons and m atomic orbitals, and where the $C_{pj}^{(1)}$ values are determined by a matrix-diagonalisation procedure from equations described below for each example.

For the polarisability $H^{(1)} = eF_e x$ where

$$x = \sum_{k=1}^{N} x_k$$

and $H^{(2)} = 0$, so the integrals are of the form

$$x_{pi} = x_{ip} = \langle \phi_p^{(0)} | x | \phi_i^{(0)} \rangle$$

The equation for the coefficients $C_{pi}^{(1)}$ of the first-order wave function is*

$$(\varepsilon_p^{(0)} - \varepsilon_i^{(0)})C_{pi}^{(1)} + x_{pi} + \sum_{j=1}^{n} \sum_{q=n+1}^{m} [4(jq \mid pi) - (ji \mid pq) - (qi \mid pj)]C_{qj}^{(1)} = 0$$

and the energy is

$$E^{(2)} = 2e^2 F_e^2 \sum_{i=1}^{n} \sum_{p=n+1}^{m} C_{pi}^{(1)} x_{pi}$$

so that the perpendicular polarisability is

$$\alpha_{xx} = -\frac{2E^{(2)}}{F_e^2} = -4e^2 \sum_{i=1}^{n} \sum_{p=n+1}^{m} C_{pi}^{(1)} x_{pi}$$

---

*Each first-order perturbation operator can be expressed as $H^{(1)} = K\xi$ where K is the perturbation parameter (a constant times the perturbing field) and ξ is x, M_x or some other simple operator. The $C_{pi}^{(1)}$ here are related to the coefficients $\hat{C}_{pi}^{(1)}$ of the first-order wave function by the equation $C_{pi}^{(1)} = K\hat{C}_{pi}^{(1)}$.

For the magnetic susceptibility of a diatomic molecule, $H^{(1)}$ and $H^{(2)}$ are as given above*, and the integrals are $(M_x)_{pi} = \langle \phi_p^{(0)} | M_x | \phi_i^{(0)} \rangle$ and $(r^2)_{ii} = \langle \phi_i^{(0)} | r^2 | \phi_i^{(0)} \rangle$, which are used in the equations

$$(\varepsilon_p^{(0)} - \varepsilon_i^{(0)})C_{pi}^{(1)} + (M_x)_{pi} + \sum_{j=1}^{n} \sum_{q=n+1}^{m} [(qi \,|\, pj) - (ji \,|\, pq)]C_{qj}^{(1)} = 0$$

$$E^{(2)} = \frac{e^2 H^2}{6mc^2} \sum_{i=1}^{n} (r^2)_{ii} - \frac{e^2 H^2}{3m^2c^2} \sum_{i=1}^{n} \sum_{p=n+1}^{m} C_{pi}^{(1)}(M_x)_{pi}$$

so that the susceptibility is†

$$\chi_{av} = -\frac{2N_0 E^{(2)}}{H^2} = -\frac{N_0 e^2}{3mc^2} \sum_{i=1}^{n} \left[(r^2)_{ii} - \frac{2}{m} \sum_{p=n+1}^{m} C_{pi}^{(1)}(M_x)_{pi} \right]$$

where N_0 is Avogadro's number.

The shielding is a cross term between μ and H. The perturbations are‡

$$H^{(1)} = \frac{e}{2mc} \boldsymbol{H} \cdot \boldsymbol{M} + \frac{e}{mc} \frac{\boldsymbol{\mu}}{r^3} \cdot \boldsymbol{M}$$

$$H^{(2)} = \frac{e^2}{2mc^2} \boldsymbol{H} \cdot (r^2 \boldsymbol{I} - \boldsymbol{rr}) \cdot \frac{\boldsymbol{\mu}}{r^3}$$

and *either*

$$(\varepsilon_p^{(0)} - \varepsilon_i^{(0)})C_{pi}^{(1)} + (M_x)_{pi} + \sum_{j=1}^{n} \sum_{q=n+1}^{m} [(qi \,|\, pj) - (ji \,|\, pq)]C_{qj}^{(1)} = 0$$

*Averaged over all molecular orientations, as indicated by a bar, these perturbing operators are

$$\overline{H^{(1)}} = \frac{eH}{3mc} M_x \text{ where } M_x = \sum_{k=1}^{N} M_{xk}, \text{ and } \overline{H^{(2)}} = \frac{e^2 H^2}{12mc^2} r^2$$

where $r^2 = \sum_{k=1}^{N} r_k^2$

†It is interesting to compare this expression with that obtained by ordinary second-order perturbation theory. The result of the latter treatment is

$$\chi_{av} = -\frac{N^0 e^2}{6mc^2} \sum_{i=1}^{N} (r_i^2)_{00} + \frac{N_0 e^2}{3m^2c^2} \sum_{m \neq 0} \frac{|(M_x)_{m0}|^2}{E_m^{(0)} - E_0^{(0)}}$$

The first term is, of course, identical (the difference of a factor of 2 arises since N = number of electrons and n = number of occupied MOs), and the 'paramagnetic' term bears a close resemblance to the summation over excited states. A similar correspondence is found in the expressions for the polarisability and shielding.

‡Averaged over all orientations $H^{(1)} = \frac{eH}{3mc} M_x + \frac{2e\mu}{3mc} \frac{M_x}{r^3}$

where $M_x = \sum_{k=1}^{N} M_{xk}$ and $M_x/r^3 = \sum_{k=1}^{N} M_{xk}/r_k^3$, and $H^{(2)} = \frac{e^2 H \mu}{3mc^2} \frac{1}{r}$ where $1/r = \sum_{k=1}^{N} (1/r_k)$.
Equations for a component are

$$\chi_{xx} = -\frac{N_0 e^2}{2mc^2} \sum_{i=1}^{n} \left[(y^2 + z^2)_{ii} - \frac{2}{m} \sum_{p=n+1}^{m} C_{pi}^{(1)}(M_x)_{pi} \right] \text{ and}$$

$$\sigma_{xx} = \frac{e^2}{2mc^2} \sum_{i=1}^{n} \left[\left(\frac{y^2 + z^2}{r^3} \right)_{ii} - \frac{2}{m} \sum_{p=n+1}^{m} C_{pi}^{(1)} \left(\frac{M_x}{r^3} \right)_{pi} \right]$$

so that shielding is

$$\sigma_{av} = \frac{E^{(2)}}{\mu H} = \frac{2e^2}{3mc^2} \sum_{i=1}^{n} \left[\left(\frac{1}{r}\right)_{ii} - \frac{2}{m} \sum_{p=n+1}^{m} C_{pi}^{(1)} \left(\frac{M_x}{r^3}\right)_{pi} \right]$$

or (equivalently)

$$(\varepsilon_p^{(0)} - \varepsilon_i^{(0)}) C_{pi}^{(1)} + (M_x/r^3)_{pi} + \sum_{j=1}^{n} \sum_{q=n+1}^{m} [(qi \mid pj) - (ji \mid pq)] C_{qj}^{(1)} = 0$$

so that the shielding is

$$\sigma_{av} = \frac{2e^2}{3mc^2} \sum_{i=1}^{n} \left(\frac{1}{r}\right)_{ii} - \frac{4e^2}{3m^2c^2} \sum_{i=1}^{n} \sum_{p=n+1}^{m} C_{pi}^{(1)} (M_x)_{pi}$$

where the average is taken over all molecular orientations.

The general procedure is to evaluate the integrals, solve the equations for the coefficients $C_{pi}^{(1)}$, and then evaluate the expressions for polarisability, susceptibility and shielding constants.

Some general comments about the theory are as follows. The total magnetic susceptibility and shielding are invariant to a change of gauge, i.e. a change of origin in our restricted exploration, in the limit of a complete basis set[5]. We find that this invariance is approached when one chooses about three times the minimum basis set of Slater orbitals, but the choices must be appropriate to the perturbations and variation of orbital exponents must be explored. In practice, we find that the calculation stabilises when the second-order energy for the total magnetic susceptibility is minimised*. Here, we recall that the zero-order energy is independent of the external field, while the second-order energy varies with the square of the field. The shielding does not show this energy minimum[5] and the polarisability is independent of the choice of origin.

The electron currents induced in a magnetic field can be found by expanding the expression for the electron probability current in powers of the magnetic field strength. Results are

$$\mathbf{J}^{(0)} = 0$$

$$\mathbf{J}_j^{(1)}(\mathbf{r}) = (\hbar^2 H/mc^2)[\phi_j^{(0)} \nabla \hat{\phi}_j^{(1)} - \hat{\phi}_j^{(1)} \nabla \phi_j^{(0)} + (z\mathbf{j} - y\mathbf{k}) \phi_j^{(0)2}]$$

where $\mathbf{J}_j^{(1)}(\mathbf{r})$ is the first-order current at $\mathbf{r}$ due to the jth molecular orbital†.

$$\hat{\phi}_j^{(1)} = \sum_{p=n+1}^{m} \hat{C}_{pj}^{(1)} \phi_p^{(0)}$$

The uniform magnetic field is along x, while the molecular axis is taken along z. We specify in each figure the direction of circulation which is diamagnetic, because conventions differ according to whether the direction of current flow corresponds to electron flow or to positive current.

---

*There are calculational advantages when the origin is taken at the centre of electronic charge where the absolute value of the diamagnetic susceptibility is a minimum, and where, in the limit of a complete basis set, the paramagnetic susceptibility is a minimum[15].

†$C_{pj}^{(1)} = \dfrac{eH\hbar}{2mci\varepsilon_0} \hat{C}_{pj}^{(1)}$ and $\phi_j^{(1)} = \dfrac{eH\hbar}{2mci\varepsilon_0} \hat{\phi}_j^{(1)}$, where $\varepsilon_0 = e^2/a_0$

5.3 RESULTS FOR INDIVIDUAL MOLECULES

5.3.1 LiH[5, 16]

Only σ orbitals are occupied. Since the operator M_x mixes only orbitals which differ in m quantum number by 1, the perturbation yielding the paramagnetic susceptibility is of pure π symmetry. In this molecule, and also in Li_2 and BH which are discussed below, the symmetry of the first-order wave function is different from that of all orbitals of the unperturbed wave function. In this circumstance, variation of the parameters of the π orbitals leaves the ground-state energy unchanged, so that the energy minimum principle can be applied rigorously to the second-order energy. Also a careful study of the dependence of the results of changes in the internuclear distances has yielded vibrational corrections to these second-order properties.

Table 5.1   Second-order properties of LiH[16]

| | Origin at | | Experimental |
	Li	H	
χ/p.p.m.	-7.63	-7.66	—
$\mu_J/J(\mu_N)$	-0.67	-0.67	-0.654 ± 0.007[17]
σ_{Li}/p.p.m.	90.2	90.2	—
C_{Li}/kHz	9.55	9.50	9.92 ± 0.04 (v = 0)[18]
σ_H/p.p.m.	26.45	26.47	—
C_H/kHz	-9.56	-9.60	-10.1 ± 1.0 (v = 0)[18]
$\alpha_\perp$/Å^3	4.00	4.00	—

Basis sets for the first-order wave function were 3π on Li and 1π on H for Li as origin, 4π on Li and 1π on H for H as origin, and 3π on Li and 2π on H for the origin-independent polarisability $\alpha_\perp$.

Agreement with available experimental data (Table 5.1) is excellent, considering the Hartree–Fock approximations. It seems likely, from a fairly extensive study of the basis sets, that agreement better than 2–5% can only be achieved by the introduction of some kind of spin-correlated wave functions, or by the use of configuration interaction theory.

A qualitative view of the electronic circulation induced by the magnetic field is shown in Figure 5.1, where the contours are in atomic units which are to be multiplied by the square of the field. The currents are mainly diamagnetic around both Li and H, but there is a small region (shaded) in which paramagnetic circulation of current occurs, centred at the node in the bonding orbital of LiH. Mixing from this region into the (mainly) Li π orbitals by the field gives rise to this paramagnetic circulation. However, in LiH, this paramagnetic current is neither very large, nor is it centred near enough to the Li nucleus to produce a large down-field chemical shift.

The perpendicular component of the polarisability was shown to be 4.00 A^3. The first-order change in the wave function (Figure 5.2) indicates that the electron density is concentrated nearer the Li atom, rather than nearer the H atom (Figure 5.1). Indeed, the optimal basis set for the susceptibility is still sufficiently incomplete to be satisfactory for the polarisability, so that a revised wave function was used for $\alpha_\perp$. Probably even this wave function

Figure 5.1 First-order electron density current calculated with the origin at lithium. (a) Contour map of the modulus taken in a half-plane with the internuclear axis at the bottom; (b) vector direction map over the identical region shown in 2(a). The shading indicates the region of paramagnetic current loops. The magnetic field is perpendicular to the plane of the paper, and points inward. Clockwise circulation corresponds to a diamagnetic current

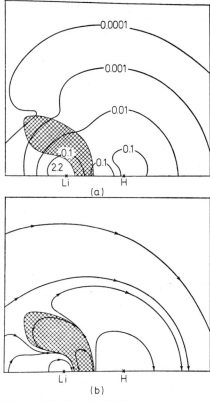

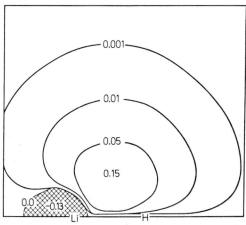

Figure 5.2 First-order change in the electron density due to an electric field perpendicular to the internuclear axis. The contours are taken over the same region as in Figures 5.1 and 5.2. The shading indicates the region of negative change in electron density. The electric field is in the plane of the paper and points from the bottom to the top of the figure

is incomplete for the calculation of $\alpha_{\parallel}$, which was shown to be 3.35 A^3, but is only to be regarded as a lower limit. The inner-shell Li orbital contributes only c. 0.2–0.3 % to $\alpha_{\perp}$, so that the major effect arises from mixing of the bonding 2σ orbitals with π orbitals by the electric field.

5.3.2 Li_2[19]

Results (Table 5.2) for the rotational magnetic moment are deceptively good, as compared with experiment, because of the large nuclear contribution. Electronic contributions are −0.144 for 1σ (inner shell), −0.144 for 2σ (inner shell) and −0.031 for 3σ (valence pair). If the nuclear contribution is divided equally, it contributes 0.143 to each, so that the resulting net contributions are −0.001, −0.001 and 0.112 to the theoretical value of

Table 5.2 Second-order properties of Li_2 [19]

	Origin at C	Experimental
χ/p.p.m.	− 28.94	—
$\mu_J/J(\mu_N)$	0.1100	0.10797 ± 0.00011[20]
σ_{Li}/p.p.m.	98.58	—
$C_{(Li^7)}$/kHz	0.240	—

Basis sets were 5σ and 3π on each Li nucleus.

0.110. Compared with experiment, then, the valence-pair contribution of −0.031 is computed from coupled Hartree–Fock theory within an error of c. 0.002 or 6%. The small contributions of the inner shells show that they rotate with the nuclei, but the valence pair follows the nuclear rotation only rather poorly. The primary effect on the valence-shell electrons is mixing with the π orbitals by the magnetic field produced when the two Li^+ cores rotate about the molecular centre.

5.3.3 BH[21-23]

The results for BH (Table 5.3) are analysed in terms of orbital and nuclear contributions in Table 5.4, and are shown in Figures 5.3–5.6. The ground state molecular orbitals in parts (a) of Figures 5.3–5.5 show that to a good approximation the 1σ orbital is a lone-pair 1s orbital on B, 2σ is the bond between B and H, and 3σ is the lone pair on B. At 0.57 a.u. from B on the opposite side from H we find the charge centroid. The anomalous properties, (a) the molecular paramagnetism in this closed-shell $^1\Sigma$ molecule, (b) the large electronic contribution to the rotational magnetic moment, μ_J/J, (c) the large anti-shielding σ_B, and (d) the exceedingly large spin–rotation constant C_B, all arise primarily from this 3σ orbital. For example, the rotational magnetic moment would be $+0.96$ nuclear magnetons (μ_N) if the electrons did not follow the rotation of the nuclei, or about $-3\mu_N$ if electrons and nuclei rotated together, as contrasted with the value of -8.3 μ_N (B as origin)

Table 5.3 Second-order properties of BH($^1\Sigma$)[21-23]

	Origin at B	Origin at H
χ/p.p.m.	18.75	18.52
$\mu_J/J(\mu_N)$	−8.30	−8.24
σ_B/p.p.m.	−262	−265
C_B/kHz	493	497
σ_H/p.p.m.	24.5	25.4
C_H/kHz	−14.9	−18.2

Basis sets were 10σ and either 3π on B and 2π on H (gauge origin at B), or 5π on B and 3π on H (gauge origin at H).

Table 5.4 Orbital and nuclear contributions* to magnetic properties of BH[23]

	1σ 1s of B†	2σ Bonding †	3σ Lone pair†	B nucleus	H nucleus
$\chi^d(B)$	−0.2	−8.3	−9.5		
$\chi^p(B)$	0.0	4.5	32.3		
$\chi_{xx}(B)$	−0.2	−3.3	37.6		
$\chi_{zz}(B)$	−0.2	−4.7	−7.0		
$\chi(B)$	−0.2	−3.8	22.8		
$\mu_J(J(B))$	−0.02	−0.94	−8.30	0.04	0.92
$\mu_J/J(H)$	0.05	−0.58	−8.67	0.04	0.92
$\sigma_B^d(B)$	166	22	22		
$\sigma_B^p(B)$	0	−35	−436		
$\sigma_{xx}^B(B)$	166	−28	−630		
$\sigma_{xx}^B(B)$	166	16	17		
$\sigma^B(B)$	166	−13	−414		
$\sigma_B^d(H)$	165	10	28		
$\sigma_B^p(H)$	1	−23	−446		
$\sigma_B(H)$	166	−13	−418		
$C_B(B)$	0	38	464		−8
$C_B(H)$	0	38	467		−8
$\sigma_H^d(B)$	0.0	10.9	8.8		
$\sigma_H^p(B)$	0.5	13.0	−8.6		
$\sigma_H(B)$	0.5	23.9	0.2		
$\sigma_H^d(H)$	15.2	28.2	14.6		
$\sigma_H^p(H)$	−14.7	−3.6	−14.2		
$\sigma_{xx}^H(H)$	0.6	24.5	−3.7		
$\sigma_{zz}^H(H)$	0.4	24.8	8.6		
$\sigma_H(H)$	0.5	24.6	0.4		
$C_H(B)$	49	14	48	−126	
$C_H(H)$	49	12	47	−126	

*A positive χ is paramagnetic; a positive σ is diamagnetic, in this table. As indicated by the values for zz components, circulation around the molecular axis is purely diamagnetic. Gauge origin is indicated in parentheses; the reader is reminded that diamagnetic (D) and paramagnetic (P) contributions are not origin-independent. In the limit of a complete basis set the contributions of individual MOs are gauge-invariant. The z-axis coincides with the molecular axis. Occupied–occupied interactions are excluded.
†To a good approximation.

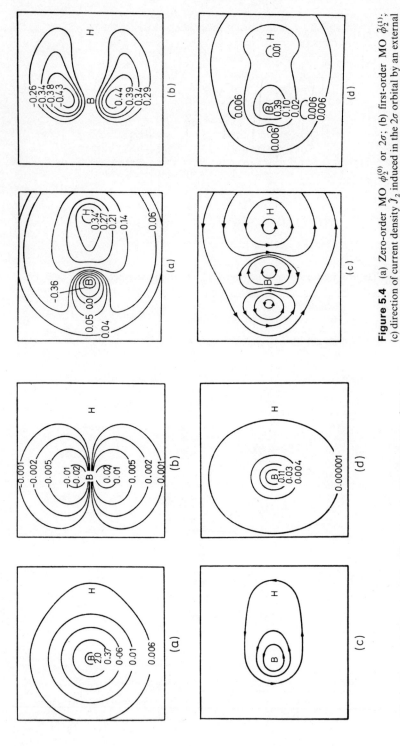

Figure 5.4 (a) Zero-order MO $\phi_2^{(0)}$ or 2σ; (b) first-order MO $\phi_2^{(1)}$; (c) direction of current density $\hat{J}_2$ induced in the 2σ orbital by an external field (note that 'paramagnetic' circulations are centred at points where the internuclear axis intersects nodal surfaces of the zero-order molecular orbital); (d) magnitude of $\hat{J}_2$

Figure 5.3 (a) Zero-order MO $\phi_1^{(0)}$ or 1σ; (b) first-order MO $\phi_1^{(1)}$; (c) direction of the current density $\hat{J}_1$ induced in the 1σ orbital by an external magnetic field; (d) magnitude of $\hat{J}_1$

shown in Table 5.3. Similarly the spin-rotation interaction constant for B, C_B, would be -8 kHz from nuclear contributions only, or c. 140 kHz if the electron distribution accompanied the molecular rotation, as compared with the value of 493 kHz (B as origin) in Table 5.3. These results show the great importance of the excitations which partially unquench the electronic angular momentum as the molecule rotates. These excitations arise from the magnetic field created by the molecular rotation.

The three orbitals show diamagnetic (1σ), intermediate (2σ) and paramagnetic (3σ) effects in a particularly clear way. The 1σ orbital (Figure 5.3a) gives a small diamagnetic current (counter-clockwise) around both B and H, and (because of the extra three inverse powers of r) a large diamagnetic shielding to B. This current arises from the Larmor precession in the ground-state wave function, rather than from excitations by the field into the first-order wave function. Indeed the only evidence of these excitations from 1σ is the very small contribution to the rotational magnetic moment (Table 5.4).

The 2σ molecular orbital has both $2s_B$ and $1s_H$, and some $2p_\sigma$, contributions, whereas the first-order wave function resembles $2p_\pi$ mostly on B but with a strong distortion towards H (Figure 5.4). The current necessarily vanishes at the two nodal points of the unperturbed orbital, and hence current loops are centred at these points. In the region of the excitations the circulation is paramagnetic (clockwise), but most of the remainder of the current is diamagnetic, particularly that about H. This current gives a net diamagnetic contribution to the total susceptibility, a small anti-shielding to the B nucleus, and a substantial shielding to the H nucleus. Currents induced in this orbital by molecular rotation give an intermediate contribution to the rotational magnetic moment and the spin–rotation constants.

The 3σ orbital, mostly $2p_{\sigma B}$ with some contribution of $2s_B$, shows a very large paramagnetic circulation about the B nucleus. This circulation is centred about the node which is 0.57 a.u. from B (away from H) and is associated with mixing by the field or molecular rotation, into the π-like first-order molecular orbital (Figure 5.5). Both the total current and the paramagnetic current (B chosen as origin; see Figure 5.6) are dominated by the paramagnetic circulation in this 3σ orbital. Excitations from 3σ to the first-order $\phi_3^{(1)}$ (Figure 5.5b) which lies 0.39 a.u. higher than 3σ account for 57% of the paramagnetic susceptibility (origin at B) when the perturbation of the Hartree–Fock potential is included, and for 88% when excitations from $\phi_3^{(0)}$ to all excited molecular orbitals are included. Qualitatively one may say that the unusual magnetic properties of BH arise primarily from the large paramagnetic current due to virtual excitations, by the external field or by molecular rotation, of the lone-pair electrons on boron into the empty p_π orbitals on this atom.

5.3.4   AlH[24]

Results for AlH (Table 5.5) indicate that the perpendicular component of the susceptibility is positive, as for BH, but that the total susceptibility may be very slightly diamagnetic. In view of the large, probably accurately computed, diamagnetic component an experimental finding of a very small

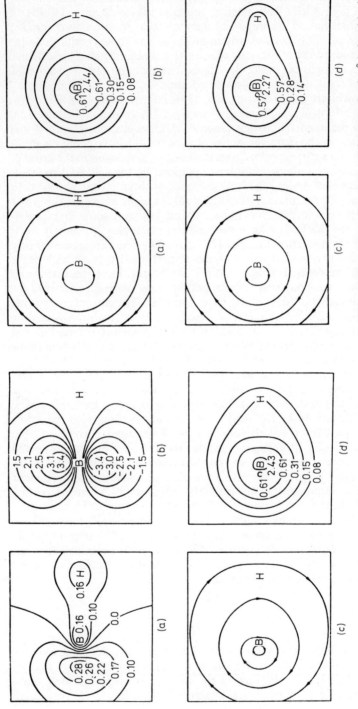

Figure 5.6 (a) Direction of the total current density $\hat{J}$ induced in the BH molecule by a uniform magnetic field; (b) magnitude of $\hat{J}$; (c) direction of the total paramagnetic current $\hat{J}^p$ induced in the BH molecule by a uniform magnetic field (the origin is at the boron nucleus); (d) magnitude of $\hat{J}^p$ (the origin is at the boron nucleus). The clockwise direction of paramagnetic current applies to Figures 5.3–5.6

Figure 5.5 (a) Zero-order MO $\phi_2^{(0)}$ or 3σ; (b) first-order MO $\hat{\phi}_3^{(1)}$; (c) direction of current density $\hat{J}_3$ induced in the 3σ orbital by an external field (note that 'paramagnetic' circulations are centred at points where the internuclear axis intersects nodal surfaces of the zero-order molecular orbital); (d) magnitude of $\hat{J}_3$

total susceptibility in AlH would provide strong evidence for the temperature-independent paramagnetism expected for the perpendicular component. Such a finding would also support the predictions for temperature-independent paramagnetism in BH, which has similar virtual excitations.

The modulus and direction of the current density are shown in Figure 5.7, in which the gauge origin is the centre of electronic charge. Contributions of filled 1σ, 2σ, 1π, 3σ, 4σ and 5σ orbitals are 0.00, 0.00, 0.00, 0.00,

Table 5.5 Second-order properties* of AlH[24]

χ/p.p.m.	-1.37
χ^{p}	32.56
χ^{d}	-33.93
$\chi_{\|\|}$	-21.87
$\chi_{\perp}$	9.17
$\mu_J/J(\mu_N)$	-3.46
σ_{Al}/p.p.m.	223
σ_{Al}^{p}	$-573\,(-573)$†
σ_{Al}^{d}	795
$\sigma_{Al,\|\|}$	786
$\sigma_{Al,\perp}$	-59
σ_{H}	26
σ_{H}^{p}	$0.3\,(-69)$†
σ_{H}^{d}	26
$\sigma_{H,\|\|}$	34
$\sigma_{H,\perp}$	22
C_{Al}/kHz	260
C_{H}	-9.34

*Basis set 16 σ and 10 π orbitals.

†Values of σ_{Al}^{p} and σ_{H}^{p} with origin at Al and H, respectively, are computed from σ^{d} evaluated at the nucleus and σ evaluated at the centre of electronic charge of the molecule.

7.38 and 25.90 to χ^{p} (gauge at Al), 0, 0, 42, -50, -67 and -498 to σ^{p} of Al (gauge at Al), and -9, -7, -2, -7, -3 and -12 to σ^{p} of H (gauge at H). It should be noted that the centre of electronic charge is almost at the Al nucleus, so that those excitations from the inner shells of Al which are almost spherically symmetric about this origin contribute essentially nothing to the paramagnetic susceptibility.

5.3.5 HF[25]

In this molecule there is no symmetry separation of unperturbed and perturbed wave functions. The general requirement for the variational method to be valid for the second-order energy is that the zero-order energy remain constant. Although we have assumed that this criterion is met if the zero-order basis set is sufficiently large that the addition and variation of further orbitals does not significantly change the zero-order energy, it is not clear how much change can be tolerated. In the study of the HF molecule changes in the zero-order energy never exceeded 0.015 a.u. for variation of exponents of π orbitals or 0.005 a.u. for σ orbitals. In preliminary work a large change did occur in the zero-order wave function when a $2p_{\pi}$ orbital with high

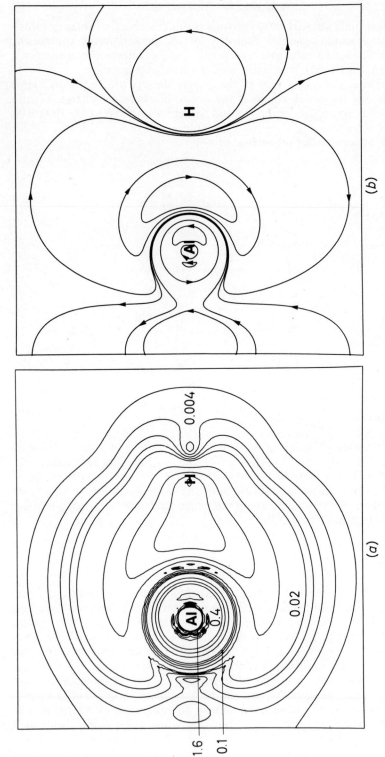

(a)

(b)

Figure 5.7 (a) Modulus of the first-order current density induced in AlH by a magnetic field perpendicular to the molecular axis and normal to the plane of the diagram. Contour levels are 0.004, (increments of 0.004), 0.02, (increments of 0.02), 0.1, (increments of 0.1), 0.4, (increments of 0.4), 1.6. (b) Direction of first-order current density induced in AlH by a magnetic field pointing into the plane of the diagram. Paramagnetic currents circulate clockwise. The gauge origin is at the Al atom for both a) and b)

exponent was added on the F atom, thus producing a large change in magnetic properties computed with the origin at the H atom.

The results for the origin taken at F are surely the more reliable (Table 5.6). Because of the excellent agreement with experiment for the rotational magnetic moment, the paramagnetic susceptibility has been predicted

Table 5.6 Second-order properties of HF[25]

	Origin at		Experimental
	F*	H	
χ/p.p.m.	-10.4	-10.9	-8.6 ± 0.1[25]
χ^d	-11.0		
χ^p	0.6		
$\mu_J/J(\mu_N)$	0.75	0.95†	0.7392 ± 0.0005[27]
$\sigma(F)$	414	437	410 ± 6[28]
$\sigma^d(F)$	482		
$\sigma^p(F)$	-68		-62.8 ± 0.3[28]
$\sigma(H)$	28.5	30.8	27.9[29], 29.2 $\pm$ 0.5[28]
$\sigma^d(H)$	15.8	108	109[30]
$\sigma^p(H)$	12.7		
C_F/kHz‡	312	189	307.6[31-33]
C_H	-71	-85§	-71.1[31, 32]
$\alpha_{||}/Å^3$	0.86	0.86	
$\alpha_\perp$	0.62	0.62	

*For the origin at F, the basis set consisted of 14 σ, 9 π and 3 δ orbitals.
†The paramagnetic term is calculated as 21.18 p.p.m., observed 21.7 p.p.m., but an error of 3 % in this term leads to an error of 30 % for the rotational magnetic moment.
‡The distance-dependent part of the vibrational corrections is predicted[25, 28] as 24.0 kHz for C_F and 0.8 kHz for C_H, and the experimental values[32] are 23.0 kHz and 1.8 kHz for C_H.
§The total shielding at H of 30.85 p.p.m. has a paramagnetic contribution of -77.52 p.p.m., which is only 3 % low, but this error produces a 20 % error in the calculated spin rotation constant of H when H is the origin. The situation for C_F is even worse.

Table 5.7 Orbital contributions to magnetic properties of HF*

	1σ	2σ	3σ	π_x	π_y
$\chi^d(F)$	-0.1	-2.1	-3.4	-2.7	-2.7
$\chi^p(F)$	0.0	0.0	0.3	0.0	0.2
$\chi(F)$	-0.1	-2.1	-3.0	-2.7	-2.5
$\chi^d(H)$	-4.8	-5.7	-7.1	-7.2	-7.2
$\chi^p(H)$	4.1	2.1	2.9	4.5	7.6
$\chi(H)$	-0.7	-3.6	-4.3	-2.7	0.3
$\sigma_F^d(F)$	306	49	39	43	43
$\sigma_F^p(F)$	0	1	46	0	-115
$\sigma_F(F)$	306	51	85	43	-71
$\sigma_H^d(F)$	0.0	0.8	7.2	3.9	3.9
$\sigma_H^p(F)$	0.0	2.3	5.2	0.4	4.8
$\sigma_H(F)$	0.0	3.1	12.4	4.3	8.7
$\alpha_{xx}/Å^3$	0.00	0.02	0.14	0.34	0.13
$\alpha_{zz}/Å^3$	0.00	0.02	0.56	0.14	0.14

*Occupied–occupied interactions are excluded

reliably. Also, because the diamagnetic susceptibility is usually given accurately by this method, the 15 % disagreement between theory and experiment for the total susceptibility raises a question about the experimental value of -8.6 p.p.m.

Orbital contributions to the magnetic and electric properties are shown in Table 5.7. When the magnetic field is along x, and the molecular axis is z, the π_y orbital (in the plane of Figure 5.8) makes a paramagnetic contribution to the shielding of F. The π_z orbital, as part of 3σ, has a sufficient displacement of its node towards H so that its contribution to the shielding is diamagnetic. Thus the weak paramagnetic currents, which modify the susceptibility very little, can produce large shielding contributions when weighted by the inverse cube of the distance from a nucleus. The change in the electron density in the xz plane for an electric field along x, shown in Figure 5.9, has a major contribution from π_x. When the electric field is along the molecular axis (z) it is the 3σ orbital which dominates the polarisability (α_{zz}) of HF. If π_x, π_y and 3σ are associated with $2p_x$, $2p_y$ and $2p_z$ of F one sees that the two orbitals not parallel to the field (E_x or E_z) are about equally polarisable, and that anisotropy arises from the fact that the bonding orbital 3σ (for α_{zz}) is more polarisable than is the π_x orbital (for α_{xx}).

5.3.6 BF[36]

Results (Table 5.8) appear to have converged quite well, if gauge dependence is taken as a criterion. No experimental quantities are as yet available for comparison. Contributions of occupied orbitals are omitted. Most of these orbitals make some contributions to the susceptibility and shielding, but the diamagnetic shieldings of the 1s orbital of F (1σ) and the 1s orbital of

Table 5.8   Second-order properties* of BF[23]

	Origin at	
	B	F
χ^P/p.p.m.	46.5	29.9
χ^d	-60.8	-44.1
$(\chi^d)_{xx}$	-82.0	-57.0†
$(\chi^P)_{xx}$	69.7	44.8
χ_{xx}	-12.4	-12.2
χ_{zz}	-18.2	-18.2
χ	-14.3	-14.2
$\mu_J/J(\mu_N)$	-0.21	-0.21
σ_F^P/p.p.m.	-213	-264
σ_F^d	470	508
$(\sigma_F^d)_{xx}$	298	520
$(\sigma_F^P)_{xx}$	-277	-397
$(\sigma_F)_{xx}$	21	123
$(\sigma_F)_{zz}$	209	484
σ_F	258	243
C_F/kHz	83	88
σ_B^P/kHz	-185	-121
σ_B^d	268	203
σ_B	84	82
C_B/kHz	15.7	15.9

*Basis set 16σ and 8π for the unperturbed wavefunction, plus 2σ(F), 3π(F), 2 (F), 2π(B), and 1δ(B) for origin at F, or plus 2σ(F), 2π(F), 2 (F), 1σ(B), 2π(B) and 1δ(B) for origin at B.
†The molecular axis is z, and the magnetic field is perpendicular to x and z.

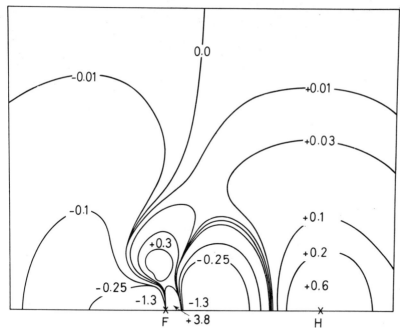

Figure 5.8 The first-order change in the electron density in hydrogen fluoride in the presence of an electric field parallel to the internuclear axis. The electric field points from the fluorine towards the hydrogen. The map shows a half-section through the internuclear axis

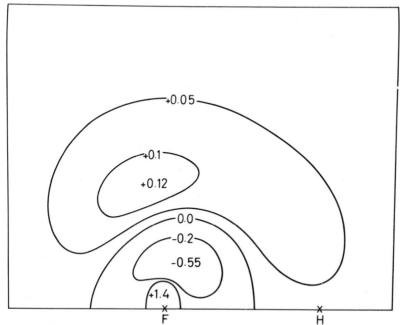

Figure 5.9 The first-order change in the electron density in hydrogen fluoride in the presence of an electric field perpendicular to the internuclear axis. The electric field points outward from the axis in the plane of this map which shows a half-section through the internuclear axis. On the opposite side of the axis the field will be pointing in and the signs of all areas in the map will be reversed

B are large: two or three times larger in magnitude than the large para-magnetic contribution to shielding of both F and B by the highest occupied orbital 5σ. The superscripts on the components xx and zz of the shielding refer explicitly in Table 5.8 to the atom chosen for the origin of the vector potential.

5.3.7 CO[34]

At each of four distances a basis set consisting of 18σ orbitals, 13π (pairs of) orbitals and 3δ (pairs of) orbitals was employed. The only experimental number available for comparison with the results of Table 5.9 is that for the rotational magnetic moment, which is in excellent agreement with the theoretical value. The sign of this quantity is also firmly established from

Table 5.9 Second-order properties of CO[34]

	Calculated*	Experimental
χ/p.p.m.	-13.26	
χ	-17.90	
χ	-10.93	
$\mu_J/J(^{13}C^{16}O)\,(\mu_N)$	-0.242	-0.2689 ± 0.0001[35]
σ_O/p.p.m.	64	
$\sigma_{O\parallel}$	411	
$\sigma_{O\perp}$	-109	
σ_C	11	
$\sigma_{C\parallel}$	271	
$\sigma_{C\perp}$	-118	
$C_O(^{12}C^{17}O)$/kHz	-23.1	
$C_C(^{13}C^{16}O)$	31.6	

*Basis set: $18\,\sigma$, $13\,\pi$ and $3\,\delta$.

these calculations, and has been shown to yield the direction of the small dipole moment ($\mu = 0.112D$) as C^-O^+, in agreement with the sign as determined by Rosenblum, Nethercot and Townes[36]. This sign is not correctly given in the Hartree–Fock approximation, but has been confirmed by extensive configuration interaction extensions of these Hartree–Fock results[37, 38].

5.3.8 N₂[24]

The nitrogen molecule shows the phenomenon of anti-shielding (total σ negative) by -101 ± 20 p.p.m. The failure of coupled Hartree–Fock theory to yield better agreement than -20 p.p.m. (Table 5.10) is not understood. However, this number is the difference between two much larger terms of σ^d (356 p.p.m.) and σ^p (-375 p.p.m.). Of these two, σ^p is far more difficult to evaluate accurately, and the loss of accuracy may be due to the u = g constraint on the σ orbitals, to the contributions of many excitations not present in the calculations, or to the need for electron correlation corrections (configurational interaction). It has been shown that a vibrational correction is not sufficient to account for this discrepancy, which depends on the wave function near the nucleus. When a program accommodating a larger basis set is available, the addition of a fourth 2p set, which made a noticeable difference in ground-state properties of NH_3 [41] could be added such that exponents of these four sets would be initially 7.677, 3.27, 1.89 and 1.222, before optimisation.

On the other hand the paramagnetic susceptibility, which depends on the wave function far from the nucleus, is given quite well, as are most other

Table 5.10   Second-order properties of N_2 [24]

	Calculated*	Experimental[39, 40]
χ/p.p.m.	−13.44	
χ^p	17.52‡	18.40
χ^d	−30.96	
$\chi_{\parallel}$	−18.44	
$\chi_{\perp}$	−10.93	
$\mu_J/J(\mu_N)$ $^{15}N_2$	−0.224	±0.2593 ± 0.005
$^{14}N_2$	−0.240	
σ/p.p.m.	−20	−101 ± 20
σ^p	−375 (−403)†	−483 ± 20
σ^d	356	
$\sigma_{\parallel}$	340	
$\sigma_{\perp}$	−200	
C_N/kHz $^{15}N_2$	−17.9	−22 ± 1
$^{14}N_2$	−13.7	

*Basis set: 16 σ and 14 π orbitals. The origin is taken at the molecular centre.
†The value of σ^p in parentheses is evaluated at the N atom as origin, from the value of σ^d calculated with N as origin and the value of σ calculated at the centre of electric charge.
‡Vibrational correction makes $\chi^p = 17.72$ and $\mu_J/J = -0.242$ for $^{14}N_2$.

second-order properties. One might risk the generalisation that coupled Hartree–Fock theory can almost give the paramagnetic susceptibility to within c. 5% or less.

Major contributions to the virtual excitations are the filled σ_g, $1\sigma_u$, $2\sigma_g$, $2\sigma_u$, $3\sigma_g$ and $1\pi_u$ orbitals, which contribute 1.67, 1.47, 0.95, 1.03, 11.23 and 1.18 to χ^p (centre of charge), and −8, −9, −31, 26, −373 and −13 to σ^p (origin at N nucleus), respectively. While the excitations from the filled $3\sigma_g$ orbitals dominate, contributions from other orbitals preclude a simple explanation of these magnetic properties.

The current maps for N_2 (Figure 5.10) show that the outer region is dominated by diamagnetic circulation (Figure 5.10a, b), while the region near the N nucleus has a strong paramagnetic dominance which arises primarily from excitations from the highest filled σ level ($3\sigma_g$).

5.3.9 F_2 [42]

Of the calculated values shown in Table 5.11, the values of $\chi = 10.2$ and $\mu_J/J = -0.108$ μ_N have been extrapolated after consideration of effects of basis-set extension. Agreement with experiment is fairly good for the pre-ferred values of $\sigma_F = -200$ p.p.m. and $C_F = 152$ kHz when the gauge origin

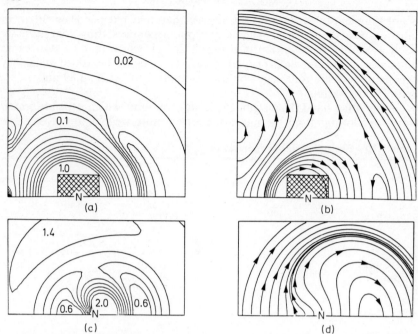

Figure 5.10 (a) Modulus of first-order current density induced in N_2 by a magnetic field perpendicular to the molecular axis, omitting the cross-hatched region. The molecular centre is at the lower left origin. The gauge origin is at the molecular centre. Contours are at 0.02, (intervals of 0.02), 0.1, (intervals of 0.1), 1.0

(b) Direction of the first-order current density induced in N_2 by a magnetic field perpendicular to and pointing inward towards the plane of the figure. The molecular centre is at the lower left, and the gauge origin is at the molecular centre. Paramagnetic currents circulate clockwise. The cross-hatched region is omitted.

(c) Modulus of first-order current density in N_2, as in (a) but with the gauge origin at the N nucleus. Intervals are 0.6, (increments of 0.2), 2.0. The area is the cross-hatched region of (a)

(d) Direction of first-order current density induced in N_2 as in (b). The gauge origin is at the N atom, and the area is that cross-hatched in (b)

Table 5.11   Second-order properties of F_2 [42]

	Origin at F	Origin at C	Experimental
χ/p.p.m.	−10.98	−10.62	—
χ^p	56.33	31.08	—
χ^d	−67.31	−41.70	—
$\mu_J/J(\mu_N)$	−0.095	−0.101	−0.121 ± 0.001 [43]
σ_F/p.p.m.	−276	−200	−210 [39, 44]
σ^p	−805	−700	—
σ^d	529	500	—
C_F/kHz	169	152	157.3 ± 0.8 [43]
$\alpha_{\parallel}$/Å^3	2.15	2.15	—
$\alpha_{\perp}$/Å^3	0.77	0.77	—

Basis sets are 8 σ, 5 π and 2 δ for magnetic properties, and 8 σ, 5 π and δ for the polarisabilities.

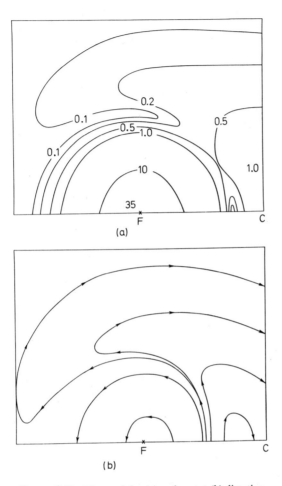

(a)

(b)

Figure 5.11 The modulus (a) and vector (b) direction of the induced electron density current in the fluorine molecule, when a magnetic field is applied perpendicular to the internuclear axis. F indicates the position of one nucleus and C the inversion centre. The magnetic field is perpendicular to the plane of the map, pointing inward. The induced current density has been calculated with the gauge origin at the fluorine nucleus F. Diamagnetic currents are clockwise

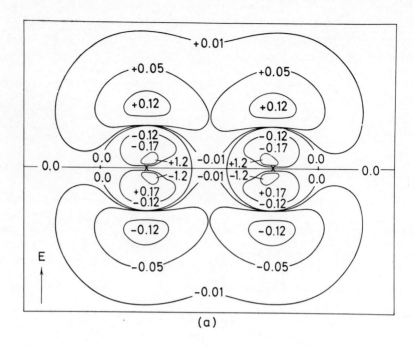

(a)

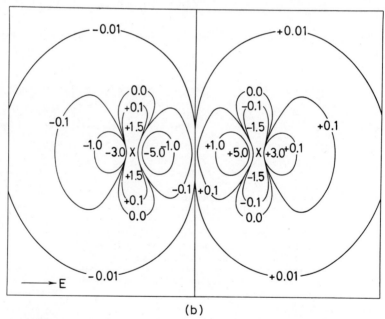

(b)

Figure 5.12 These maps show the first-order change in the electron density in the fluorine molecule in the presence of an electric field perpendicular (a) or parallel (b) to the internuclear axis. The maps show a cross-section through the entire molecule, the Xs indicating the positions of the fluorine nuclei. The direction of the electric field is indicated by the vector in each map

is chosen at the molecular centre. However, the rather poor gauge-dependence of shielding and spin–rotation constants is an indication that the basis sets are still incomplete.

The anti-shielding of the F atom is clearly associated with excitations from the $1\pi_{uy}$ level (magnetic field along x, molecular axis along z). This orbital also is the largest contributor to the paramagnetic part of the magnetic susceptibility, whether the origin is taken at the F atom or at the molecular centre (C). When interactions among occupied orbitals are neglected, this orbital contributes -732 p.p.m. (origin at F) or -724 p.p.m. (origin at C) to the shielding. An error of only 3% in this contribution would result in an error of about 10% in the total shielding. The strong paramagnetic circulation in the neighbourhood of the F nucleus is shown in Figure 5.11.

The polarisability along the molecular axis is dominated by excitation from the filled $3\sigma_g$ to the lowest unoccupied π_u orbital (Figure 5.12). This parallel component is considerably greater in F_2 than in HF or the F atom, as may be expected from the modification of excitation energies by homopolar bonding.

5.3.10 ClF[45]

Full results are yet to appear on an extensive study of the second-order properties of ClF (Table 5.12). The anomalously large diamagnetic shielding is due to a dominant mixing, with a small energy gap, between the highest filled π level to the lowest unfilled σ level. Thus, the symmetries are just opposite to those for the dominant mixing in BH.

Table 5.12   Second-order properties of ClF [45]

	Calculated	Experimental
χ/p.p.m.	-20.28	—
$\mu_J/J(\mu_N)$		
^{35}Cl	-0.103	—
^{37}Cl	-0.101	—
$\alpha_{Cl}/\text{Å}^3$	-408	
$C_{(^{35}Cl)}/kHz$	21.5	21.7[46]
$C_{(^{37}Cl)}$	17.5	—
σ_F	700	667[47]
C_F/kHz	-30	-22.5[46]

5.3.11   More complex molecules

Arrighini, Moccia and Maestro[48] have computed magnetic susceptibilities for H_2O, NH_3, CH_4 and H_2O_2 by coupled Hartree–Fock methods. Values for shielding constants in H_2O, NH_3, CH_4 and CH_3F have been computed by these same authors. The results for the largest basis sets are shown in Tables 5.13 and 5.14. Although some variation of exponents was explored, full optimisation would have involved extensive calculations. Consequently,

Table 5.14 Proton magnetic shielding components (p.p.m.) for H_2O, NH_3, CH_4 and CH_3F[48]

	H_2O		NH_3		CH_4		CH_3F	
	CA*	H	CA*	H	CA*	H	CA*	H
σ^d_{xx}	32	75	13	119			75	94
σ^d_{yy}	9	130	43	57			46	158
σ^d_{zz}	25	103	20	112			20	114
σ^d	22	103	25	96 (98)¶	28	87 (87)[30]	47	122 (121)[30]
σ^p_{xx}	4 (7)[55]	−26 (−37)[55]	15 (1)[58]	−55 (−105)[58]			−24	−36 (−98 av)[30]
σ^p_{yy}	11 (15)[55]	−79 (−107)[55]	−2 (−9)[58]	−12 (−5)[58]			3	−93
σ^p_{zz}	2 (7)[55]	−52 (−72)[55]	9 (8)[58]	−61 (−84)[58]			10	−70 (−84)[30]
σ^p_{av}	6 (10)[55]	−52 (−72)[55]	7 (6)[58]	−43 { (−65)[58] (−67)[59] }	4 (3)[60]	−54 (−56)[60]	−4	−66
σ_{av}	28 (30)†	50 (30)†	32 (31)‡	53 (31)‡	32 { (31)§ (32)[61] }	33 { (31)§ (32)[61] }	44 (27)‖	56 (27)‖

*CA refers to the choice of gauge origin at the central atom. H to the hydrogen atom as origin. Experimental results are in parentheses.
†From $\sigma_{H2O} = (3.60 + \sigma_{H2})$ p.p.m.[56] and $\sigma_{H2} = 26.43 \pm 0.60$ p.p.m.[57].
‡Using $\sigma_{NH3} = (0.65 + \sigma_{H2O})$ p.p.m. and footnote †.
§Using $\sigma_H - \sigma_{CH4} = -4.20$ p.p.m. and footnote †.
‖Using $\sigma_{CH3F} - \sigma_{CH4} = -4.00$ p.p.m.[62] and footnote §.
¶The experimental value of 94 p.p.m.[30] is corrected to 98 p.p.m. in Ref. 41.

some gauge-dependence of total susceptibility and shielding remains. Choices of gauge origins which maximise the total susceptibility for each of these molecules are 0, 0, 0.106 for H_2O; 0, 0, -0.0133 for NH_3; 0, 0, 0 for CH_4; and 0, 0, 2.388 for CH_3F (all in a.u.). Average proton shielding constants for these 'best' gauges are 29 for H_2O, 32 for NH_3, 32 for CH_4

Table 5.13 Magnetic susceptibilities (p.p.m.) for H_2O, NH_3, CH_4 and H_2O_2[48]. Experimental values are in parentheses

	H_2O*	NH_3†	CH_4‡	H_2O_2§
χ_{xx}^p	0.4 (0.77)[49]	2.9 (3.00)[51]	3.3 (9.3)[54]	26.7
χ_{xy}^p	2.1 (2.49)[49]			2.5
χ_{yy}^p	2.1 (2.49)[49]			2.2
χ_{zz}^p	0.8 (1.47)[49]	5.7 (5.77)[51]		29.7
χ_p	1.1 (1.57)[49]	3.8 (3.92)[51]	3.3 (9.3)[54]	19.5
χ_{xx}^d	-14.7	-19.7	$-28.7\,(-21.5)$	-64.4
χ_{xy}^d				-2.6
χ_{yy}^d	-16.3			-25.3
χ_{zz}^d	-15.5	-21.9		-66.1
χ^d	$-15.5\,(-14.6)$[50]	-20.5	$-28.7\,(-21.5)$	-52.0
χ	$-14.4\,(-13.0)$[50]	$-16.6\begin{cases}(-18.0)^{52}\\(-16.3)^{53}\end{cases}$	$-25.3\begin{cases}(-12.2)^{52}\\(-17.4)^{53}\end{cases}$	-32.3

*H_2O in xz plane with C_2 along z, O at origin. 27 basis functions.
†N and H in xz plane with C_3 along z, N at origin. 32 basis functions.
‡S_4 axes along x, y and z, C at origin. 18 basis functions.
§C_2 along y, z axis bisecting the O—O bond, centre of O—O bond as origin. 30 basis functions.

and 27 for CH_3F, all in excellent agreement with experiment. Since proton shielding constants are usually dominated by the diamagnetic contribution, the paramagnetic term need not be calculated with equivalent accuracy, particularly if it is minimised by suitable choice of origin.

References

1. Van Vleck, J. H. (1932). *Electric and Magnetic Susceptibilities*, 186. (London and New York: Oxford University Press)
2. Ramsey, N. F. (1956). *Molecular Beams*, 163 and 207. (London and New York: Oxford University Press)
3. Schlichter, C. P. (1963). *Principles of Magnetic Resonance*, Chapter 4. (New York: Harper and Row)
4. Hegstrom, R. and Lipscomb, W. N. (1968). *Rev. Mod. Phys.*, **40**, 354
5. Stevens, R. M., Pitzer, R. M. and Lipscomb, W. N. (1963). *J. Chem. Phys.*, **38**, 550
6. Hegstrom, R. and Lipscomb, W. N. (1967). *J. Chem. Phys.*, **46**, 1594
7. Karplus, M. and Das, T. P. (1961). *J. Chem. Phys.*, **34**, 1683
8. Pople, J. A. (1963–1964). *Mol. Phys.*, **7**, 301
9. Cornwell, C. D. (1966). *J. Chem. Phys.*, **44**, 874
10. Langhoff, P. W., Karplus, M. and Hurst, R. P. (1966). *J. Chem. Phys.*, **44**, 505
11. Caves, T. C. and Karplus, M. (1969). *J. Chem. Phys.*, **50**, 3649
12. Lipscomb, W. N. (1966). *Advances in Magnetic Resonance*, Ed. by J. T. Waugh, Vol. 2, 137. (New York: Academic Press)
13. Roothaan, C. C. J. (1951). *Rev. Mod. Phys.*, **23**, 69
14. Hall, G. G. (1951). *Proc. Roy. Soc. (London)*, **A205**, 541
15. Chan, S. I. and Das, T. P. (1962). *J. Chem. Phys.*, **37**, 1527
16. Stevens, R. M. and Lipscomb, W. N. (1964). *J. Chem. Phys.*, **40**, 2238

17. Lawrence, T., Anderson, C. H. and Ramsey, N. F. (1963). *Phys. Rev.*, **130**, 1865
18. Klemperer, W., Gold, L. P. and Wharton, L. (1962). *J. Chem. Phys.*, **37**, 2149, and private communication
19. Stevens, R. M. and Lipscomb, W. N. (1965). *J. Chem. Phys.*, **42**, 4302
20. Brooks, R. A., Anderson, C. H. and Ramsey, N. F. (1964). *Phys. Rev.*, **136**, A62
21. Stevens, R. M. and Lipscomb, W. N. (1965). *J. Chem. Phys.*, **42**, 3666
22. Hegstrom, R. A. and Lipscomb, W. N. (1966). *J. Chem. Phys.*, **45**, 2378
23. Hegstrom, R. A. and Lipscomb, W. N. (1968). *J. Chem. Phys.*, **48**, 809
24. Laws, E. A., Stevens, R. M. and Lipscomb, W. N. (1971). *J. Chem. Phys.*, **54**, 4269
25. Stevens, R. M. and Lipscomb, W. N. (1964). *J. Chem. Phys.*, **41**, 184
26. Ehrlich, P. (1942). *Z. Anorg. Allgem. Chem.*, **249**, 219
27. Baker, M. R., Anderson, C. H., Pinkerton, J. and Ramsey, N. F. (1961). *Bull. Amer. Phys. Soc.*, **6**, 19
28. Hindermann, D. K. and Cornwell, C. D. (1968). *J. Chem. Phys.*, **48**, 4148, 2017
29. Schneider, W. G., Bernstein, H. J. and Pople, J. A. (1958). *J. Chem. Phys.*, **28**, 601
30. Wofsy, S. C., Muenter, J. S. and Klemperer, W. (1971). *J. Chem. Phys.*, **55**, 2014
31. Baker, M. R., Nelson, H. M., Leavitt, J. A. and Ramsey, N. F. (1961). *Phys. Rev.*, **121**, 807
32. Muenter, J. S. and Klemperer, W. (1970). *J. Chem. Phys.*, **52**, 6033
33. Weiss, R. (1963). *Phys. Rev.*, **131**, 659
34. Stevens, R. M. and Karplus, M. (1968). *J. Chem. Phys.*, **49**, 1094
35. Ozier, I., Yi, P., Khocla, A. and Ramsey, N. F. (1967). *J. Chem. Phys.*, **46**, 1530
36. Rosenblum, B., Nethercot, Jr., A. H. and Townes, C. H. (1958). *Phys. Rev.*, **109**, 400
37. Grimaldi, F., Lecourt, A. and Moser, C. (1967). *Int. J. Quantum Chem.*, **1S**, 153
38. Green, S. (1971). *J. Chem. Phys.*, **54**, 827
39. Baker, M. R., Anderson, C. H. and Ramsey, N. F. (1964). *Phys. Rev.*, **133**, A1533
40. Chan, S. I., Baker, M. R. and Ramsey, N. F. (1964). *Phys. Rev.*, **136**, A1224
41. Laws, E. A., Stevens, R. M. and Lipscomb, W. N. (1972). *J. Chem. Phys.*, **56**, 2029
42. Stevens, R. M. and Lipscomb, W. N. (1964). *J. Chem. Phys.*, **41**, 3710
43. Ozier, I., Crapo, L. M., Cederberg, J. W. and Ramsey, N. F. (1964). *Phys. Rev. Lett.*, **13**, 482
44. Gutowsky, H. S. and Hoffman, C. J. (1951). *J. Chem. Phys.*, **19**, 1259
45. Stevens, R. M., to be published
46. Muenter, J., private communication of preliminary experimental values for the $v = 0$, $J = 1$ electric resonance spectra.
47. Cornwell, C. D. (1966). *J. Chem. Phys.*, **44**, 874
48. Arrighini, G. P., Maestro, M. and Moccia, R. (1968). *J. Chem. Phys.*, **49**, 882; (1970) **52**, 6411; (1971) **54**, 825. For polarisabilities see (1967). *Chem. Phys. Lett.*, **1**, 242
49. Battaglia, A., Januzzi, M. and Polacco, E. (1963). *Ric. Sci. Rend.*, **A3**, 385
50. Landolt-Börnstein. (1951). *Zahlwerte Funktionen*, Ed. by J. Bartels *et al.*, Vol. 10, Part 3, Sec. 2, p. 354. (Berlin: Springer-Verlag)
51. Eshbach, J. R. and Strandberg, M. W. P. (1952). *Phys. Rev.*, **85**, 24
52. *Handbook of Chemistry and Physics* (45th Ed.), Ed. by C. D. Hodgman *et al.*, (1964). (Cleveland, Ohio: The Chemical Rubber Co.)
53. Barter, C., Meisenheimer, R. G. and Stevenson, D. P. (1960). *J. Phys. Chem.*, **64**, 1312
54. Anderson, C. H. and Ramsey, N. F. (1966). *Phys. Rev.*, **149**, 14
55. Bluyssen, H., Dymanus, A., Reuss, J. and Verhoeven, J. (1967). *Phys. Lett.*, **25A**, 584
56. Pople, J. A., Schneider, W. G. and Bernstein, H. J. (1959). *High Resolution Nuclear Magnetic Resonance.* (New York: McGraw-Hill)
57. Myint, T., Kleppner, D., Ramsey, N. F. and Robinson, H. G. (1966). *Phys. Rev. Lett.*, **17**, 405
58. Gordon, J. P. (1955). *Phys. Rev.*, **99**, 1253
59. Kukolich, S. G. and Wofsy, S. C. (1970). *J. Chem. Phys.*, **52**, 5477
60. Hegstrom, R. and Lipscomb, W. N. (1967). *J. Chem. Phys.*, **46**, 4538
61. Flygare, W. H. and Goodisman, J. (1968). *J. Chem. Phys.*, **49**, 3122
62. Emsley, J. W., Feeney, J. and Sutcliffe, L. H. (1966). *High Resolution Nuclear Magnetic Resonance Spectroscopy*, Vol. 2. (Oxford, England: Pergamon Press)

6
Statistical Thermodynamics
of Liquids

D. D. FITTS
University of Pennsylvania

6.1 INTRODUCTION

The statistical-thermodynamic theories of dilute gases and of crystalline solids were formulated many years ago and are now well understood. The elementary aspects of such theories are presented in the typical textbook on 'statistical mechanics' and in most introductory 'physical chemistry' textbooks. The molecular theory of the liquid state, on the other hand, has been formulated in a satisfactory form more recently and is at the present time in a stage of rapid development.

Since the liquid–gas co-existence curve in the pressure–temperature (p,T) phase diagram for a typical substance ends at a critical point, the distinction between a liquid and a gas disappears at temperatures above the critical temperature. As a result, it is possible to change continuously the state of a substance from one which is clearly gaseous to one which is clearly liquid without having the substance undergo a first-order phase transition. For this reason the two fluid states are treated together and a satisfactory theory of liquids must of necessity predict the equilibrium properties of both phases.

The scope of this review is limited to theories for the equilibrium properties of simple liquids. Simple liquids are composed of spherical, non-polar molecules which obey, at least to a first approximation, the laws of classical mechanics. Where necessary, corrections for quantum-mechanical behaviour can be added. Thus, typical simple liquids include the inert elements (neon, argon, krypton, xenon), nitrogen, methane, and to some extent liquid metals. Helium, on the other hand, must be treated by purely quantum-mechanical means and will not be considered here. In addition to the *real* simple liquids just mentioned, there are *hypothetical* simple liquids whose molecules interact according to a specified mathematical expression. Examples of such systems are the hard-sphere fluid, the square-well fluid, and the Lennard-Jones fluid.

Research work in the statistical-thermodynamic theory of liquids has expanded so much in the past 15 or so years that it would not be possible in this short review to discuss in meaningful detail all the significant contributions to the understanding of liquid structure and properties. Fortunately, a recent book entitled *Physics of Simple Liquids*[1] surveys progress in this field up to about 1967. This excellent book contains 16 chapters on various aspects of liquids, by authors who are working actively on that particular aspect. Thus, this review will be concerned largely with developments made since those reported in *Physics of Simple Liquids*.

More recently, Neece and Widom[2] have written a shorter review on the theory of liquids which includes, among many other items, a list of books dealing with the fluid state. We wish to add to that list the excellent two-volume treatise of Münster[3] and a forthcoming two-volume set edited by Henderson[4].

6.2 STATISTICAL MECHANICS

6.2.1 Intermolecular potentials

The main objective of statistical thermodynamics is to determine the bulk, macroscopic properties of a substance from a knowledge of the forces which each molecule in the system exerts on every other molecule. Statistical-mechanical theories of classical liquids are based on the evaluation of either the canonical ensemble partition function $Q_N(V,T)$ for a closed system of N particles at temperature T in a fixed volume V or the grand-canonical ensemble partition function $\Xi(\mu,V,T)$ for an open system with chemical potential μ, temperature T, and volume V:

$$Q_N(V,T) = \frac{1}{N!\Lambda^{3N}} \int \exp(-\beta\Phi_N)\, d1 \ldots dN \qquad (6.1)$$

$$\Xi(\mu,V,T) = \sum_{N=0}^{\infty} \frac{z^N}{N!} \int \exp(-\beta\Phi_N)\, d1 \ldots dN \qquad (6.2)$$

where $\Lambda = h/(2\pi mkT)^{\frac{1}{2}}$, h is Planck's constant, m is the molecular mass, k is Boltzmann's constant, $\beta = 1/kT$, Φ_N is the total potential energy, and z is the activity, $z = \exp(\beta\mu/\Lambda^3)$. The symbol d$j$ denotes the volume element at position j. The book by Hill[5] is a standard reference for this basic material.

For one-component systems composed of N molecules which are or may be treated as spherically symmetric, the potential energy $\Phi(1,\ldots,N)$ is a function of the positions $1,\ldots,N$ of those molecules and is given by

$$\Phi(1,\ldots,N) = \sum_{i<j=1}^{N} u_{ij} + \sum_{i<j<k=1}^{N} w_{ijk} + \ldots \qquad (6.3)$$

where u_{ij} is the potential of intermolecular force between any two molecules i and j and depends only on the distance r_{ij} between them. The summation is taken over each pair of particles. The triplet potential w_{ijk} depends only on the three distances r_{ij}, r_{jk}, and r_{ik} and the sum is over all molecular triplets. The pair potential u_{ij} has been studied extensively, but at the present time only the asymptotic form of w_{ijk} is known. Thus, it is pointless to consider any higher-order terms in the expansion (6.3) of the potential energy.

Calculations of the thermodynamic properties of fluids have been made primarily for model systems in which the pair potential $u(r)$ is given by some simple analytic form and the triple and higher-order potentials are neglected. Such simple forms considered in this review include the following:

hard-sphere potential

$$\begin{aligned} u(r) &= \infty & r &< d \\ &= 0 & r &\geqslant d \end{aligned} \qquad (6.4)$$

where d is the hard-sphere diameter;

square-well potential

$$\begin{aligned} u(r) &= \infty & r &< d \\ &= -\varepsilon & d &\leqslant r \leqslant Kd \\ &= 0 & r &> Kd \end{aligned} \qquad (6.5)$$

where $K > 1$ and ε is the depth of the well;

Lennard-Jones (LJ) 12-6 potential

$$u(r) = \varepsilon\left\{\left(\frac{r_m}{r}\right)^{12} - 2\left(\frac{r_m}{r}\right)^{6}\right\} = 4\varepsilon\left\{\left(\frac{\sigma}{r}\right)^{12} - \left(\frac{\sigma}{r}\right)^{6}\right\} \qquad (6.6)$$

where ε is the well depth, r_m is the value of r at the potential minimum, and $\mu(\sigma) = 0; r_m = 2^{\frac{1}{6}}\sigma;$

inverse-power potential

$$u(r) = \varepsilon(\sigma/r)^n \tag{6.7}$$

where σ and ε are scaling factors and n is a positive integer.

Details regarding these and other simple potential forms are reviewed by Fitts[6]. Except for Section 6.6, where the influence of the triplet contributions w_{ijk} are considered, we review here calculations involving only pair potentials.

Thermodynamic functions calculated by statistical-mechanical means for these model systems are often compared with experimental results for real systems, most commonly argon. However, the major value of calculations on model systems is to compare results for the same pair potential obtained by different approximations to the statistical-thermodynamic theory. Most of the work on liquids reported here deals with this latter consideration.

6.2.2 Correlation functions

The n-particle correlation function $g(1,...,n)$ in the canonical ensemble is[5]

$$\rho^n g(1,...,n) = \frac{\Lambda^{-3N}}{(N-n)!Q_N} \int \exp(-\beta\Phi_N)\mathrm{d}(n+1)...\mathrm{d}N \tag{6.8}$$

where $\rho = N/V$ is the number density, and in the grand-canonical ensemble is[5]

$$\rho^n g(1,...n) = \frac{1}{\Xi} \sum_{N \geqslant n} \frac{z^N}{(N-n)!} \int \exp(-\beta\Phi_N)\mathrm{d}(n+1)...\mathrm{d}N \tag{6.9}$$

The two-particle correlation function $g(1,2)$ in a homogeneous fluid is a function only of r_{12} and is known as the *radial distribution function* (RDF).

It is often convenient to define the *pair correlation function* $h(r)$ as

$$h(r) = g(r) - 1 \tag{6.10}$$

the so-called *Mayer f-function* as

$$f(r) = \exp[-\beta u(r)] - 1 \tag{6.11}$$

and an unnamed function $y(r)$ as

$$g(r) = \exp[-\beta u(r)]y(r) \tag{6.12}$$

Since $g(r)$ tends to $\exp[-\beta u(r)]$ as the density ρ tends to zero, the function $y(r)$ approaches unity at low densities. The *direct correlation function* $c(r)$ is defined by the Ornstein–Zernike equation,

$$h_{12} = c_{12} + \rho \int c_{13}h_{23}\mathrm{d}3 \tag{6.13}$$

For systems in which the potential energy Φ is a sum of pair potentials only, the internal energy E and the pressure p are given by

$$E = \frac{3}{2}NkT + 2\pi\rho N \int_0^\infty u(r)g(r)r^2\,dr \tag{6.14}$$

$$\frac{pV}{NkT} = 1 - \frac{2\pi\rho}{3kT} \int_0^\infty \frac{du(r)}{dr} g(r)r^3\,dr \tag{6.15}$$

Equation (6.15) is called the *pressure equation* to distinguish it from an alternative expression for the equation of state called the *compressibility equation*, which can be derived from the fluctuations in the grand-canonical ensemble,

$$kT\left(\frac{\partial\rho}{\partial p}\right)_T = 1 + 4\pi\rho \int_0^\infty h(r)r^2\,dr = \left[1 - 4\pi\rho \int_0^\infty c(r)r^2\,dr\right]^{-1} \tag{6.16}$$

For a hard-sphere fluid, with $u(r)$ given by equation (6.4), the pressure equation (6.15) has the simple form

$$\frac{pV}{NkT} = 1 + \frac{2}{3}\pi\rho d^3 g(d) \tag{6.17}$$

Chen, Henderson and Barker[7] have introduced a third procedure for calculating the equation of state. In their procedure the Helmholtz free energy A is obtained by integrating equation (6.14) for E and the equation of state is obtained from the thermodynamic relation $p = -(\partial A/\partial V)_T$. The resulting equation is

$$\left(\frac{pV}{NkT}\right) - \left(\frac{pV}{NkT}\right)_0 = \frac{1}{2}\rho \int_{\beta_0}^\beta \left[\mathscr{G}(\beta,\rho) + \rho\left(\frac{\partial\mathscr{G}(\beta,\rho)}{\partial\rho}\right)_T\right]d\beta \tag{6.18}$$

where the subscript 0 refers to some reference state and $\mathscr{G}$ is defined as

$$\mathscr{G}(\beta,\rho) = 4\pi \int_0^\infty u(r)g)(r)r^2\,dr \tag{6.19}$$

This procedure is equivalent to equation (6.15) or (6.16), but gives a more satisfactory equation of state when an approximate RDF is used[7].

6.2.3 Significant structure theory

An alternative approach to the study of liquids is the *significant structure theory* due to Eyring and his collaborators. The features of this approach, which relates the structure of a liquid to the solid and gaseous phases, are summarised in a recent book by Eyring and Jhon[8]. In addition, Henderson, Barker, and Kim[9] provide a brief, but meaningful summary. However, a book review[10] of Eyring's and Jhon's monograph[8] points out that the development of the significant structure theory is based on the use of erroneous experimental data for the heat capacity of liquid argon at constant volume. The correct data appears to destroy the very basis of the theory. Accordingly, it does not seem while in this review to consider this approach further.

6.3 COMPUTER STUDIES

One procedure for obtaining the thermodynamic properties of a model system is a direct computer calculation. There are two computer-simulation techniques: the method of molecular dynamics, due to Alder and Wainright[11-14], and the Monte Carlo method, devised by Metropolis *et al.*[15] and developed by Wood *et al.*[16-18].

In the molecular dynamics (MD) method, the classical trajectories of N particles, where N is less than 1000, are determined step-wise by solving numerically the Newtonian equations of motion from an initial choice of velocities and positions. This initial configuration may not correspond to thermodynamic equilibrium, in which case the MD method can be used to follow a system from a non-equilibrium configuration to an equilibrium state. To compensate for the small size of the system, periodic boundary conditions are introduced; thus, the system is surrounded by replicas of itself so that when a particle leaves the system through one wall, it enters through the opposite wall. Once the system has reached an equilibrium state, thermodynamic quantities can be calculated as the time average of the appropriate dynamical variable over a large number (say 1500) of microscopic states.

In the Monte Carlo (MC) method, an initial static configuration for 32–1000 particles is selected and the particles are randomly displaced one at a time. By accepting all such displacements which lower the potential energy Φ and accepting only a fraction $\exp(-\delta\Phi/kT)$ of those displacements which increase the potential energy by $\delta\Phi$, the MC method generates a series of configurations in which each individual configuration appears with a frequency proportional to the Boltzmann factor $\exp(-\Phi/kT)$. Again, periodic boundary conditions are used to compensate for the finite size of the system. A thermodynamic quantity is then obtained by averaging the appropriate dynamical variable over many (say 10^6) such configurations.

The RDF for the system of N particles can also be calculated by the MD and MC methods. If $n(r)$ particles are situated in the averaging process at a distance between r and $r+\Delta r$ from a given particle, then the RDF $g(r)$ is given by

$$g(r) = n(r)/4\pi\rho r^2 \Delta r \qquad (6.20)$$

Wood[19] provides more details of the actual computational process.

The molecular dynamics and Monte Carlo methods differ in the principles on which they are based, namely, classical kinetic theory and classical statistical mechanics, respectively. In practice, for the same model system the two techniques produce practically identical results, thereby supporting one of the fundamental assumptions of statistical mechanics that the time average is equal to the ensemble average. In terms of their demands on the computer, the MD and MC methods are about the same. Typically, the MD method requires more particles, but fewer configurations, than the MC method for the same degree of reliability in the calculation of a thermodynamic quantity. The MD method has the advantage that non-equilibrium systems may be studied. On the other hand, the MC method has been extended to ensembles other than the canonical (NVT) ensemble, in particular to the isobaric–isothermal (NpT) ensemble[19].

6.3.1   Hard-sphere fluid

The hard-sphere fluid is a collection of N particles in a volume V (hence, a number density $\rho = N/V$) with a pairwise potential of interaction $u(r)$ given by equation (6.4). Although the hard-sphere potential differs markedly from that for a pair of real atoms or molecules, a real liquid at high densities and high temperatures behaves much like a hard-sphere fluid. Thus, a hard-sphere fluid may be regarded as an 'ideal' liquid and used as a first approximation to real liquid behaviour in much the same way that the ideal gas and the harmonic-oscillator model of a crystal are used to treat gases and crystalline solids. In fact, the newly developed perturbation theories of liquids discussed in Section 6.5 are based on just such considerations.

The one-dimensional case (hard rods) and the two-dimensional case (hard disks) of a hard-sphere system are of interest in the study of fluids in so far as they reflect on the validity and accuracy of the numerical methods used for calculating the properties of the three-dimensional system. For this same purpose, the properties of a hard-square system have been investigated[20, 21]. Since hard rods, disks, and spheres are treated in great detail in the review by Wood[19], we need comment here only briefly on some of the older results and concentrate on work published since that review.

6.3.1.1   Fluid phase

The density of a system of hard-core particles is often expressed relative to the close-packed volume V_0. For hard disks of diameter d, the close-packed volume (actually an area for the two-dimensional case) for N particles is $V_0 = (\sqrt{3}/2)Nd^2$; the density is then the ratio V_0/V. An alternative designation is the reduced density ρ^* defined as $\rho^* = \rho d^2$, where $\rho = N/V$ is the number density. Sometimes the reduced volume $V^* = V/V_0 = 2/\sqrt{3}\rho^*)$ is used. For hard spheres of diameter d, the close-packed volume V_0 is given by $V_0 = Nd^3/\sqrt{2}$, the density relative to V_0 is again the ratio V_0/V, the reduced density $\rho^* = \rho d^3$, and the reduced volume V^* is related to V_0 and ρ^* by $V^* = V/V_0 = \sqrt{2}/\rho^*$.

The equation of state pV_0/NkT as a function of V^* for both hard disks and hard spheres has been studied extensively by MD and MC techniques and has in each case two branches, one of higher density and one of lower density. These branches are interpreted as representing, respectively, solid and fluid phases (see Wood[19] for a thorough discussion). Thus, a hard-particle system has an important feature in common with a real molecular system. We consider the fluid branch in this section, the solid branch in the next section, and the melting curve between these two phases in the following section.

The second, third[22], and fourth[23, 24] virial coefficients for a hard-disk fluid are known exactly and the fourth[25], fifth[25], sixth[25], and seventh[26] have been calculated by Monte Carlo integrations. From these virial coefficients an approximate analytic expression for the equation of state can be obtained[25] by representing pV/NkT as a Padé approximant $P(m, n)$. The approximant $P(m, n)$ is a ratio of two polynomials in density with degrees m and n; the coefficients are chosen so that $P(m, n)$ reproduces the first $m+n$ virial coefficients. Ree and Hoover[26] have obtained $P(3, 3)$, $P(3, 4)$, and $P(4, 3)$ for the equation of state of the hard-disk fluid and found good agreement ($\leqslant 5\%$) with the MD calculations of Hoover and Alder[27] over the entire fluid range ($V_0/V < 0.762$, $V^* > 1.312$, $\rho^* < 0.880$).

Chae, Ree and Ree[28] have recently tabulated the RDF and the equation of state for a hard-disk fluid at densities V_0/V of 0.4, 0.5, and 0.6 ($V^* = 2.5$, 2.0, 1.677; $\rho^* = 0.4619$, 0.5774, 0.6928) as obtained by MC NVT-ensemble calculations with 90 and 208 particles averaged over more than 10^6 configurations. Wood has formulated[19, 29] the MC method for the NpT ensemble and has calculated[30] the equation of state and the RDF for the hard-disk fluid. The RDF is tabulated[30] for $pV_0/NkT = 1.25, 2.0, 3.0, 4.0, 5.0, 6.0$. Where a comparison is possible, Wood's equation of state and RDF agree with the NVT-ensemble calculations of Chae, Ree and Ree[28].

For the hard-sphere fluid, the equation of state has been determined by MD calculations by Alder and Wainwright[31] with their later, unpublished work tabulated by Ree and Hoover[25]. Monte Carlo calculations of the equation of state for a hard-sphere fluid are reviewed by Wood[19].

The second, third and fourth virial coefficients for a system of hard spheres are known exactly (see Rowlinson's review[32] for details), while the fifth[25, 33–35], sixth[25] and seventh[26] have been calculated by a combination of analytical, numerical, and Monte Carlo integrations. From these virial coefficients Ree and Hoover[25, 26] have determined several Padé approximants for the equation of state, the approximant $P(3, 3)$ agreeing best with the later MD calculations of Alder and Wainwright. However, Carnahan and Starling[36] have discovered a much simpler form, namely,

$$\frac{pV}{NkT} = \frac{1+\eta+\eta^2-\eta^3}{(1-\eta)^3} \tag{6.21}$$

where $\eta = \pi\rho d^3/6 = \pi\rho^*/6 = b\rho/4$ and $b = 2\pi d^3/3$ is the second virial coefficient. Equation (6.21) fits the MD results of Alder and Wainwright even better than the Ree and Hoover form and is quite accurate over the entire fluid range ($\eta < 0.50$; $V^* > 1.50$; $\rho^* < 0.95$).

Carnahan and Starling[37] have also related the internal energy ΔE, enthalpy ΔH, entropy ΔS, Gibbs free energy ΔG, and fugacity f, all relative to the ideal gas, to the equation of state for the hard-sphere fluid and then have used equation (6.21) to calculate these properties. For $\eta = 0.00 (0.02) 0.74$ ($\rho^* = 0.00–1.41$) they have tabulated pV/RT, $\Delta H/RT$, $\Delta S/R$, $\Delta G/RT$ and f/p.

In the last few years the RDF for the hard-sphere fluid has been calculated for a wide range of densities. Alder and Hecht[38] have tabulated the RDF as determined by the MD method for $r/d = 1.00 (0.04) 2.80$ at reduced volumes $V^* = 3.0, 2.0, 1.7, 1.6$ ($\rho^* = 0.4714, 0.7071, 0.8319, 0.8839$). Their RDF at $V^* = 2.0$ was previously tabulated by Ree, Keeler and McCarthy[39]. Barker and Henderson[40] have tabulated the mean RDF as calculated by the MC technique with 108 particles for $r/d = 1.0173–2.2727$ at reduced volumes $V^* = 7.071–1.571$ [$\rho^* = 0.20 (0.10) 0.90$]. They extrapolated their results to $r/d = 1$ and thereby reported $g(d)$ as well. Verlet and Weis[41] have expressed the RDF for the hard-sphere fluid in an analytical form.

$$g_{HS}(r/d;\eta) = g_{PY}(r/d_W;\eta_W) + \delta g_1(r) \tag{6.22}$$

where $g_{PY}(r)$ is the RDF given by the Percus–Yevick theory (see Section 6.4), $\delta g_1(r)$ is an analytical correction term, and d_W, η_W are scaling factors which are empirical functions of η.

There have been no new computer simulation results on mixtures of hard spheres reported since Neece's and Widom's review[2].

6.3.1.2   Solid phase

It is desirable to have a rigorous series expansion of the high-density solid-phase thermodynamic properties, analogous to the low-density virial expansion for a gas. Such expansions are useful, for example, for representing the solid phase in the determination of the solid–fluid phase transition (see Section 6.3.1.3). On the basis of computer calculations, Alder, Hoover and Wainwright[42] suggested $\alpha \equiv (V - V_0)/V_0$ as an appropriate expansion parameter. In terms of α, the Helmholtz free energy A_N of the high-density solid phase of N v-dimensional rigid particles of diameter d is given by the asymptotic expansion

$$\frac{A_N}{NkT} \underset{\alpha \to 0}{\sim} v \ln \frac{\Lambda}{d} - v \ln \alpha + C + D\alpha + E\alpha^2 + \ldots \qquad (6.23)$$

where Λ is defined in Section 6.2.1 and $C, D, E, \ldots$ are constants dependent on the lattice structure. Although this expansion was proposed earlier, Rudd et al.[43] have recently presented a formal derivation and obtained explicit expressions for C, D and E. They also report numerical values of C and D for the hexagonal-close-packed and face-centred-cubic packings of hard spheres and for the triangular lattice of hard disks. Since Rudd et al.[43] refer to earlier work, we omit these references here. The corresponding expression for a system of hard squares at high densities has also been studied and the first coefficient C evaluated numerically[21]. The entropy S can be obtained from equation (6.23) by the usual relation

$$S_N = -(\partial A_N/\partial T)_{V,N} \qquad (6.24)$$

The pressure of the high-density solid phase for rigid particles may also be expanded in terms of α, with the result

$$\frac{pV}{NkT} = \frac{v}{\alpha} + C_0 + C_1\alpha + C_2\alpha^2 + \ldots \qquad (6.25)$$

where again v is the dimensionality of the system. Alder, Hoover and Young[44] have calculated by the MD technique pV/NkT as a function of V^* for hard disks $(1.01 \leqslant V^* \leqslant 1.25)$ and for hard spheres $(1.005 \leqslant V^* \leqslant 1.42)$. The pressure difference and hence the entropy difference between the hexagonal and face-centred-cubic packings of hard spheres could not be detected and thus the relative stability of these two phases could not be determined. By fitting the MD results for hard disks and for hard spheres to equation (6.25), Alder, Hoover and Young determined C_0 and C_1 for each case.

6.3.1.3 Phase transition

For pure phases, either fluid or solid, the MD and MC techniques can calculate thermodynamic properties quite accurately. However, present computers are not capable of accommodating enough particles to treat co-existing fluid and solid phases. Accordingly, an accurate determination of the melting transition can be obtained by treating the fluid and solid phases separately and then locating the solid-phase density and fluid-phase density which satisfy the thermodynamic equilibrium conditions of equal pressure and chemical potential at constant temperature. Since the entropy, which is part of the chemical potential, is not a dynamical variable and therefore cannot be calculated by the usual MD and MC averaging techniques, Hoover and Ree[45] have developed a computer 'experiment' for obtaining the entropy and chemical potential by integrating along a reversible path from a 'standard state' to the state of interest and have applied[46] this technique to confirm the existence of a first-order melting transition for a classical-mechanical system of hard spheres. By extrapolating their hard-sphere results to the thermodynamic limit (i.e., infinite number of particles), they found that at $pV_0/NkT = 8.27 \pm 0.13$ the solid phase with density $V_0/V = 0.736 \pm 0.003$ ($V^* = 1.359 \pm 0.006$, $\rho^* = 1.041 \pm 0.004$) and the fluid phase with density $V_0/V = 0.667 \pm 0.003$ ($V^* = 1.500 \pm 0.007$, $\rho^* = 0.943 \pm 0.004$) have equal values for p and chemical potential at constant T.

For a system of hard disks, Alder and Wainwright[47] located a phase transition at $pV_0/NkT = 7.72$ between a solid phase with density $V_0/V = 0.790$ ($V^* = 1.266$, $\rho^* = 0.912$) and a fluid phase with density $V_0/V = 0.762$ ($V^* = 1.312$, $\rho^* = 0.880$). Hoover and Ree[46] have verified that these two states do indeed have the same chemical potential and are therefore in thermodynamic equilibrium.

Recently Meeron[48] purported to show via rigorous theory that a rigid particle system has no phase transitions. Wood[49] pointed out, however, that Meeron's theorem excludes phase transitions only in the density range where

$$0 \leqslant 2b\rho \exp(\beta\mu^e) < 1 \qquad (6.26)$$

(μ^e is the chemical potential in *excess* of the ideal gas value) and thus only for $V_0/V < 0.147$ for hard disks and $V_0/V < 0.091$ for hard spheres. These densities are well below the respective densities where the phase transitions are observed.

6.3.2 Square-well fluid

The lower-order terms in the density expansions of the equation of state and the RDF for a square-well fluid with the potential given by equation (6.5) have been reviewed and studied further by Barker and Henderson[50]. The second and third virial coefficients are known analytically for arbitrary K, the fourth and fifth are known either analytically or numerically for $K = 1.5$ and 2. The first two terms in the density expansion of the RDF are known analytically[51, 52] for $K = 2$ and partly analytically and partly numerically[50] for $K = 1.5$. The one-dimensional square-well fluid has been recently studied by Katsura and Tago[53], who review earlier work.

Rotenberg[54] has calculated by the MC method the equation of state for $K = 1.5$ at $T^* = kT/\varepsilon = 0.5$, 1.0 and 3.333 with V^* ranging from 1.1 to 4.0. Lado and Wood[55] also calculated the equation of state for $K = 1.5$ with the MC method at $T^* = 0.5$ and $V^* = 1.185$ for 256, 500 and 864 particles. Their results show a strong dependence on the number of particles in the calculation. Alder has some unpublished MD equation-of-state calculations, quoted by Barker and Henderson[56], for $K = 1.5$ over a range of densities at $T^* = 1.115$, 1.7, 2.09 and 3.02.

6.3.3   Inverse-power fluid

Hoover et al.[57, 58] have studied the thermodynamic properties of a fluid whose molecules interact according to the inverse-power law, equation (6.7), with $n = 4$, 6, 9, and 12. With $n = \infty$, the inverse-power fluid becomes a hard-sphere fluid. It is easy to show[57] for an inverse-power fluid that the *excess* thermodynamic properties (with respect to an ideal gas at the same density and temperature) are functions of $(\rho\sigma^3)(\varepsilon/kT)^{3/n} = \rho^*T^{*-3/n}$ only. For this reason a single isotherm or isochore can describe the entire range of thermodynamic states.

Using the MC method to calculate the properties of the fluid state and the lattice dynamics method for the solid state, Hoover et al.[57, 58] determined the equation of state and entropy as a function of $\rho^*T^{*-3/n}$ and located the solid–fluid phase transition for each value of n. The effect of the 'softness' of the pair potential on the melting transition is discussed in great detail[58].

6.3.4   Lennard–Jones 12-6 fluid

The properties of a system of N particles in a volume V with pairwise inter-molecular interactions of the LJ 12-6 form, equation (6.6), have been extensively studied by the MD and MC methods. In addition to Wood[19], McDonald and Singer[59] have also reviewed such calculations. Hoover and Ross[60] have written a review article on statistical theories of melting with an emphasis on computer simulation results. An LJ 12-6 fluid possesses a critical temperature, above which there is only one fluid phase and below which two fluid phases can co-exist. Such behaviour is the result of an attractive term in the potential function. Moreover, the LJ 12-6 fluid possesses a solid–liquid phase transition. Thus, the LJ 12-6 fluid can be reasonably used as a model for a real fluid. The thermodynamic properties of the fluid phases are reviewed first and the phase transitions are discussed in Section 6.3.4.2.

6.3.4.1   Fluid phases

The thermodynamic properties of an LJ 12-6 system of particles depend on two independent variables, the temperature and the density being a convenient pair. The temperature may be expressed in reduced form as $T^* = kT/\varepsilon$, where ε is the well depth. There are many arbitrary choices for expressing the density in reduced form. A commonly used convention, which we shall adopt in this review, is the reduced density $\rho^* = \rho\sigma^3$, where ρ is the number density N/V and σ is the finite value of intermolecular distance r where the potential $u(r)$ is zero; i.e., $u(\sigma) = 0$. In the discussion which follows, all reported densities are converted, if necessary, to ρ^*.

The virial coefficients through the fifth B_5 for an LJ 12-6 gas have been calculated with the results summarised by Rowlinson[61]. Kim, Henderson and Oden[62] have since reported more extensive and more accurate values for B_5. Earlier MC calculations of thermodynamic properties and the RDF are summarised by Wood[19]. We review here more recent work.

Rahman[63] was the first to apply the MD technique to a system of LJ 12-6 particles. His RDF at $T^* = 0.787$ and $\rho^* = 0.807$ is presented in graphical form. Verlet[64] has calculated by the MD method the equation of state and the internal energy at various T^* from 0.591 to 4.625 and various ρ^* from 0.35 to 0.88, most of these states corresponding to the liquid region. In an appendix to a report on another aspect of liquids, Levesque and Verlet[65] presented some new MD and MC results for the equation of state and the internal energy at various T^* and ρ^*, as well as some corrections to the MD data of Verlet[64]. Their tabulation of the equation of state and internal energy at $T^* = 1.35$ supersedes the earlier MC calculations of Verlet and Levesque[66]. Using the MD method with 864 particles, Verlet[67] has also tabulated the RDF for $r/\sigma = 0.84$ (0.04) 2.40 (0.20) 5.00 at 25 T^*, ρ^* pairs with T^* ranging from 0.591 to 3.669 and ρ^* from 0.450 to 0.880. This compilation is the most useful set of RDFs for the LJ 12-6 system currently available. Fehder[68, 69] has studied the two-dimensional system of LJ 12-6 disks by the MD method; such calculations are an aid for visualising the dynamic behaviour of a system of LJ 12-6 particles.

Recent MC calculations on the LJ 12-6 fluid are largely due to McDonald and Singer[70-72]. At supercritical temperatures ($T^* = 1.446-3.533$) and at reduced densities $\rho^* = 0.17, 0.34, 0.51$ and 0.68, they calculated the equation of state, configurational internal energy and heat capacity C_V, and the excess entropy for 32 particles by an MC procedure[73] somewhat different from the one described above. Using the conventional MC procedure for 108 particles, they studied[71] the thermodynamic properties of the liquid region for $T^* = 0.721-1.237$ and $\rho^* = 0.5948-0.9600$. In a later paper[72], McDonald and Singer report improved MC calculations for the liquid region over essentially the same range of T^* and ρ^*. They observe that the pressures obtained by Verlet[64] using the MD method appear to be some 20 atm higher than their pressures obtained by the MC technique.

The results of computer calculations for the LJ 12-6 fluid may be used to approximate the thermodynamic properties of argon. It turns out (see Wood[19]) that the results of the MD and MC calculations for the LJ 12-6 fluid with values of the interaction parameters $\varepsilon/k = 119.76$ K and $\sigma = 3.405$ Å (chosen to fit the second virial coefficient of argon at high temperatures[74]) are in generally good agreement with the properties of liquid argon, except near the critical point. Thus, the LJ 12-6 potential is a good *effective* pair potential for argon at high densities, representing the many-body interactions as a *quasi*-pair potential. With these parameter values, the triple-point temperature 83.81 K for argon corresponds to $T^* = 0.700$ and the critical temperature 150.86 K corresponds to $T^* = 1.260$.

The discrepancies between computer calculations and argon data near the critical point (see Reference 71, for example) are due to features in the MD and MC methods. The critical point is characterised by large fluctuations in density, which are suppressed in the computer calculations by the small value of N and the periodic boundary conditions. Thus, the computed thermodynamic properties near the critical point are estimated incorrectly.

McDonald and Singer[70-72] have made extensive comparisons of their MC calculations with experimental data for argon using $\varepsilon/k = 119.76$ K and $\sigma = 3.405$ Å. At supercritical temperatures[70] the agreement between MC results and argon data is very good for the equation of state and the excess entropy, fair for the configurational internal energy, and generally poorer for the configurational heat capacity C_V. When the liquid-state MC calculations[71, 72] are compared with argon data, the MC configurational internal energy is too low by 1.1–4.2%, the discrepancy increasing with temperature, and the MC pressure is on the average 23 atm too low[72]. McDonald and Singer[72] found that a small change in the value of ε/k to 117.2 K with no change in σ results in the best overall fit between the MC calculations and the argon experimental data. Since the pressures obtained by Verlet[64] with the MD method are about 20 atm above these MC results, the value $\varepsilon/k = 119.76$ K relates the MD and argon pressures very well on the average. Levesque and Vieillard–Baron[75] and McDonald and Singer[72] have carried out a series of calculations for other potential forms, such as the LJ 18-6, LJ 9-6, Kihara, and exp-6 potentials, and conclude that the LJ 12-6 potential provides the best correlation between computed and experimental properties of argon.

McDonald[76] has applied the MC method to the LJ 12-6 fluid in the NpT ensemble. In this ensemble, the temperature and pressure are fixed and the density is calculated. With $T^* = 0.967$, corresponding to 115.8 K for argon with $\varepsilon/k = 119.76$ K, and with the pressure fixed at 0.6 atm, the calculated density ρ^* is 0.7097 ± 0.0017, which compares favourably with the argon reduced value of 0.7108.

6.3.4.2 Phase transitions

Hansen and Verlet[77] have recently studied by the MC technique both the liquid–gas and solid–fluid phase transitions for an LJ 12-6 system of 864 particles. In their study of the liquid-gas co-existence region, they calculated the equation of state and internal energy along two isotherms, $T^* = 0.75$ and $T^* = 1.15$, fitted the pressure with a polynomial in ρ^*, and then calculated the Helmholtz free energy A by means of the relation

$$A = NkT \int_0^\rho [(p/\rho kT) - 1]\, d\rho/\rho \qquad (6.27)$$

By suppressing large fluctuations in the density, they were able to calculate the equation of state along the unstable isotherm on which the pressure data exhibit van der Waals loops. The liquid–gas co-existence line was then located by means of the double-tangent construction[78] made on the A v. V curve.

The resulting heat of vaporisation, pressure and liquid density at $T^* = 0.75$ and 1.15 compare favourably with those for argon with the usual values of $\varepsilon/k = 119.8$ K and $\sigma = 3.405$ Å. The densities of the gas phase for the LJ 12-6 fluid and for argon at these two temperatures do not agree at all well, largely because the LJ 12-6 potential is not a good representation of the true pair potential for argon at low densities[6].

Using the technique devised by Hoover and Ree[45] for the study of solid–fluid phase transitions, Hansen and Verlet[77] determined the melting transition for the LJ 12-6 system at $T^* = 0.75$, 1.15, 1.35 and 2.74. The resulting pressures, fluid and solid densities, and latent heats of melting agree well with the experimental data for argon, which has been conveniently summarised by Crafton[79].

With their MC calculations and the previously determined[66] critical point for an LJ 12-6 fluid ($T_c^* = 1.36$, $\rho_c^* = 0.36$), Hansen and Verlett[77] were able to construct a fairly complete T–ρ phase diagram for the LJ 12-6 system. From the T–ρ phase diagram they located the triple point at $T_t^* = 0.68 \pm 0.02$ and $\rho_t^* = 0.85 \pm 0.01$. The triple point for argon in reduced units is $T_t^* = 0.70$ and $\rho_t^* = 0.841$.

Hansen[80] has calculated the pressure, solid density and fluid density of the LJ 12-6 melting transition at high temperatures ($T^* = 2.74$, 5, 10, 20, 50 and 100). At these temperatures it is feasible to consider an inverse-power fluid (with potential $4\varepsilon(\sigma/r)^{12}$) as a zeroth-order approximation to the behaviour of the LJ 12-6 system and to treat the attractive part $[-4\varepsilon(\sigma/r)^6]$ of the potential as a perturbation. As mentioned in Section 6.3.3, one isotherm determines the properties of the unperturbed system at all T^* and ρ^*. Moreover, at these temperatures only the first-order term (see Section 6.5.1) in the perturbation is needed. Thus, the separate treatment of the two terms in the LJ 12-6 potential results in a considerable saving of computer time. Except at $T^* = 2.74$, there are no argon data for comparison.

Hansen and Weis[81] have included quantum corrections in a treatment of the liquid–gas and solid–fluid phase transitions parallel to the calculations of Hansen and Verlet[77] and have compared the results with neon data. With $\varepsilon/k = 36.76$ K and $\sigma = 2.786$ Å in the LJ 12-6 potential[82], the agreement between the MC calculations with quantum corrections and experimental data on neon is similar to the agreement reported between MC and argon results. Quantum corrections for argon were shown to be negligible.

6.3.4.3 Mixtures

The MC method has been applied to calculate the thermodynamic properties of binary mixtures of LJ 12-6 fluids by McDonald[76] and by Singer and Singer[83, 84]. McDonald and Singer[59] provide a general discussion of such calculations. In this application the MC method is preferred over the MD technique because the temperature of the system can be predetermined more precisely. In the mixture of two LJ 12-6 fluids, A and B, with interaction parameters ε_{AA}, σ_{AA} and ε_{BB}, σ_{BB}, the potential between an A–B pair was described by the LJ 12-6 form with parameters

$$\varepsilon_{AB} = (\varepsilon_{AA}\varepsilon_{BB})^{\frac{1}{2}}, \ \sigma_{AB} = \tfrac{1}{2}(\sigma_{AA}+\sigma_{BB}) \tag{6.28}$$

The only calculated properties currently available are the excess Gibbs free energy, enthalpy, and volume for a series of equimolar mixtures with $\sigma_{AA}/\sigma_{BB} = 1.00\text{--}1.12$ and $\varepsilon_{AA}/\varepsilon_{BB} = 0.810, 0.900, 1.000, 1.111$ and 1.235. More extensive calculations of this type would greatly accelerate the development of theories for mixtures.

6.4 INTEGRAL EQUATIONS FOR THE RDF

Since the RDF plays an important role in determining the bulk thermodynamic properties of a classical fluid, a theoretical expression relating $g(r;\rho,T)$ and $u(r)$ is highly desirable. Such an integral equation can be derived from equation (6.8) or (6.9) and can be expressed in several, equivalent forms[3]. However, none of these forms can be solved by known techniques unless drastic approximations are made.

Three widely-used approximations for relating $g(r)$ and $u(r)$ are the Born–Green–Yvon (BGY) integrodifferential equation, which includes the Kirkwood superposition approximation[85], and the hypernetted-chain (HNC) and Percus–Yevick (PY) integral equations. Hill[5], Rushbrooke[86], Rowlinson[32, 61], and Rice and Gray[87] discuss these theories in detail and provide references to the original articles and to numerous applications. The HNC and PY theories have also been studied in terms of a graph–theoretical analysis of the density expansion of the RDF[87]. Fitts and Smith[88] have recently analysed the BGY theory on the same basis.

The applications of these theories, as reviewed by Rowlinson[32, 61], for example, lead to the conclusion that the PY theory is a better approximation than either the HNC or the BGY theories. For this reason, few applications of the HNC theory have been published in the last few years, a notable exception being an article by Watts[89]. The BGY theory appears to predict a liquid–solid phase transition[90], although many are skeptical regarding this interpretation. If it were not for this particular unresolved feature, the BGY theory would probably receive less attention than it now does. Accordingly, this section on integral equations emphasises results from the PY theory.

6.4.1 PY integral equation

The PY equation for the hard-sphere RDF has been solved exactly by Wertheim[91, 92] and by Thiele[93]. Wertheim obtained the Laplace transform of $g(r)$ and an explicit expression for $g(r)$ itself in the region $d < r < 2d$, where d is the hard-sphere diameter. Bearman et al.[94, 95] performed a numerical inversion of the transform in the region $d < r < 6d$ and presented tables of $g(r)$ at several densities. Smith and Henderson[96] have obtained explicit analytical expressions, written in terms of complex variables, which are suitable for rapid computer evaluation of the PY hard-sphere RDF for $r < 5d$. They also investigated the effect of truncating the PY hard-sphere RDF to unity past $r = 5d$ and found that essentially no error is introduced into the compressibility integral for $\rho^* < 0.5$, but that the error becomes increasingly larger as ρ^* increases.

The PY equation of state for the hard-sphere fluid from the pressure equation (6.15) is

$$\left(\frac{pV}{NkT}\right)_p = \frac{1+2\eta+3\eta^2}{(1-\eta)^2} = \frac{1+\eta+\eta^2-3\eta^3}{(1-\eta)^3} \qquad (6.29)$$

and from the compressibility equation (6.16) is

$$kT\left(\frac{\partial\rho}{\partial p}\right)_T = \frac{(1-\eta)^4}{(1+2\eta)^2} \qquad (6.30)$$

or equivalently,

$$\left(\frac{pV}{NkT}\right)_c = \frac{1+\eta+\eta^2}{(1-\eta)^3} \qquad (6.31)$$

where, as in Section 6.3.1.1, $\eta = \pi\rho^*/6$. Both PY equations of state are in excellent agreement with MC and MD results at low densities and are in reasonably good agreement (especially that obtained from the compressibility equation) at higher densities (see Rowlinson[61] for a graphical comparison). The Carnahan–Starling (CS) equation of state (6.21) is a weighted average of equations (6.29) and (6.31)

$$\left(\frac{pV}{NkT}\right)_{CS} = \frac{1}{3}\left(\frac{pV}{NkT}\right)_p + \frac{2}{3}\left(\frac{pV}{NkT}\right)_c \qquad (6.32)$$

The PY theory does not predict the hard-sphere solid–fluid transition since both equations (6.29) and (6.31) are continuous for η less than its close-packed value of 0.7405. However, for $\eta > 0.618$ ($\rho^* > 1.18$), the PY RDF takes on negative values and is therefore physically unacceptable.

For the LJ 12-6 potential, the PY integral equation must be solved numerically, usually by a procedure due to Broyles[97]. In the liquid region an equation of state is readily obtained from the pressure equation (6.15). However, until Baxter[98] demonstrated how to make the calculation, no method was known for calculating the PY equation of state below the critical temperature from the compressibility equation (6.16). Extensive calculations of the RDF and thermodynamic quantities have been made recently by Levesque[99], Verlet and Levesque[66], Throop and Bearman[100], and Mandel, Bearman and Bearman[95]. Older work is summarised by Rowlinson[61]. Levesque and Verlet[66, 99] tabulate only thermodynamic functions, while Throop and Bearman[100] tabulate in addition the RDF for $r/\sigma = 0.90\,(0.05)\,2.75\,(0.25)\,6.00$ at densities ρ^* from 0.10 to 0.60 and temperatures T^* from 1.2 to 1.5. Mandel et al.[95] have solved the PY equation for a wide ρ^* range (0.10–0.85) at high values of T^* (1.40–2.63) and a wide T^* range (0.75–2.63) at high values of ρ^* (0.60–0.925); these solutions, in a form suitable for calculating either the RDFs or direct correlation functions $c(r)$, are available from the ASIS National Auxiliary Publications Service.

For $T^* > 1.32$, the PY equation has solutions at all densities. However, for $T^* \leqslant 1.32$ there is a range of densities where the PY equation has no solutions[99]. Thus, the PY theory gives a gas–liquid co-existence region ending at a critical point[66] where $T_c^* = 1.32$ and $\rho_c^* = 0.28$. At the critical point, the PY isothermal compressibility from equation (6.16) is infinite, a feature characteristic of real fluids. The isothermal compressibility obtained from the pressure equation (6.15) is finite everywhere. The PY critical temperature $T_c^* = 1.32$ compares favourably with the MC result[66] of 1.36 for the LJ 12-6 potential, but since computer simulation results are unreliable near the critical point (see Section 6.3.4.1), this close agreement is not meaningful.

The adequacy of the PY theory for an LJ 12-6 fluid is best ascertained by a comparison with MC and MD calculations. Mandel et al.[95] compared their PY RDF, equations of state, and internal energy with MD calculations over a range of temperatures at $\rho^* = 0.85$ and found that the RDFs and internal energies are in reasonably good agreement, but that the PY equation is incapable of yielding quantitative pressures in the liquid region. Both PY equations of state differ from the MD values by more than 100%. Watts[101] came to similar conclusions. Bearman et al.[102] have compared at different temperatures the maxima and minima in the density dependence of the heat capacity C_V as calculated from the PY theory and as determined experimentally[103] for krypton and xenon; the agreement is surprisingly close. Encouraged by this favourable comparison, Theeuwes and Bearman[104] calculated the internal pressure by differentiating the internal energy from the PY theory and found that the theoretical values agreed within computational errors with results obtained from MC calculations. From the comparison of the internal pressures of the PY theory for the LJ 12-6 potential with the internal pressures from the equation of state of liquid krypton and xenon, they concluded that the theoretical values $T_c^* = 1.285$ and $\rho_c^* = 0.285$ are suitable scaling parameters for these two substances. The theoretical and experimental pressures agree within about 10%.

Watts[105, 106] solved the Baxter[98] formulation of the PY equation for an LJ 12-6 potential truncated to zero at some predetermined values of r. His results agree qualitatively with previous PY calculations[66, 99] in the vicinity of the critical point, but the location of his critical point depends strongly on the value of r at which the potential is truncated. Moreover, his calculated compressibilities[105] differ markedly from those of Throop and Bearman[100] and he ascribed the discrepancies to Throop and Bearman's approximation in their iterative solution that $g(r) = 1$ for $r \geqslant 6\sigma$. However, Mandel and Bearman[107] pointed out and Watts[108] concurred that the major portion of the discrepancy arose because Watts used a truncated LJ 12-6 potential whereas Throop and Bearman used analytical end corrections for $r > 6\sigma$ and that the additional discrepancy may be the result of approximations in Watts' numerical integration procedure.

The equation of state from the PY theory has also been obtained from the energy equation by means of equations (6.18) and (6.19). Thus, the fourth and fifth virial coefficients for the square-well potential[7] and the LJ 12-6 potential[109] have been calculated and agree with the exact values better than those obtained from either equation (6.15) or (6.16). Barker, Henderson and Watts[110, 111] also applied this procedure with the PY RDF to calculate pV/NkT at $T^* = 0.72, 1.325, 1.35$ and 2.74 as a function of ρ^* for the LJ 12-6 potential. The agreement with the computer simulations[17, 64, 65, 71] is excellent and is much better than the PY and the PY2 (see Section 6.4.2) results obtained from either the pressure or compressibility equations.

The PY theory has also been recently applied to mixtures of hard spheres[112–116] and LJ 12-6 particles[117, 118] with some success. Since space does not permit a detailed discussion, the reader is referred to the literature.

6.4.2 Higher-order approximations

The PY and HNC theories can be extended to 'second-order' theories[86, 87] through a functional derivative approach. Although the PY2 integral equation gives more accurate results than the PY theory[61], the numerical computation is time-consuming and approaches the expense of a more accurate MD or MC calculation. For this reason, little new work has been reported since Rowlinson's review[61].

Lee, Ree and Ree[119, 120] have extended the BGY theory to 'second order' by applying the superposition approximation[85] to $g(1,2,3,4)$ in the second equation of the BGY hierarchy[5]. They have applied this BGY2 theory to calculated the virial coefficients[119] and RDF[120] for a hard-sphere fluid and have found this approach to be more accurate than the BGY or PY theories when compared with MD and MC results. It should be noted, however, that except for the original application of the Kirkwood superposition approximation to the first equation, truncation of the BGY hierarchy of integro-differential equations with a superposition approximation yields expressions for the mean force on a molecule which are not conservative[121].

6.4.3 Pressure-consistent integral equations

Since the equations of state obtained from the pressure equation (6.15) and the compressibility equation (6.16) differ in both the PY and HNC theories, Hurst[122], Carley and Lado[123], Rowlinson[124], Henderson[125], Stell[126], Morita[127], and Hutchinson and Conkie[128] have suggested approximate theories based on the PY and HNC integral equations with parameters which may be selected to achieve consistency between equations (6.15) and (6.16). Such pressure-consistent approximations have been applied to the hard-sphere fluid[122–125, 127–133] and the LJ 12-6 fluid[123, 134, 135]. The results for the hard-sphere fluid have been especially encouraging. For example, Watts and Henderson[131] have tabulated the RDF obtained by a self-consistent procedure based on the PY theory for $\rho^* = 0.10\ (0.10)\ 1.00$ at $r/d = 1.00\ (0.05)$ $2.00\ (0.10)\ 5.00$; these RDFs agree excellently with the corresponding MC results[40], Mandel and Bearman[136] have recently shown why pressure-consistent approximations should work better for the hard-sphere fluid than for the LJ 12-6 fluid.

6.5 PERTURBATION AND VARIATIONAL THEORIES

In perturbation and variational approaches to the study of liquid properties, the pair potential $u(r)$ is written as the sum of a reference potential $u_0(r)$ and a perturbing term $u_1(r)$,

$$u(r) = u_0(r) + u_1(r) \qquad (6.33)$$

The properties of the system whose molecules interact according to $u_0(r)$

are presumed to be known. The perturbation or variational theory must then determine the modification of these properties for the unperturbed system by the perturbation $u_1(r)$ in the pair potential.

General reviews of perturbation and variational theories are those of Mansoori and Canfield[137], who give a historical presentation, and of Stell[138]. More limited reviews are due to Henderson, Barker and Kim[9], and Barker and Henderson[139].

The concept of evaluating the configurational partition function by a perturbation technique was first suggested by Peierls[140]. Zwanzig[141] was the first to carry out an actual calculation. After demonstrating that the gaseous equation-of-state data for argon and for nitrogen over a wide range of temperature and density follows the form

$$\frac{pV}{NkT} = a_0(\rho) + \frac{a_1(\rho)}{NkT} + \frac{a_2(\rho)}{(NkT)^2} + \dots \qquad (6.34)$$

Zwanzig proposed β as an expansion parameter. He selected the LJ 12-6 potential for $u(r)$ and the hard-sphere potential $u_{HS}(r;d)$ as the unperturbed potential $u_0(r)$ and then expanded the ratio of the canonical-ensemble partition functions for the perturbed and unperturbed systems in powers of β, keeping only the first term. From the ratio of partition functions, the perturbation to the hard-sphere equation of state could be determined. Smith and Alder[142] and Frisch et al.[143] improved on Zwanzig's[141] calculation.

These calculations[141-143] on the gas phase suggested that the expansion in inverse powers of temperature is useful only at high temperatures. Moreover, they were not particularly satisfactory, partly because accurate results for the hard-sphere system were not available at the time and partly because the results are strongly dependent on the value of the hard-sphere diameter d, for the selection of which the procedure provides no prescription.

Realising that the failure at low temperatures of the calculations with hard spheres as the reference system could result from inadequacies either in the perturbational treatment of the attractive forces or in the treatment of the finite steepness of the repulsive potential, Barker and Henderson[56] applied Zwanzig's method to the square-well fluid. In this case the repulsive forces are identical with those for the hard-sphere reference system. Moreover, Barker and Henderson developed approximate expressions for the second term in the perturbation expansion and so were able to include one more term than Zwanzig. The perturbation calculation of the equation of state for the square-well fluid with $K = 1.5$ agrees excellently with the MC calculations of Rotenberg[54] and the unpublished MD calculations of Alder at all temperatures in the high-density region corresponding to liquid densities. Thus, the attractive forces can be adequately accounted for by a perturbation expansion and the hard-sphere potential is not suitable as an unperturbed potential in treating a realistic system with finite repulsive forces.

6.5.1 Perturbation expansion

Since the LJ 12-6 potential is an excellent effective pair potential for argon at high densities, most perturbation calculations have been done on the

LJ 12-6 system. Moreover, the properties of the hard-sphere fluid have been studied extensively by MD and MC techniques (principally because the hard-sphere potential is the simplest), making that fluid the most sensible choice for an unperturbed system. Thus, the simplest decomposition (6.33) of the LJ 12-6 potential $u(r)$ is to set $u_0(r) = u_{HS}(r;d)$, but, as discussed above, this choice leads to poor results in liquids. Since the decomposition (6.33) of $u(r)$ is quite arbitrary, other selections have been made. Ultimately, however, the unperturbed or reference system with pair potential $u_0(r)$ must be related to the hard-sphere fluid.

One possible decomposition is

$$u_0(r) = 4\varepsilon(\sigma/r)^{12}$$
$$u_1(r) = -4\varepsilon(\sigma/r)^6$$

(6.35)

which was studied by McQuarrie and Katz[144]. In this case the perturbation becomes infinitely large as $r \to 0$, thereby limiting the applicability of a perturbation treatment with equation (6.35) to high temperatures. Hansen[80] used the decomposition (6.35) in his treatment of the LJ 12-6 melting transition at high temperatures (see Section 6.3.4.2). Another possibility is to divide $u(r)$ into two parts

$$u_0(r) = u(r), u_1(r) = 0, \qquad \text{for } r < \zeta$$
$$u_0(r) = 0, \qquad u_1(r) = u(r), \qquad \text{for } r > \zeta$$

(6.36)

Usually the value σ is selected[145] for the division point ζ. Still another alternative is to divide the potential $u(r)$ into its repulsive $(du(r)/dr < 0)$ and attractive $(du(r)/dr > 0)$ parts

$$u_0(r) = u(r)+\varepsilon, u_1(r) = -\varepsilon, \qquad \text{for } r < r_m$$
$$u_0(r) = 0, \qquad u_1(r) = u(r), \qquad \text{for } r > r_m$$

(6.37)

as used by Weeks, Chandler and Andersen[146, 147]. The addition of ε to $u_0(r)$ in equation (6.37) insures that the reference system pair potential obeys the thermodynamic requirement $u_0(r) \to 0$ as $r \to \infty$.

With the selection of $u_0(r)$ and $u_1(r)$, the calculation of the properties of the liquid under consideration consists of two parts, the tasks of relating the properties of the reference system to the hard-sphere fluid and of determining the perturbation due to $u_1(r)$ on the reference system. A systematic approach to this two-part problem lies in the introduction of a family of potentials $v(r;\alpha,\lambda)$ characterised by the two parameters α and λ. The potentials $v(r;\alpha,\lambda)$ have the following properties

$$v(r;0,0) = u_{HS}(r;d)$$
$$v(r;1,0) = u_0(r)$$
$$v(r;1,\lambda) = u_0(r)+\lambda u_1(r)$$
$$v(r;1,1) = u_0(r)+u_1(r) = u(r)$$

(6.38)

Thus, the parameter α is associated with the repulsive part of the pair potential $u(r)$ and λ with the attractive tail.

A convenient computational starting point is to expand the RDF $g(r;\rho,T;\alpha,\lambda)$ for a system with pair potential $v(r;\alpha,\lambda)$ as a double power series in α and λ about the hard-sphere ($\alpha = \lambda = 0$) RDF $g_{HS}(r;\rho;d)$. However, since $y_{HS}(r;d)$ is a continuous function at $r = d$ whereas $g_{HS}(r;d)$ is not, it is preferable to expand the function $y(r)$ in powers of α about its hard-sphere value rather than $g(r)$. To first order the expansion is

$$g(r;\rho,T;\alpha,\lambda) = g(r;\alpha,0) + \lambda\left(\frac{\partial g}{\partial \lambda}\right)_{\rho,T,\alpha;\,\alpha=\lambda=0} + \ldots$$

$$= \exp\left[-\beta v(r;\alpha,0)\right]\left[y_{HS}(r;d) + \alpha\left(\frac{\partial y}{\partial \alpha}\right)_{\rho,T,\lambda;\,\alpha=\lambda=0}\right]$$

$$+ \lambda\left(\frac{\partial g}{\partial \lambda}\right)_{\rho,T,\alpha;\,\alpha=\lambda=0} + \ldots \tag{6.39}$$

Buff and Schindler[148] have pointed out that the use of the canonical ensemble in perturbation expansions with subsequent passage to an infinite system may give incorrect results for fluid systems unless due care is exercised. Such difficulties are avoided with the use of the grand-canonical ensemble. Accordingly, following the procedures of Buff and Schindler[148] and of Schofield[149], the derivative $(\partial g/\partial \lambda)_{\rho,T,\alpha}$ is obtained[150] from equation (6.9),

$$\left(\frac{\partial g}{\partial \lambda}\right)_{\rho,T,\alpha} = -\beta g(1,2)\frac{\partial v_{12}}{\partial \lambda} - 2\beta\rho\int\frac{\partial v_{13}}{\partial \lambda}g(1,2,3)\,d3$$

$$-\tfrac{1}{2}\beta\rho^2\int\frac{\partial v_{34}}{\partial \lambda}\left[g(1,2,3,4) - g(1,2)g(3,4)\right]d3\,d4$$

$$+\frac{1}{2\rho}\left(\frac{\partial \rho}{\partial p}\right)_T\left(\frac{\partial \rho^2 g(1,2)}{\partial \rho}\right)_T\frac{\partial}{\partial \rho}\left(\rho^2\int\frac{\partial v_{34}}{\partial \lambda}g(3,4)\,d3\right) \tag{6.40}$$

For the moment we leave the term $(\partial y/\partial \alpha)_{\alpha=\lambda=0}$ unevaluated and defer its discussion until Section 6.5.3.

Evaluating equation (6.40) at $\lambda = 0$, substituting the result into equation (6.39), and then setting $\alpha = 0$ and $\lambda = 1$, we obtained the RDF $g(r)$ with the hard-sphere fluid as the reference system,

$$g(1,2) = g_{HS}(1,2) - \beta u_1(1,2)g_{HS}(1,2) - 2\beta\rho\int u_1(1,3)g_{HS}(1,2,3)\,d3$$

$$-\tfrac{1}{2}\beta\rho^2\int u_1(3,4)\left[g_{HS}(1,2,3,4) - g_{HS}(1,2)g_{HS}(3,4)\right]d3\,d4$$

$$+\frac{1}{2\rho}\left(\frac{\partial \rho}{\partial p}\right)_{HS}\left(\frac{\partial \rho^2 g_{HS}(1,2)}{\partial \rho}\right)\frac{\partial}{\partial \rho}\left(\rho^2\int u_1(3,4)g_{HS}(3,4)\,d3\right)$$

$$+ \ldots \tag{6.41}$$

Smith, Henderson and Barker[151] used equation (6.41) to calculate the RDF for the square-well fluid.

To obtain the equation of state, one can substitute equation (6.41) into either equation (6.15) or (6.16). However, as discussed in Section 6.2.2, the determination of the Helmholtz free energy A by integrating equation (6.14) and then calculating the equation of state from $p = -(\partial A/\partial V)_T$ generally gives more satisfactory results when an approximate RDF is used. The resulting expression for A in terms of the hard-sphere fluid as a reference state is[7]

$$A - A_{HS} = \tfrac{1}{2}\rho NkT \int_0^\beta \int_V u_1(1, 2)g(1, 2; \rho, T)\, d2\, d\beta \qquad (6.42)$$

where the observation that a real fluid at high temperatures ($\beta = 0$) behaves like a hard-sphere fluid has been used. Substitution of equation (6.41) into equation (6.42) gives

$$A = A_{HS} + \sum_{n=1}^{\infty} A_n \beta^n$$

$$\frac{A_1}{NkT} = \frac{1}{2}\rho \int u_1(1, 2)g_{HS}(1, 2)\, d2$$

$$\frac{A_2}{NkT} = -\frac{1}{4}\rho \int [u_1(1, 2)]^2\, g_{HS}(1, 2)\, d2$$

$$-\frac{1}{2}\rho^2 \int u_1(1, 2)u_1(1, 3)g_{HS}(1, 2, 3)\, d2\, d3 \qquad (6.43)$$

$$-\frac{1}{8}\rho^3 \int u_1(1, 2)u_1(3, 4)\left[g_{HS}(1, 2, 3, 4) - g_{HS}(1, 2)g_{HS}(3, 4)\right] d2\, d3\, d4$$

$$+\frac{1}{8}kT\left(\frac{\partial \rho}{\partial p}\right)_{HS}\left\{\frac{\partial}{\partial \rho}\left[\rho^2 \int u_1(1, 2)g_{HS}(1, 2)\, d2\right]\right\}^2$$

. . . .

The last term in the expression for A_2 in equation (6.43) does not appear in Zwanzig's[141] expansion obtained from the petit-canonical ensemble partition function and was first presented, but not derived, by Smith, Henderson and Barker[152]. Equation (6.43) may also be obtained by expanding $g(r; \rho, T)$ in powers of β about $\beta = 0$ using the first derivative $(\partial g/\partial \beta)_\rho$ as obtained by Schofield[149].

6.5.2 Approximations to A_2

Since the three- and four-particle correlation functions for a hard-sphere fluid are not known and are too complicated to obtain by computer simulation methods, the term A_2 must be approximated if it is to be included at all in a calculation. Barker and Henderson[56] derived an expression for A_2 which is formally equivalent to equation (6.43), namely,

$$\frac{A_2}{NkT} = -\tfrac{1}{2}\sum_{i,j}\left\{\langle v_i v_j\rangle - \langle v_i\rangle\langle v_j\rangle\right\}u_1(r_i)u_1(r_j) \qquad (6.44)$$

where the brackets $\langle \ldots \rangle$ indicate averages over the hard-sphere reference system of the numbers $v_1, \ldots, v_i, \ldots$ of intermolecular distances lying in the intervals r_0 to $r_1, \ldots, r_{i-1}$ to $r_i, \ldots$, the intervals being sufficiently narrow that $u_1(r)$ is effectively constant within each interval. From equation (6.44) Barker and Henderson[56] derived an approximation for A_2 which they called the *macroscopic compressibility approximation* (MCA),

$$\frac{A_2}{NkT} \approx -\pi\rho kT \left(\frac{\partial\rho}{\partial p}\right)_{HS} \int_\sigma^\infty [u_1(r)]^2 g_{HS}(r; d) r^2 \, dr \qquad (6.45)$$

and proposed without derivation another expression which they called the *local compressibility approximation* (LCA),

$$\frac{A_2}{NkT} \approx -\pi\rho kT \left(\frac{\partial\rho}{\partial p}\right)_{HS} \frac{\partial}{\partial\rho} \left\{ \rho \int^\infty [u_1(r)]^2 g_{HS}(r; d) r^2 \, dr \right\} \qquad (6.46)$$

Using the potential decomposition (6.36), Barker, Henderson and Smith[56, 145, 153–155] applied these approximations to the square-well and LJ 12-6 fluids. They found that the inclusion of an approximate A_2 improved the agreement between perturbation calculations and computer simulation results and that the LCA is somewhat better than the MCA in this respect. They also found that these approximations agree poorly with A_2 calculated from equation (6.44) by the Monte Carlo method. From an exact expression for A_2 similar to equation (6.44), Verlet *et al.*[41, 65] studied the convergence of the expansion (6.43) and found the first-order term 17 times larger than the second-order term for the potential decomposition (6.36), but 200 times larger for the potential decomposition (6.37) for a state near the triple point of an LJ 12-6 fluid. Thus, in order to minimise the errors resulting from the approximation of A_2, the selection (6.37) is to be preferred.

Praestgaard and Toxvaerd[156] have presented a derivation of the LCA expression by showing what approximation in the exact A_2 leads to the LCA for that term. When they applied the same approximation to each of the higher-order terms in the expansion of A, they were able to sum the entire expansion to obtain

$$\frac{A - A_{HS}}{NkT} = -2\pi\beta \int_0^\infty r^2 \, dr \int_{\mu_0}^{\mu_1} \rho_{HS}(\mu) g_{HS}(r; \rho_{HS}(\mu)) \, d\mu \qquad (6.47)$$

where μ_0 is the chemical potential of the hard-sphere reference system with density $\rho = N/V$, $\rho_{HS}(\mu)$ is the number density for a hard-sphere system with chemical potential μ, and $\mu_1(r) = \mu_0 - u_1(r)$. They applied this approximation to the square-well fluid[156] and to the LJ 12-6 fluid[157] and found little difference between their results and the LCA results at higher temperatures. Only at very low temperatures did equation (6.47) with its approximate inclusion of all higher-order terms differ significantly from the LCA results. In the case of the one-dimensional square-well fluid, where exact values of A and p are known, equation (6.47) is far superior to the LCA expression at very low temperatures.

Moffat and Kozak[158] have analysed the LCA in terms of a density expansion with coefficients expressed in graph-theoretic form. By comparing the density expansion of the equation of state in the LCA with a similar, but exact expansion of Kozak and Rice[159], Moffat and Kozak[158] found that the LCA neglects many diagrams in the coefficient of each power of ρ and that numerically each coefficient (except the first) is poorly approximated. Moreover, the sum of the first few terms in the exact expansion does not describe, even approximately, the results predicted by the complete LCA expression.

Their conclusion is that the LCA for A_2 is successful because it introduces a closed expression which effectively approximates the sum of all coefficients in the density expansion.

Smith, Henderson, and Barker[152] have also evaluated A_2 in equation (6.43) using the Kirkwood superposition approximation[85] (SA)

$$g(1,2,3) = g(1,2)g(1,3)g(2,3)$$
$$g(1,2,3,4) = g(1,2)g(1,3)g(1,4)g(2,3)g(2,4)g(3,4) \qquad (6.48)$$

They substituted equations (6.48) into A_2, justified the omission of all reducible cluster integrals, and calculated the irreducible cluster integrals numerically using PY results for the hard-sphere reference system. For the square-well fluid, they found that although the SA results agree better with exact results than do LCA calculations, the SA results are poor at high densities. For the LJ 12-6 fluid, SA calculations are in good agreement with exact results for A whereas LCA results are poor; SA and LCA pressures are equally good. For both the square-well fluid and the LJ 12-6 fluid, the equation of state obtained from either the SA or LCA expression is in much better agreement with computer simulation results than is the PY equation of state from either equation (6.15) or (6.16).

6.5.3 Reference system in terms of hard-sphere fluid

To account for the finite steepness of the repulsive part of $u(r)$, we consider the expansion of $v(r; \alpha, \lambda)$ in α. That this approach is appropriate for treating the repulsive forces was first shown by Rowlinson[160], who considered the case where $v(r; \alpha, 0) = 4\varepsilon(\sigma/r)^{12/\alpha}$, thereby making $u_0(r)$ that of equation (6.35). When $\alpha = 0$, Rowlinson's $v(r; \alpha, 0)$ becomes $u_{HS}(r; d)$.

Generalising Rowlinson's treatment, Barker and Henderson[145] used the potential decomposition of equation (6.36) and considered the parametrisation

$$
\begin{aligned}
v(r; \alpha, \lambda) &= u\left(d + \frac{r-d}{\alpha}\right), & r < d + \alpha(\zeta - d) \\
&= 0, & d + \alpha(\zeta - d) < r < \zeta \\
&= \lambda u(r) & \zeta < r
\end{aligned}
\qquad (6.49)
$$

which satisfies the properties (6.38). In addition to α and λ, the Barker–Henderson potential introduces two other parameters, the point ζ and the hard-sphere diameter d. Actually, Barker and Henderson[145] set

$$
\exp(-\beta v) = \left[1 - H\left(d + \frac{r-d}{\alpha} - \zeta\right)\right]\exp\left[-\beta u\left(d + \frac{r-d}{\alpha}\right)\right]
$$
$$
+ H\left(d + \frac{r-d}{\alpha} - \zeta\right) + H(r - \zeta)\{\exp[-\lambda\beta u(r)] - 1\} \qquad (6.50)
$$

where $H(x)$ is the Heaviside step function, which is 0 for $x < 0$ and 1 for $x > 0$. Equation (6.50) is not precisely equivalent to equation (6.49).

Since first-order terms are given correctly in the canonical ensemble, Barker and Henderson[145] substituted equation (6.50) into equation (6.1) and expanded $A(= -kT \ln Q_N)$ in a double Taylor series in α and λ to obtain

$$\frac{A - A_{HS}}{NkT} = -\alpha 2\pi\rho d^2 g_{HS}(d)\left(d - \int_0^\zeta \{1 - \exp[-\beta u(z)]\}\, dz\right)$$
$$+ \lambda 2\pi\rho\beta \int_\zeta^\infty g_{HS}(r)u(r)r^2\, dr + \dots \qquad (6.51)$$

They then set $\zeta = \sigma$, $\lambda = 1$, and chose d to make the first-order term in α vanish,

$$d = \int_0^\zeta \{1 - \exp[-\beta u(z)]\}\, dz \qquad (6.52)$$

Thus, the diameter d depends on T, but not on ρ. With this choice for d, equation (6.51) becomes identical with equation (6.43) when both are carried to first order only. Higher-order terms in λ can be obtained by the procedure that led to equation (6.43), but higher-order terms in α and cross terms in $\alpha^l\lambda^n$ have not been considered. For applications to systems at very high temperatures Henderson and Barker[161] found that the condition that A be insensitive to small variations in ζ yields better results than letting $\zeta = \sigma$. At lower temperatures this condition gives $\zeta = \sigma$. In addition to applications to the fluid state of square-well and LJ 12-6 fluids, Henderson and Barker[162] successfully applied equations (6.51) and (6.52) along with a cell model for the solid phase to study the melting curve of argon.

Another possibility for $v(r;\alpha,0)$ is

$$\exp[-\beta v(r;\alpha,0)] = \alpha \exp[-\beta u_0(r)] + (1-\alpha)\exp[-\beta u_{HS}(r;d)] \qquad (6.53)$$

Gubbins et al.[150] used equation (6.53) in equation (6.1), expanded A in a Taylor series in α, and found that if

$$\int_0^\infty \{\exp[-\beta u_0(r)] - \exp[-\beta u_{HS}(r;d)]\}\, y_{HS}(r;\rho)r^2\, dr = 0 \qquad (6.54)$$

then the term linear in α vanishes. Andersen, Weeks, and Chandler[163] have published an equivalent derivation. Both derivations[150, 163] are accompanied by expressions for the neglected higher-order terms in α. Equation (6.54) may be used to determine the diameter d of the hard-sphere system which approximates the unperturbed system and gives a value which is dependent on both ρ and T. The derivation quoted here applies to any decomposition of $u(r)$. If desired, the perturbation $u_1(r)$ may also be included in equation (6.53) by means of the parameter λ.

Weeks, Chandler and Andersen[146, 147] observed in the MD calculations of Verlet[67] that at high densities ($\rho^* \geqslant 0.65$) the RDF for an LJ 12-6 fluid is closely approximated by the RDF for the potential $u_0(r)$ of equation (6.37). Accordingly, they concluded that the decomposition (6.37) of $u(r)$ into $u_0(r)$ and $u_1(r)$ should be the most favourable for the perturbation expansion. Using empirical arguments, they were the first to propose equation (6.54) with $u_0(r)$ given by equation (6.37). To test these proposals, Weeks, Chandler and Andersen[147] calculated the thermodynamic properties of the LJ 12-6 fluid using for A the expression

$$A - A_{HS} = 2\pi\rho N \int u_1(r) y_{HS}(r;d) \exp[-\beta u_0(r)] r^2 \, dr \qquad (6.55)$$

with $d(\rho,T)$ determined by equation (6.54). Their initial use of PY values for the hard-sphere system introduced some errors[41, 164, 165]. Therefore, they repeated these calculations[166] using the MC $g_{HS}(r;d)$ as presented by Verlet and Weis[41]. The agreement between their perturbation calculations and computer simulation results is excellent. Verlet and Weis[41] and Andersen, Weeks and Chandler[163] have studied in further detail the use of $u_0(r)$ as given in equation (6.37) and of d as given by equation (6.54).

Using equation (6.37) to define $u_0(r)$, Gubbins et al.[150] calculated the RDF for the LJ 12-6 fluid. They considered three expansions of $y(r)$ about $y_{HS}(r;d)$ in equation (6.39). The first expansion used essentially the Barker–Henderson parameterisation (6.50) so that $d(T)$ is given by equation (6.52) with $\zeta = r_m$. The second expansion used equation (6.53) with $d(\rho, T)$ given by equation (6.54), while the third expansion involved setting the reference system equation of state obtained from equation (6.15) equal to the exact result. Their calculated RDFs agree well with MD results, the third expansion procedure being the best.

Perturbation theory has been reformulated to apply to binary mixtures of hard spheres by Henderson and Barker[167] and of LJ 12-6 particles by Leonard, Henderson and Barker[168] and by Weres and Rice[169]. Recent applications of perturbation theory to hard-sphere mixtures include those of Smith[170] and of Henderson and Leonard[171]. As with the applications of PY theory to mixtures, we omit a detailed discussion.

6.5.4 The γ expansion

An alternative expansion scheme for the Helmholtz free energy A is known as the γ expansion and was developed by Hemmer and Hauge[172, 173] and by Lebowitz, Stell and Baer[174], all of whom refer to earlier work. In this perturbation approach the parameterised pair potential is

$$v(r;\gamma) = u_0(r) + \gamma^3 u_1(\gamma r) \qquad (6.56)$$

For $\gamma = 1$, the potential $v(r;\gamma)$ becomes the pair potential $u(r)$ for the system under consideration. In the limit $\gamma \to 0$, one has an infinitely weak and long-ranged attractive potential and the equation of state is determined by applying the Maxwell equal-area construction[78] to the van der Waals type equation

$$\lim_{\lambda \to 0} p(\rho;\lambda) = p_0(\rho) - a\rho^2 \qquad (6.57)$$

where $p_0(\rho)$ is the pressure of the reference system at number density ρ and

$$a = -2\pi \int u_1(r) r^2 \, dr \qquad (6.58)$$

Corrections to the van der Waals equation, in the form of expansions in powers of γ, have been explicitly obtained in three-dimensions to order γ^3 by Hemmer[172], to order γ^5 by Zittartz[175], and to order γ^6 by Lebowitz, Stell and Baer[174], who also developed a method for obtaining the expansion to all orders of γ. Jalickee, Siegert and Vezzetti[176, 177] have considered corrections in the form of an expansion in the fugacity in the grand-canonical ensemble and have related their results to the γ expansion. Little in the way of numerical calculations for explicit potentials has been done. Although Stell[138] alludes to applications to the LJ 12-6 fluid, none have been published as yet.

Keeping only terms to order γ^3, Stell[138] has derived for the RDF two approximations, which are too involved to discuss here. Numerical applications of these approximations have not been reported.

6.5.5 Variational procedure

It can be readily shown[178] that the first integral in the expansion (6.43) provides a rigorous upper bound on the Helmholtz free energy

$$A \leqslant A_{HS} + \beta A_1 \qquad (6.59)$$

Minimisation of the right-hand side of equation (6.59) can be used as a criterion for determining the diameter $d(\rho, T)$ of the hard-sphere reference system. Mansoori and Canfield[179] (see also Mansoori and Canfield's review[137]) have carried out an elaborate variational calculation on the LJ 12-6 fluid using the PY theory for the unperturbed hard-sphere system; their results agree well with MC and MD calculations. Rasaiah and Stell[180, 181] also considered upper bounds on A and made calculations[180] similar to those of Mansoori and Canfield. The variational approach has been extended to binary mixtures of LJ 12-6 fluids by Mansoori and Leland[182] (see also Reference 137). Rudd and Frisch[183] have applied the procedure to binary mixtures of hard-sphere and square-well fluids.

In their original treatment Mansoori and Canfield[179] proposed that higher-order inequalities could be obtained by truncating the expansion (6.43) after terms other than A_1. However, Ashurst and Hoover[184] pointed out that such inequalities do not necessarily bound the true free energy. In a reply to Ashurst and Hoover[184], Mansoori and Canfield[185] derived a new higher-order inequality, but this inequality has not yet been applied to the LJ 12-6 fluid.

Finally, we note that Henderson and Barker[186] have criticised the variational procedure, claiming that in general the computations are more difficult and the results poorer than a perturbation treatment.

6.6 THREE-BODY INTERMOLECULAR FORCES

Although most of the published work on the statistical-thermodynamic theory of liquids is based on the assumption of a pairwise-additive potential of intermolecular forces, there is now substantial evidence that many-body forces are important in dense systems and should not be neglected. In a consideration of non-additive intermolecular potentials, the present state of knowledge allows only the inclusion of three-body forces. Thus, the potential Φ in equation (6.3) is truncated after the last term shown.

The triplet contribution w_{123} to the London dispersion forces was first obtained by Axilrod and Teller[187] using third-order quantum-mechanical perturbation theory. The triple–dipole contribution to this three-body interaction is

$$w_{123} = v(1 + 3 \cos \theta_1 \cos \theta_2 \cos \theta_3)/r_{12}^3 r_{23}^3 r_{13}^3 \qquad (6.60)$$

where θ_1, θ_2, θ_3 are the interior angles of the three-atom triangle and v is a positive constant, which for argon is 73.2×10^{-109} J m^9 as determined, but not published, by Leonard and quoted by Barker, Henderson and Smith[188]. Although equation (6.60) is the asymptotic form for w_{123} and does not properly represent the triplet potential at close distances (see Reference 6), this form appears to be adequate for the study of liquid argon. Thus, Barker, Henderson and Smith[188, 189] calculated the thermodynamic properties of argon by means of the Barker–Henderson perturbation method (equations (6.51) and (6.52)) using the Barker–Pompe pair potential (see below) and the triplet potential (6.60) and found good agreement with experiment.

Coupled with the inclusion of the triplet potential in a statistical-thermo-dynamic calculation is the necessity for obtaining the *true* pair potential $u(r)$ for a pair of interacting atoms. It is well established[6] that the LJ 12-6 potential and many other similar expressions are not *true* pair potentials. Accordingly, Barker and Pompe[190] proposed an analytic form for the true $u(r)$ of argon with 11 parameters, which were selected to reproduce numerous data which depend only on pairwise interactions. Bobetic and Barker[191] (see also Barker, Bobetic and Pompe[192]) added two more parameters and re-evaluated the pair potential using additional data. Barker, Fisher and Watts[193] calculated the thermodynamic properties of liquid argon by MD and MC techniques including the triplet potential (6.60) and found that a weighted linear combination of the Barker–Pompe and Bobetic–Barker potentials for $u(r)$ give excellent agreement with experiment.

A number of recent studies have been directed to the determination of an *effective* pair potential from the true pair potential and the triplet potential. Such an effective pair potential is not unique and its formulation depends on the physical property that is represented by the statistical-mechanical formula. Thus, Sinanoğlŭ[194, 195] has considered an effective pair potential based on the internal energy equation (6.14) and Chen and Present[196] one based on the RDF. The use of the LJ 12-6 potential as an effective pair potential is based on a comparison of computer simulation calculations with experiment.

Rushbrooke and Silbert[197], Rowlinson[198], Casanova et al.[199], and Dulla et al.[200] have considered the inclusion of three-body forces in the HNC and PY theories. Such considerations become quite involved and are beyond the scope of this review.

References

1. Temperley, H. N. V., Rowlinson, J. S. and Rushbrooke, G. S., eds. (1968). *Physics of Simple Liquids*. (Amsterdam: North-Holland Publishing Co.)
2. Neece, G. A. and Widom, B. (1969). *Ann. Rev. Phys. Chem.*, **20**, 167
3. Münster, A. (1968, 1972). *Statistical Thermodynamics*, Vol. I and II. (Heidelberg: Springer-Verlag Berlin)
4. Henderson, D., ed. (1971). *Physical Chemistry, An Advanced Treatise*, Vol. VIII, parts A and B, *Liquid State*. (New York: Academic Press)
5. Hill, T. L. (1956). *Statistical Mechanics*. (New York: McGraw-Hill Book Co.)
6. Fitts, D. D. (1966). *Ann. Rev. Phys. Chem.*, **17**, 59
7. Chen, M., Henderson, D. and Barker, J. A. (1969). *Can. J. Phys.*, **47**, 2009
8. Eyring, H. and Jhon, M. S. (1969). *Significant Liquid Structures*. (New York: John Wiley and Sons)
9. Henderson, D., Barker, J. A. and Kim, S. (1969). *Int. J. Quantum Chem.*, **3S**, 265
10. Rowlinson, J. S. (1971). *Trans. Faraday Soc.*, **67**, 576
11. Alder, B. J. and Wainwright, T. (1958). *Proceedings of the International Symposium on Transport Processes in Statistical Mechanics, Brussels*, 97. (New York: Interscience Publishers, Inc.)
12. Alder, B. J. and Wainwright, T. E. (1957). *J. Chem. Phys.*, **27**, 1208
13. Wainwright, T. E. and Alder, B. J. (1958). *Nuovo Cimento Suppl.*, **9**, 116
14. Alder, B. J. and Wainwright, T. E. (1959). *J. Chem. Phys.*, **31**, 459
15. Metropolis, N., Rosenbluth, A. W., Rosenbluth, M. N., Teller, A. H. and Teller, E. (1953). *J. Chem. Phys.*, **21**, 1087
16. Wood, W. W. and Jacobson, J. D. (1957). *J. Chem. Phys.*, **27**, 1207
17. Wood, W. W. and Parker, F. R. (1957). *J. Chem. Phys.*, **27**, 720
18. Wood, W. W., Parker, F. R. and Jacobson, J. D. (1958). *Nuovo Cimento Suppl.*, **9**, 133
19. Wood, W. W. (1968). Reference 1, 115
20. Hoover, W. G. and De Rocco, A. G. (1962). *J. Chem. Phys.*, **36**, 3141
21. Beyerlein, A. L., Rudd, W. G., Salsburg, Z. W. and Buynoski, M. (1970). *J. Chem. Phys.*, **53**, 1532
22. Tonks, L. (1936). *Phys. Rev.*, **50**, 955
23. Rowlinson, J. S. (1964). *Mol. Phys.*, **7**, 593
24. Hemmer, P. C. (1965). *J. Chem. Phys.*, **42**, 1116
25. Ree, F. H. and Hoover, W. G. (1964). *J. Chem. Phys.*, **40**, 939
26. Ree, F. H. and Hoover, W. G. (1967). *J. Chem. Phys.*, **46**, 4181
27. Hoover, W. G. and Alder, B. J. (1967). *J. Chem. Phys.*, **46**, 686
28. Chae, D. G., Ree, F. H. and Ree, T. (1969). *J. Chem. Phys.*, **50**, 1581
29. Wood, W. W. (1968). *J. Chem. Phys.*, **48**, 415
30. Wood, W. W. (1970). *J. Chem. Phys.*, **52**, 729
31. Alder, B. J. and Wainwright, T. E. (1960). *J. Chem. Phys.*, **33**, 1439
32. Rowlinson, J. S. (1965). *Rep. Progr. Phys.*, **28**, 169
33. Katsura, S. and Abe, Y. (1963). *J. Chem. Phys.*, **39**, 2068
34. Rowlinson, J. S. (1964). *Proc. Roy. Soc. (London)*, **A279**, 147
35. Kilpatrick, J. E. and Katsura, S. (1966). *J. Chem. Phys.*, **45**, 1866
36. Carnahan, N. F. and Starling, K. E. (1969). *J. Chem. Phys.*, **51**, 635
37. Carnahan, N. F. and Starling, K. E. (1970). *J. Chem. Phys.*, **53**, 600
38. Alder, B. J, and Hecht, C. E. (1969). *J. Chem. Phys.*, **50**, 2032
39. Ree, F. H., Keeler, R. N. and McCarthy, S. L. (1966). *J. Chem. Phys.*, **44**, 3407
40. Barker, J. A. and Henderson, D. (1971). *Mol. Phys.*, **21**, 187
41. Verlet, L. and Weis, J. J. (1972). *Phys. Rev.*, **5**, A939
42. Alder, B. J., Hoover, W. G. and Wainwright, T. E. (1963). *Phys. Rev. Lett.*, **11**, 241
43. Rudd, W. G., Salsburg, Z. W., Yu, A. P. and Stillinger, F. H., Jr. (1968). *J. Chem. Phys.*, **49**, 4857
44. Alder, B. J., Hoover, W. G. and Young, D. A. (1968). *J. Chem. Phys.*, **49**, 3688
45. Hoover, W. G. and Ree, F. H. (1967). *J. Chem. Phys.*, **47**, 4873
46. Hoover, W. G. and Ree, F. H. (1968). *J. Chem. Phys.*, **49**, 3609
47. Alder, B. J. and Wainwright, T. E. (1962). *Phys. Rev.*, **127**, 359
48. Meeron, E. (1970). *Phys. Rev. Lett.*, **25**, 152

49. Wood, W. W. (1971). *Phys. Rev. Lett.*, **26**, 225
50. Barker, J. A. and Henderson, D. (1967). *Can. J. Phys.*, **45**, 3959
51. McQuarrie, D. A. (1964). *J. Chem. Phys.*, **40**, 3455
52. Katsura, S. and Nishihara, K. (1969). *J. Chem. Phys.*, **50**, 3579
53. Katsura, S. and Tago, Y. (1968). *J. Chem. Phys.*, **48**, 4246
54. Rotenberg, A. (1965). *J. Chem. Phys.*, **43**, 1198
55. Lado, F. and Wood, W. W. (1968). *J. Chem. Phys.*, **49**, 4244
56. Barker, J. A. and Henderson, D. (1967). *J. Chem. Phys.*, **47**, 2856
57. Hoover, W. G., Ross, M., Johnson, K. W., Henderson, D., Barker, J. A. and Brown, B.C. (1970). *J. Chem. Phys.*, **52**, 4931
58. Hoover, W. G., Gray, S. G. and Johnson, K. W. (1971). *J. Chem. Phys.*, **55**, 1128
59. McDonald, I. R. and Singer, K. (1970). *Quart. Rev. Chem. Soc.*, **24**, 238
60. Hoover, W. G. and Ross, M. (1971). *Contemp. Phys.*, **12**, 339
61. Rowlinson, J. S. (1968). Reference 1, 59
62. Kim, S., Henderson, D. and Oden, L. (1969). *Trans. Faraday Soc.*, **65**, 2308
63. Rahman, A. (1964). *Phys. Rev.*, **136**, A405
64. Verlet, L. (1967). *Phys. Rev.*, **159**, 98
65. Levesque, D. and Verlet, L. (1969). *Phys. Rev.*, **182**, 307
66. Verlet, L. and Levesque, D. (1967). *Physica*, **36**, 254
67. Verlet, L. (1968). *Phys. Rev.*, **165**, 201
68. Fehder, P. L. (1969). *J. Chem. Phys.*, **50**, 2617
69. Fehder, P. L. (1970). *J. Chem. Phys.*, **52**, 791
70. McDonald, I. R. and Singer, K. (1967). *J. Chem. Phys.*, **47**, 4766
71. McDonald, I. R. and Singer, K. (1967). *Discuss. Faraday Soc.*, **43**, 40
72. McDonald, I. R. and Singer, K. (1969). *J. Chem. Phys.*, **50**, 2308; **52**, 2166
73. Singer, K. (1966). *Nature, (London)*, **212**, 1449
74. Michels, A., Wijker, H. and Wijker, H. (1949). *Physica*, **15**, 627
75. Levesque, D. and Vieillard-Baron, J. (1969). *Physica*, **44**, 345
76. McDonald, I. R. (1969). *Chem. Phys. Lett.*, **3**, 241
77. Hansen, J.-P. and Verlet, L. (1969). *Phys. Rev.*, **184**, 151
78. Rowlinson, J. S. (1969). *Liquids and Liquid Mixtures*, 2nd edn., 77. (London: Butterworth and Co.)
79. Crafton, R. G. (1971). *Phys. Lett.*, **36A**, 121
80. Hansen, J.-P. (1970). *Phys. Rev.*, **2**, A221
81. Hansen, J.-P. and Weis, J. J. (1969). *Phys. Rev.*, **188**, 314
82. Brown, J. S. (1966). *Proc. Phys. Soc. London*, **89**, 987
83. Singer, K. (1969). *Chem. Phys. Lett.*, **3**, 164
84. Singer, J. V. L. and Singer, K. (1970). *Mol. Phys.*, **19**, 279
85. Kirkwood, J. G. (1935). *J. Chem. Phys.*, **3**, 300
86. Rushbrooke, G. S. (1968). Reference 1, 25
87. Rice, S. A. and Gray, P. (1965). *The Statistical Mechanics of Simple Liquids*. (New York: Interscience Publishers)
88. Fitts, D. D. and Smith, W. R. (1971). *Mol. Phys.*, **22**, 625
89. Watts, R. O. (1969). *J. Chem. Phys.*, **50**, 1358
90. Kirkwood, J. G. and Monroe, E. (1942). *J. Chem. Phys.*, **10**, 394
91. Wertheim, M. S. (1963). *Phys. Rev. Lett.*, **10**, 321
92. Wertheim, M. S. (1964). *J. Math. Phys.*, **5**, 643
93. Thiele, E. (1963). *J. Chem. Phys.*, **39**, 474
94. Throop, G. J. and Bearman, R. J. (1965). *J. Chem. Phys.*, **42**, 2408
95. Mandel, F., Bearman, R. J. and Bearman, M. Y. (1970). *J. Chem. Phys.*, **52**, 3315
96. Smith, W. R. and Henderson, D. (1970). *Mol. Phys.*, **19**, 411
97. Broyles, A. A. (1961). *J. Chem. Phys.*, **35**, 493
98. Baxter, R. J. (1967). *J. Chem. Phys.*, **47**, 4855
99. Levesque, D. (1966). *Physica*, **32**, 1985
100. Throop, G. J. and Bearman, R. J. (1966). *Physica*, **32**, 1298
101. Watts, R. O. (1969). *Can. J. Phys.*, **47**, 2709
102. Bearman, R. J., Theeuwes, F., Bearman, M. Y., Mandel, F. and Throop, G. J. (1970). *J. Chem. Phys.*, **52**, 5486
103. Theeuwes, F. and Bearman, R. J. (1970). *J. Chem. Thermodyn.*, **2**, 513
104. Theeuwes, F. and Bearman, R. J. (1970). *J. Chem. Phys.*, **53**, 3114

105. Watts, R. O. (1968). *J. Chem. Phys.*, **48**, 50
106. Watts, R. O. (1969). *J. Chem. Phys.*, **50**, 984
107. Mandel, F. and Bearman, R. J. (1969). *J. Chem. Phys.*, **50**, 4121
108. Watts, R. O. (1969). *J. Chem. Phys.*, **50**, 4122
109. Henderson, D. and Chen, M. (1970). *Can. J. Phys.*, **48**, 634
110. Barker, J. A., Henderson, D. and Watts, R. O. (1970). *Phys. Lett.*, **31A**, 48
111. Henderson, D., Barker, J. A. and Watts, R. O. (1970). *IBM J. Res. Develop.*, **14**, 668
112. Lebowitz, J. L. (1964). *Phys. Rev.*, **133**, A895
113. Lebowitz, J. L. and Zomick, D. (1971). *J. Chem. Phys.*, **54**, 3335
114. Baxter, R. J. (1970). *J. Chem. Phys.*, **52**, 4559
115. Leonard, P. J., Henderson, D. and Barker, J. A. (1971). *Mol. Phys.*, **21**, 107
116. Mansoori, G. A., Carnahan, N. F., Starling, K. E. and Leland, T. W., Jr. (1971). *J. Chem. Phys.*, **54**, 1523
117. Throop, G. J. and Bearman, R. J. (1966, 1967). *J. Chem. Phys.*, **44**, 1423; **47**, 3036
118. Grundke, E. W., Henderson, D. and Murphy, R. D. (1971). *Can. J. Phys.*, **49**, 1593
119. Lee, Y. T., Ree, F. H. and Ree, T. (1968). *J. Chem. Phys.*, **48**, 3506
120. Ree, F. H., Lee, Y. T. and Ree, T. (1971). *J. Chem. Phys.*, **55**, 234
121. Raveche, H. J. and Green, M. S. (1969). *J. Chem. Phys.*, **50**, 5334
122. Hurst, C. (1965, 1966). *Phys. Lett.*, **14**, 192; *Proc. Phys. Soc. London*, **86**, 193; **88**, 533
123. Carley, D. D. and Lado, F. (1965). *Phys. Rev.*, **137**, A42
124. Rowlinson, J. S. (1965, 1966). *Mol. Phys.*, **9**, 217; **10**, 533
125. Henderson, D. (1966). *Proc. Phys. Soc. London*, **87**, 592
126. Stell, G. (1969). *Mol. Phys.*, **16**, 209
127. Morita, T. (1969). *Progr. Theor. Phys.*, **41**, 339
128. Hutchinson, P. and Conkie, W. R. (1971). *Mol. Phys.*, **21**, 881
129. Kim, S., Henderson, D. and Oden, L. (1966). *J. Chem. Phys.*, **45**, 4030
130. Lado, F. (1967). *J. Chem. Phys.*, **47**, 4828
131. Watts, R. O. and Henderson, D. (1969). *Mol. Phys.*, **16**, 217
132. Henderson, D. and Watts, R. O. (1970). *Mol. Phys.*, **18**, 429
133. Oden, L. E. and Henderson, D. (1971). *Chem. Phys. Lett.*, **8**, 85
134. Throop, G. J. and Bearman, R. J. (1966). *Proc. Phys. Soc. London*, **88**, 539
135. Henderson, D. and Oden, L. (1966). *Mol. Phys.*, **10**, 405
136. Mandel, F. and Bearman, R. J. (1971). *J. Chem. Phys.*, **55**, 4762
137. Mansoori, G. A. and Canfield, F. B. (1970). *Ind. Eng. Chem.*, **62**, No. 8, 12
138. Stell, G. (1970). SUNY at Stony Brook College of Engineering Report 182
139. Barker, J. A. and Henderson, D. (1971). *Accounts Chem. Res.*, **4**, 303
140. Peierls, R. E. (1932). Quoted by Landau, L. D. and Lifshitz, E. M. (1969). *Statistical Physics*, 2nd edn. (Oxford: Pergamon)
141. Zwanzig, R. W. (1954). *J. Chem. Phys.*, **22**, 1420
142. Smith, E. B. and Alder, B. J. (1959). *J. Chem. Phys.*, **30**, 1190
143. Frisch, H. L., Katz, J. L., Praestgaard, E. and Lebowitz, J. L. (1966). *J. Phys. Chem.*, **70**, 2016
144. McQuarrie, D. A. and Katz, J. L. (1966). *J. Chem. Phys.*, **44**, 2393
145. Barker, J. A. and Henderson, D. (1967). *J. Chem. Phys.*, **47**, 4714
146. Chandler, D. and Weeks, J. D. (1970). *Phys. Rev. Lett.*, **25**, 149
147. Weeks, J. D., Chandler, D. and Andersen, H. C. (1971). *J. Chem. Phys.*, **54**, 5237
148. Buff, F. P. and Schindler, F. M. (1958). *J. Chem. Phys.*, **29**, 1075
149. Schofield, P. (1966). *Proc. Phys. Soc. London*, **88**, 149
150. Gubbins, K. E., Smith, W. R., Tham, M. K. and Tiepel, E. W. (1971). *Mol. Phys.*, **22**, 1089
151. Smith, W. R., Henderson, D. and Barker, J. A. (1971). *J. Chem. Phys.*, **55**, 4027
152. Smith, W. R., Henderson, D. and Barker, J. A. (1970). *J. Chem. Phys.*, **53**, 508
153. Smith, W. R., Henderson, D. and Barker, J. A. (1968). *Can. J. Phys.*, **46**, 1725
154. Barker, J. A. and Henderson, D. (1968). *Proceedings of the Fourth Symposium on Thermophysical Properties*, 30. (New York: American Association of Mechanical Engineers)
155. Barker, J. A., Henderson, D. and Smith, W. R. (1969). *J. Phys. Soc. Jap. Suppl.*, **26**, 284
156. Praestgaard, E. and Toxvaerd, S. (1969). *J. Chem. Phys.*, **51**, 1895
157. Toxvaerd, S. and Praestgaard, E. (1970). *J. Chem. Phys.*, **53**, 2389
158. Moffat, M. J. and Kozak, J. J. (1971). *J. Chem. Phys.*, **55**, 3794
159. Kozak, J. J. and Rice, S. A. (1968). *J. Chem. Phys.*, **48**, 1226

160. Rowlinson, J. S. (1964). *Mol. Phys.,* **7,** 349; **8,** 107
161. Henderson, D. and Barker, J. A. (1970). *Phys. Rev.,* **1,** A1266
162. Henderson, D. and Barker, J. A. (1968). *Mol. Phys.,* **14,** 587
163. Andersen, H. C., Weeks, J. D. and Chandler, D. (1971). *Phys. Rev.,* **4,** A1597
164. Barker, J. A. and Henderson, D. (1971). *Phys. Rev.,* **4,** A806
165. Henderson, D. (1971). *Mol. Phys.,* **21,** 841
166. Weeks, J. D., Chandler, D. and Andersen, H. C. (1971). *J. Chem. Phys.,* **55,** 5422
167. Henderson, D. and Barker, J. A. (1968). *J. Chem. Phys.,* **49,** 3377
168. Leonard, P. J., Henderson, D. and Barker, J. A. (1970). *Trans. Faraday Soc.,* **66,** 2439
169. Weres, O. and Rice, S. A. (1970). *J. Chem. Phys.,* **52,** 4475
170. Smith, W. R. (1971). *Mol. Phys.,* **22,** 105
171. Henderson, D. and Leonard, P. J. (1971). *Proc. Nat. Acad. Sci. U.S.,* **68,** 2354
172. Hemmer, P. C. (1964). *J. Math. Phys.,* **5,** 75
173. Hauge, E. H. and Hemmer, P. C. (1966). *J. Chem. Phys.,* **45,** 323
174. Lebowitz, J. L., Stell, G. and Baer, S. (1965). *J. Math. Phys.,* **6,** 1282
175. Zittartz, J. (1964). *Z. Physik.,* **180,** 219
176. Jalickee, J. B., Siegert, A. J. F. and Vezzetti, D. J. (1969). *J. Math. Phys.,* **10,** 1442
177. Jalickee, J. B., Siegert, A. J. F. and Vezzetti, D. J. (1970). *J. Math. Phys.,* **11,** 3168
178. Isihara, A. (1968). *J. Phys. A,* **1,** 539
179. Mansoori, G. A. and Canfield, F. B. (1969). *J. Chem. Phys.,* **51,** 4958
180. Rasaiah, J. and Stell, G. (1970). *Mol. Phys.,* **18,** 249
181. Stell, G. (1970). *Chem. Phys. Lett.,* **4,** 651
182. Mansoori, G. A. and Leland, T. W., Jr. (1970). *J. Chem. Phys.,* **53,** 1931
183. Rudd, W. G. and Frisch, H. L. (1971). *J. Chem. Phys.,* **54,** 3479
184. Ashurst, W. T. and Hoover, W. G. (1970). *J. Chem. Phys.,* **53,** 1617
185. Mansoori, G. A. and Canfield, F. B. (1970). *J. Chem. Phys.,* **53,** 1618
186. Henderson, D. and Barker, J. A. (1970), *J. Chem. Phys.,* **52,** 2315
187. Axilrod, B. M. and Teller, E. (1943). *J. Chem. Phys.,* **11,** 299
188. Barker, J. A., Henderson, D. and Smith, W. R. (1969). *Mol. Phys.,* **17,** 579
189. Barker, J. A., Henderson, D. and Smith, W. R. (1968). *Phys. Rev. Lett.,* **21,** 134
190. Barker, J. A. and Pompe, A. (1968). *Aust. J. Chem.,* **21,** 1683
191. Bobetic, M. V. and Barker, J. A. (1970). *Phys. Rev.,* **2,** B4176
192. Barker, J. A., Bobetic, M. V. and Pompe, A. (1971). *Mol. Phys.,* **20,** 347
193. Barker, J. A., Fisher, R. A. and Watts, R. O. (1971). *Mol. Phys.,* **21,** 657
194. Sinanoglŭ, O. (1967). *Chem. Phys. Lett.,* **1,** 340
195. Sinanoglŭ, O. (1967). *Advan. Chem. Phys.,* **12,** 283
196. Chen, C. T. and Present, R. D. (1970). *J. Chem. Phys.,* **53,** 1585
197. Rushbrooke, G. S. and Silbert, M. (1967). *Mol. Phys.,* **12,** 505
198. Rowlinson, J. S. (1967). *Mol. Phys.,* **12,** 513
199. Casanova, G., Dulla, R. J., Jonah, D. A., Rowlinson, J. S. and Saville, G. (1970). *Mol. Phys.,* **18,** 589
200. Dulla, R. J., Rowlinson, J. S. and Smith, W. R. (1971). *Mol. Phys.,* **21,** 299

7
Molecular Collisions and Reactive Scattering

R. D. LEVINE
The Hebrew University, Jerusalem

7.1   INTRODUCTION

There has been considerable progress in the theory of molecular dynamics over the last 5 years. New methods have been developed and existing techniques have been refined to a point where one is on the threshold of providing both a quantitative and a qualitative interpretation of the detailed dynamics of simple, binary, molecular (and atomic) collisions. This development paralleled a considerable growth and refinement of experimental methods and results on the dynamics of gas-phase collisions[1,2] and the theoretical developments on potential energy surfaces[3,4].

A large fraction of the previous theoretical work has been concerned, however, with tooling-up and examination of methods rather than with applications. In my opinion, the theorist in this field has yet to demonstrate the utility of the fundamental theory as a laboratory tool for the prediction and correlation of experimental results, particularly for reactive collisions. The gap between the 'basic' theory as practised by the theoretician and the 'working' theory as used by the experimentalist has not been completely closed. It is partly for this reason that many theoretical developments in this field can be traced to scientists who label themselves experimentalists but, needing theoretical guidance in order to design and perform their experiments, developed the appropriate theoretical treatments. It is, however, a safe assumption that the most important future development in this field would be the increasing role of the theory as a method of study of realistic systems, used as an adjunct to the experimental investigation.

This review considers the period between the Faraday Society discussion[5] on the Molecular Dynamics of Chemical Reactions in Gases in 1967 and the VIIth International Conference[6] on the Physics of Electronic and Atomic Collisions in 1971. During that period, there have appeared several books[7-20], volumes of advances series[21-24] and conference proceedings[4,26-30] dealing partly or wholly with collision theory. The bibliographical lists issued by the Atomic and Molecular Processes Information Center[31] provide an important source. In what follows, I shall not attempt to review (or even mention) every development in the field. Rather I have selected some areas of activity that received considerable recent attention. Even in those areas, the list of references is not exhaustive but only (hopefully) extensive enough to serve as a guide to the literature.

7.2   MOLECULAR COLLISION THEORY

7.2.1   Elastic scattering and the inversion problem

The recent experimental progress in the study of angular distributions and absolute total cross-sections for collisions of 'structureless' particles has stimulated two related developments. Both are aimed at providing the missing link – the potential – in an otherwise well-understood[32-34] route between the Schrödinger equation and the experimental scattering behaviour.

On the one hand, it is now becoming possible to obtain the long-range potential either through improved formulations of the traditional per-turbation-theoretic approach[3, 35] or, in a direct, *ab initio* solution of the Schrödinger equation for the electronic motion[3, 36, 37]. The entire potential can thus be obtained pointwise, without any expansion in inverse powers of R. Of course, real atoms can be described as structureless particles interacting via a central potential only in the Born–Oppenheimer approxi-mation. That this approximation fails for the collision of excited atoms is clear on both experimental and theoretical grounds (cf. Section 7.2.6.5). For ground-state atoms at low collision velocities (below the threshold for electronic excitation) the approximation is expected to be valid pro-vided that there are no avoided potential energy curve crossings at (or below) the energy region under study. More work on this point is, however, clearly indicated[38, 39].

The alternative, 'experimental', route to the potential is by an inversion of the scattering data. Traditionally, this has been carried out in an indirect way by selecting a functional form for the potential, with a number of free parameters, which are then determined from a comparison of the experimental and calculated results. With low-resolution data only a few (two or three) free parameters could be optimised, and only the more promi-nent features of the data could be reproduced[40]. With improved resolution, more flexibility can be introduced into the trial potential[41, 42], but the ultimate aim is an inversion procedure and not a fitting method. The complete solu-tion is not simple, as a measurement of the angular distribution at a single energy is known not to define the potential uniquely[43] and experience has shown[44] that this is not an idle worry. Progress towards an inversion which is unique has however been reported[45]. In practice, methods based on the WKB approximation for the phase shift have been reported[46-51] and applied[51]. In the WKB approximation there exists a unique inversion from the de-flection function $\Theta(b)$ to the Firsov–Sabatier transformed potential[46-49] $Q(R)$:

$$V = E\{1 - \exp[-2Q(R)]\} \tag{7.1}$$

Here b is the impact parameter and

$$Q(R) = \pi^{-1} \int_0^\infty \frac{\Theta((R^2 + x^2)^{\frac{1}{2}})}{(R^2 + x^2)^{\frac{1}{2}}} \, dx \tag{7.2}$$

For a monotonic central potential where $\Theta(b)$ is a single-valued function of b, the angular distribution is

$$\frac{d\sigma}{d\theta} = b \bigg/ \left| \frac{d\Theta}{db} \right| \tag{7.3}$$

where $\theta = |\Theta|$, and equation (7.2) provides an inversion in the semi-classical limit. For realistic potentials, which contain a well, deflection into a given angle θ can be due to collisions with different impact parameters. While the classical angular distribution is simply the sum of terms of type

(7.3) over all values of b that lead to a given $\theta = |\,\Theta(b)\,|$, the quantal result contains interference terms[32] between the different trajectories that lead to the same deflection. Buck[51] has discussed a procedure for estimating these terms using experimental information on the maxima and minima due to the interferences and thereby was able to determine $\Theta(b)$ and hence the potential.

7.2.2 Total scattering: the size of molecules

The total scattering cross-section is a measure of the 'size' of the colliding atoms (or molecules). When energy transfer (or chemical reaction) also takes place, this total can be partitioned as the sum of the cross-sections for the different processes. The relative contributions from the various processes depend, in a complicated way, on the details of the intermolecular potential, particularly on the short-range forces. It appears however that, at low velocities, the sum, i.e. the total cross-section, is determined primarily by the long-range potential and is rather insensitive to the way in which it is partitioned[52, 53]. The only currently observed manifestation of non-elastic events in the total scattering is the 'quenching of the glories'[54-57], as is discussed below.

To illustrate these considerations we assume an idealised model of a structureless atom colliding with a rigid-rotor diatom[9, 52, 58]. For an aniso-tropic potential, rotational energy transfer can take place. In the absence of rotational state selection, the total cross-section for an initial rotor angular momentum j, averaged over the initial projections (m_j) of the rotor orientation, is given by[9, 58]

$$\pi^{-1}k_j^2\sigma_T(j) = (2j+1)^{-1}\sum_J(2J+1)2\sum_{l=|J-j|}^{J+j}[1-ReS^J(j,l;j,l)]$$

$$= \sum_l(2l+1)\sum_{J=|l-j|}^{l+j}\frac{2J+1}{(2l+1)(2j+1)}2[1-ReS^J(j,l;j,l)] \qquad (7.4)$$

Here J is the total angular momentum $(\boldsymbol{J} = \boldsymbol{l}+\boldsymbol{j})$, where l is the orbital angular momentum and $S^J(j,l;j',l')$ is an element of the scattering matrix for the transition $j,l \to j',l'$ at a given (conserved) value of J.

To see the general features of equation (7.4), it has proved convenient to express the scattering matrix (at a given J) as[59]

$$S = \exp(i\boldsymbol{\delta})S'\exp(i\boldsymbol{\delta}) \qquad (7.5)$$

where $\boldsymbol{\delta}$ (and hence $\exp(i\boldsymbol{\delta})$) is a diagonal matrix whose elements are the phase shifts due to the distortion potential[59] (i.e. the part of the potential that cannot induce inelastic transitions). For high l values, the centrifugal barrier prevents a close approach of the molecules and (with few exceptions, say dipole–dipole scattering) only elastic transitions are possible, $S' = I$. As l decreases, inelastic transitions begin to contribute, but the phase shifts also increase and vary rapidly with energy. Hence one can put[52, 60, 61]

$$\langle S^J(j,l;j',l')\rangle = 0 \qquad (7.6)$$

where the angular brackets denote an average over a small energy interval. One expects (7.6) to hold for $l \leqslant L$, subject to a Massey–Mohr-type[32, 62] criterion,

$$\delta_J(j, L) = \pi/4, (\text{or } \tfrac{1}{2}) \tag{7.7}$$

where L is the highest value of l that satisfies equation (7.7). Exceptions to equation (7.7) are noted below. For low velocities L is determined by the long-range part of the potential.

Using equation (7.6) with (7.4) one sees that the contribution of the region $l < L$ to the (energy-averaged) total cross-section is $2\pi(L/k)^2$, irrespective of the extent of inelasticity. In general, the small[32, 52, 53] ($\sim 10\%$) contribution from $l > L$ is expected only to modify the numerical coefficient of L^2 but not to change the dependence of L^2 on the physical parameters (e.g. velocity) as determined from equation (7.7). The conclusion that the total cross-section is of the order of $2\pi(L/k)^2$ irrespective of the partitioning of this total into the contributions of the different elastic and inelastic cross-sections is sometimes referred to as the 'conservation' of σ_T.

Exceptions to the assumption that $S^J(j, l; j', l')$ changes rapidly occur near the so-called 'glory' region where[32] $(\mathrm{d}\delta_J(j, l)/\mathrm{d}l) \simeq 0$. This adds an oscillatory contribution[32] to the velocity dependence of σ_T, which is observed in good velocity resolution atom–atom collisions but which may be quenched in atom–molecule collisions[55, 56, 63]. There are two sources of this quenching that can be identified in (7.4):

(a) Even in the absence of any inelasticity, the degeneracy averaging over m_j (which leads to the inner summation in equation (7.4)) implies[57] that the different terms will reach the glory condition at somewhat different l values and the averaging will reduce the glory oscillation of each term.

(b) Moreover, due to the inelastic (and, if any, reactive) collisions, $|S^J(j, l; j, l)| \leqslant 1$, and hence the importance of the glory regime is quenched as compared to the pure elastic ($|S^J(j, l; j, l)| = 1$) case[54–56, 64–69].

It is clearly of interest to avoid the necessity of degeneracy averaging by using state-selected molecules which are initially in a specified j, m_j state. Such measurements, which would provide a characterisation of the anisotropic part of the potential, have been the subject of considerable theoretical and experimental activity[69–74].

7.2.3   Potential energy surfaces and classical trajectories

Potential energy surfaces from *ab initio*, Hartree–Fock (and improved Configuration Interaction) type calculations are rapidly becoming available[4, 75–78] for non-reactive atom–molecule collisions. In the past, in the absence of any *a priori* results (except for the He–H_2 Krauss–Mies potential[79]), the coupling between the internal degrees of freedom and the relative motion was described by 'reasonable' functional forms. For vibrational excitation[80] a long-time favourite was the Landau–Teller (sometimes called Born–Mayer) exponential repulsion between the colliding atom and an atom of the diatomic molecule. Occasionally, this was replaced[81] by exponential repulsion by both atoms of the diatomic molecule. Similarly, for rotational excitation[52, 58, 82] of a rigid rotor a Legendre expansion of the potential was used:

$$V(R, \theta) = \sum_n V_n(R)P_n(\cos \theta) \tag{7.8}$$

R is the atom–rotor separation and θ is the angle between R and the axis of the rotor. The coefficients $V_n(R)$ were chosen on realistic grounds, with guidance from perturbation theory[83-86] for their long-range form, but the short-range anisotropy was only available from fitting to experimental data[69, 71, 72, 74].

Figure 7.1a shows the contours of the Lester[76] Li^+–H_2 potential with the H_2 separation fixed at the equilibrium position. The first three even terms ($n = 0, 2, 4$) in (7.8) contribute significantly to this potential; however, most of the anisotropy is from the P_2 term. The short-range He–H_2 potential was recently reported by Gordon and Secrest[75] in the form

$$V(R, r, \theta) = C \exp\left\{-\alpha_0 R[1 - \alpha_1(r - r_e)]\right\}$$
$$\left\{1 + aP_2(\cos\theta) - b(r - r_e)[1 - CP_2(\cos\theta)]\right\} \tag{7.9}$$

with $C = 11.143$ Å, $\alpha_0 = 1.94$ Å^{-1}, $\alpha_0\alpha_1 = 0.25$ Å^{-2}, and r_e, (Å), the equilibrium separation of H_2. It is also possible to make such fits to potential energy surfaces for reactive collisions. The H–H_2 potential was so analysed by Tang and Karplus[87].

The most direct way of surveying the main features of a surface is by performing a classical trajectory calculation[82, 88]. While such a method will only be accurate for averaged quantities[89, 90], such as average energy transfer[80], the total inelasticity[82], or (in general) the moments of the distribution of the final states[90, 91], it does provide a comparatively rapid means of identifying the main effects to be expected. As an example, Figure 7.1b shows the total inelastic cross-section for Li^+–H_2 and Li^+–D_2 collisions, as functions of the translational energy[82]. The smaller rotational spacings in D_2 imply a less adiabatic collision, and the maximum of the cross-section (expected to occur at an energy where the duration of the collision ($\propto E^{-\frac{1}{2}}$) approximates the rotational period ($\propto I$)) occurs at a lower energy. In general, however, classical trajectory computations would be most appropriate for systems where the level spacings are sufficiently small so that only highly averaged information is required. The validity criterion for classical mechanics for inelastic transitions is that the action[89, 90] imparted to the internal degrees of freedom during the collision be large compared to $\hbar$.

Classical scattering theory was discussed by Garrido[92], Miles and Dahler[93], Marcus[94], Zeleznik and Svehla[95] and Eu[96]. Systematic studies of various aspects of inelastic processes have been reported. Marcus and co-workers have studied both rotational[97] and collinear vibrational[98] excitation, with special attention to approximation schemes and the use of action-angle variables. Rotational excitation was also considered by Bjerre[99] and by Van de Ree[100]. Kelley and Wolfsberg[81] and Feldman et al.[101], reported results on three-dimensional collisions, while the impulsive limit was discussed by Mahan[102]. An interesting application is the trajectory study[103] for the quasi-diatomic model of photo-dissociation of triatomics. Here, following light absorption, the atom is located at the classical turning point of a repulsive potential. As the atom and the diatom separate, the exponential repulsion between the atom and the near end of the diatom leads to exchange of energy between the translation and the vibration of the diatom (i.e. a 'half-collision' event).

It is to be expected that, with the better understanding of the limitations and strong points of the method and with the realisation that classical

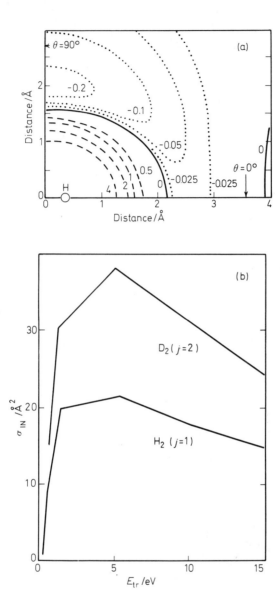

Figure 7.1 Potential energy surface[76] (a), and inelastic cross-sections obtained in a classical trajectory calculation[82] (b), for the rotational excitation of $H_2(j = 1)$ and $D_2(j = 2)$ by Li^+. The potential energy surface is shown on a polar plot where the radius is the centre-to-centre distance R and the angle with the horizontal axis is θ, the angle between R and the axis of the rotor. Note that (b) simulates the diatom as a rigid rotor. In a physical situation, vibrational excitation is also possible over most of the energy interval

trajectory calculations contain most of the essential information needed for semi-classical calculations[89, 90, 104–107], considerable activity will be reported in the near future.

7.2.4 Quantal computational methods

Practical algorithms for the numerical solution of the Schrödinger equation describing the motion of the nuclei on a potential energy surface have become a reality over the last 5 years[108–130]. Very broadly, the methods can be divided into two classes: those that treat the Schrödinger equation as a differential equation, whose solution (with arbitrary constants) is fitted to the proper scattering boundary conditions, and those which incorporate the boundary conditions explicitly. The latter invariably involve (although this is sometimes done implicitly) re-writing the differential equation as an integral equation, usually of the Volterra type. With the exception of the finite-difference boundary value method[115], which may become cumbersome for the three-dimensional case (due to the required number of grid points), the solution is usually obtained in the form of a close-coupling expansion[9, 11, 15, 116, 117]. Here, in a given sub-space of conserved quantum numbers (e.g. total angular momentum, parity, etc.), the total wave function $\Psi(R,r)$ is expanded in some complete basis $\{\phi_n(r \mid R)\}$. r is the set of internal coordinates and the basis functions may depend parametrically on R. In practice, the basis size has to be finite, and some attention should always be given to insuring that the basis size is sufficient for convergence[114, 116, 131–133]. If there are N basis functions, one can construct N independent solutions of the Schrödinger equation. The radial part of these solutions is determined by the matrix Schrödinger equation (obtained as usual by substituting the close-coupling expansion into the Schrödinger partial differential equation, pre-multiplying by ϕ_m^*, $m = 1, 2, \ldots$, and integrating over r). For a static (R-independent) basis[134]

$$\frac{d^2\psi}{dR^2} + (2\mu/\hbar^2)[E - l(l+1)/R^2 - V(R)]\psi(R) = 0 \qquad (7.10)$$

Here E and l are diagonal matrices representing the energy and orbital angular momentum and $V(R)$ is the matrix representation of the potential in the basis ϕ_n. The ordinary matrix differential equation (7.10) replaces the Schrödinger partial differential equation.

The Hamiltonian of the problem is taken as $H = H_0(R) + V(R,r)$ and, in terms of the incoming ($-$) and outgoing ($+$) solutions of $H_0(R)$,

$$H_n^{\pm}(R) \xrightarrow{R \to \infty} (\mu/k_n)^{\frac{1}{2}} \exp\left[\pm i(k_n R - l_n\pi/2 + \delta_{l_n})\right] \qquad (7.11)$$

one requires that

$$\psi(R) \xrightarrow{R \to \infty} i[-H^+(R)F + H^-(R)F^*] \qquad (7.12)$$

The scattering matrix S is given by $S = FF^{*-1}$. The solution of equation (7.10), subject to the boundary conditions expressed in equation (7.12), can be effected either as a differential equations problem[108–110, 113, 114, 117, 118, 128] or by a piecewise approximate integration procedure[119] (which is accurate for small steps).

Most newer methods can be derived from a representation of $\psi(R)$ as

$$\psi(R) = i[-H^+(R)F(R) + H^-(R)F^*(R)] \qquad (7.13)$$

where, to conform to equation (7.12), $F \equiv F(R \to \infty)$. Here the functions $F(R)$ (or related matrices) are to be determined. Again, either a differential or an integral equation approach can be used. By a suitable choice of $H_0(R)$, the functions $H^\pm(R)$ provide an accurate representation of the oscillations (or of the classically forbidden region) of $\psi(R)$. The functions $F(R)$ are then slowly varying with R and large integration steps can be taken[125, 135, 136] in solving the differential equation for $F(R)$, whereas much closer spacings would be required if the differential equation (7.10) is solved directly. In the Gordon procedure[125] $H_0(R)$ is an approximation to the potential by a series of straight-line segments, whereas the Chan–Light[119] method is essentially equivalent to using for $H_0(R)$ a series of steps-type approximations to the potential. A similar approach was discussed by Wilson and co-workers[127, 130]. In the integral equations approaches[112, 116, 123, 124] one is essentially solving for $F(R)$ (or for related matrices) using a Volterra-type integral equation. One can either solve directly for $\psi(R)$ or for $F(R)$, or convert the integral equation to a first-order, non-linear Riccati-type differential equation for $F(R)$ and derived matrices. By a judicious choice[136] of $H_0(R)$, and by using only real arithmetic, these can be solved very efficiently[116].

The unifying concept behind most newer methods is the idea of imbedding[134, 137, 138], which can be traced back to Stokes[139]. Here one compares the unknown solution of a linear second-order differential equation with the known solutions of a companion equation, as in equation (7.13). In principle, our aim is to determine the asymptotic $(R \to \infty)$ form of ψ in order to obtain the S matrix. Consider, however, a modified problem where the potential is $V_{\bar{R}}(R) \equiv V(R)\theta(\bar{R} - R)$, where $\theta(\bar{R} - R)$ is the step function that vanishes for negative values of its argument. $V_{\bar{R}}(R)$ is thus identical to $V(R)$ for $R < \bar{R}$ and vanishes for $R > \bar{R}$. The S matrix for the modified potential $V_{\bar{R}}$, denoted $S(\bar{R})$, is obtained directly from $\psi(\bar{R})$ as $F(\bar{R})F^*(\bar{R})^{-1}$, since for $R > \bar{R}$ there is no potential and the functions F will not change beyond the point $\bar{R}$. With $S(\bar{R})$ assumed known, one can solve for an 'additional' potential $V(\bar{R} + \Delta R) - V(\bar{R})$, where ΔR is a small increment. This leads to a small change in F and in S, and in the limit of $\Delta R \to 0$ one obtains[116, 134, 140] an equation for the scattering matrix density

$$\frac{dS(R)}{dR} = i[S(R)H^+(R) - H^-(R)]V(R)[H^+(R)S(R) - H^-(R)] \qquad (7.14)$$

By imbedding the actual problem of calculating $S \equiv S(\bar{R} \to \infty)$ in a family of problems of obtaining $S(\bar{R})$, one obtains the Riccati equation (7.14) together with the boundary condition of no scattering in the absence of a potential, $S(0) = I$. In practice, besides the freedom of choice of $H_0(R)$, one can also, using the K matrix, convert equation (7.14) to an equation in real variables[116]. Instead of solving equation (7.14) or its finite-difference analogues[116], one can also imbed the Volterra integral equation for $\psi(R)$, thereby simplifying its solution[123, 124]†.

---

†Addendum: Volume 10 of *Methods of Computational Physics: Atomic and Molecular Scattering* has just been published.

7.2.5   Perturbative and semi-classical methods

The available methods (Section 7.2.4) for a full quantal solution of the Schrödinger equation are limited in their ability to handle a large number of coupled states. The limiting factor is the time spent on matrix manipulations, and the present limit is about 100 coupled states. Yet the $(2j+1)$-fold degeneracy of the jth rotational state of a diatom implies that a close coupling

Table 7.1   Computational methods for molecular collisions
(Following Gordon[177])

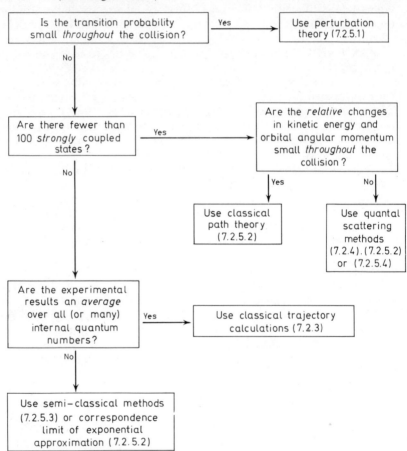

in an atom–diatom collision of all the $j = 0$ to 9 states is the largest that can be so handled. This rules out immediately all systems except hydrides, and is even much worse for diatom–diatom collisions. There are, of course, approximation procedures that can be invoked within the fully quantal treatment. (They can, however, be employed also in semi-classical or even classical solutions.) These include the use of variational principles[131–133, 141–143], the introduction of equivalent (sometimes called 'optical') potentials[144–147], and the use of perturbation theory, both in its low-order form[148–162] and in various partial summations to all orders[59, 107, 162–176] ('exponential approximations' and other types).

When the internal states are densely spaced, there are experimental problems as well, as the individual states cannot be resolved. Thus, if the experiment does not involve any resolution of quantum states, and these are densely spaced, classical trajectory calculations (Section 7.2.3) would be sufficient to reproduce the main features of the results. A fairly common case is that of partial resolution of quantum numbers (say, the total inelasticity out of a given state, as in pressure broadening and other relaxation processes). This is best treated by the appropriate version of a semi-classical procedure. The various possibilities are summarised schematically in the flow chart, adopted from a talk by Gordon[177]. Discussions of the hierarchies of approximations techniques and the interrelations between different schemes can be found in the papers of Bernstein, Curtiss, and co-workers[157, 159]. The following sub-sections summarise the previous work on the various procedures.

7.2.5.1   Perturbation theory

The coupled matrix equation (7.10) can be written

$$L\psi = V\psi \tag{7.15}$$

where L is the diagonal operator. In a first-order perturbation approximation for transitions out of the state n, one replaces V by V_n, where V_n is a non-symmetric matrix, in which only the nth column (identical to the nth column of V) is non-vanishing. The complete S matrix is

$$S = I + iA \tag{7.16}$$

where, with the normalisation adopted in equation (7.11),

$$A = -4 \int_0^\infty \psi^{(0)}(R) V(R) \psi^{(0)}(R) \mathrm{d}R \tag{7.17}$$

and $L\psi^{(0)} = 0$. Since V_n is non-symmetric, S is not unitary. Methods of removing this deficiency are discussed in Section 7.2.5.2. The distorted wave version of first-order perturbation theory is obtained by including the diagonal elements of V in L so that V in equations (7.15) and (7.17) is purely non-diagonal. In second-order perturbation theory[168, 173] one uses the solutions $L\psi^{(1)} = V\psi^{(0)}$ in (7.17) by writing $\psi^{(1)} = \psi^{(0)} + L^{-1} V\psi^{(0)}$. The computation of L^{-1} is not trivial since it is a non-local operator. Therefore second-order perturbation theory was used mostly in a classical-path approximation[173] (Section 7.2.5.2) or with additional simplifications[178, 179].

Perturbation theory is useful when the transition probability is small throughout the collision (i.e., when the *integrand* in equation (7.17) is small). The fact that A itself is small is not sufficient since the internal states may be strongly coupled *adiabatically*. When this is the case, one can usually use perturbation theory by changing the basis set $\{\phi_n\}$ (Section 7.2.4). If one uses the basis $C\phi$ where C is a unitary matrix, the coupled equations have the form $CLC^\dagger \psi' = CVC^\dagger \psi'$, where $\psi' = C\psi$ is the new set of radial wave functions. The new operator $CLC^\dagger$ need not now be diagonal, but one can reduce (or even eliminate) the non-diagonal elements of V by this procedure[180, 181]. In the sudden limit[182] L and V can be simultaneously diagonalised[170, 183, 184].

Perturbation theory is also often useful for the computation of such averaged quantities as the total inelasticity. This is based on the observation[17, 52, 149, 185] that at high l, where the centrifugal barrier is high, and tends to keep the molecules apart, perturbation theory is valid. If, for low l, the coupling is strong, $|S_{nn}|^2$ will be small compared to $\Sigma'_m |S_{mn}|^2$, which can then be approximated by its unitary limit, 1. Hence the total inelasticity is

$$\sigma_{IN} = (\pi/k^2) \sum_{l=0}^{\infty} (2l+1)P_l$$

$$= (\pi/k^2)\left[\sum_{e=0}^{L} (2l+1)1 + \sum_{e=L+1}^{\infty} (2l+1)P_e^{(0)} \right] \qquad (7.18)$$

where $P_l^{(0)} = \Sigma'_m |S_{mn}^{(0)}|^2$ is determined by perturbation theory and L is the highest value of l for which

$$P_L^{(0)} = 1 \qquad (7.19)$$

7.2.5.2 Exponential approximations and classical path theory

The S matrix obtained in perturbation theory is not unitary. This is not just an aesthetic defect, but also a practical worry, when individual transition probabilities are required, since the transition probability diverges as $b \to 0$ and the unitarity cut-off equation, (7.19), is not useful for individual transitions. Various procedures for extending the range of perturbation theory to low l values have been described[17, 52, 64, 107, 162–176, 182–188]. Most of these correspond to a partial summation of the perturbation series for S and can be summarised by the exponential formula $S = \exp[(i/\hbar)f(A)]$ where $f(x) = x + O(x^2)$ is an analytic function of x such that, for small A, the results reduce to equation (7.16). Two particular favourites are $f(x) = x$ and $f(x) = 2tn^{-1}(x/2)$ which lead to the exponential[59, 107, 162–165] and Heitler[167–169] forms

$$S = \exp[(i/\hbar)A] \quad \text{and} \quad S = [I+(i/2\hbar)A][I-(i/2\hbar)A]^{-1} \qquad (7.20)$$

Both theoretical considerations[107, 163–165] and computational experience[172] suggest that, for molecular collision theory, in the near-classical limit, the exponential form has several advantages. Particular note should be given to the exponential form obtained from distorted-wave, first-order perturbation theory[59, 107, 172],

$$S = \exp(i\delta) \exp[(i/\hbar)A'] \exp(i\delta) \qquad (7.21)$$

while the first-order result is obtained by expanding $\exp[(i/\hbar)A'] \simeq I + (i/\hbar)A'$. Equation (7.21) has been found to provide an accurate approximation procedure[172].

A particularly important development has been the introduction of the

classical-path method into the exponential form[59, 107, 159–165, 174, 176, 182–191].
Here, a classical path $R = R(t \mid b)$ is assumed known and is used to replace
$V(R)$ by $V(t \mid b)$. This is not a self-consistent procedure (unless one uses
special methods[192] to obtain $R(t \mid b)$) since the kinetic energy and orbital
angular momentum are changing in an inelastic collision. A simple classical
path should not be used unless the relative changes in kinetic energy and
orbital angular momentum are small. A partial remedy is to switch over,
at the classical turning point, from a path suitable for the initial channel
to a path suitable for the final channel. In the classical-path method

$$A = \int_{-\infty}^{\infty} e^*(t)\, V(t) \mid b) e(t) dt \qquad (7.22)$$

where $e(t)$ is a diagonal matrix with $e_n(t) = \exp(-iE_n t/\hbar)$. Further simplifications occur[105–107, 165, 170, 176, 182–184, 186] when the internal states are sufficiently
dense so that they too can be approximated semi-classically. In this, correspondence principle, limit one can obtain directly individual elements of
S by integration[107, 176, 186, 199].

7.2.5.3   Semi-classical methods

If $\mid S_{mn} \mid^2$ is the quantal transition probability for the $m \to n$ transition, then
in the classical limit[9, 105, 106]

$$S_{mn} = \sum_j (P_{mn}^{(j)})^{\frac{1}{2}} \exp \left[i\Delta_j(m, n) \right] \qquad (7.23)$$

where $P_{mn}^{(j)}$ is the 'classical' transition probability for the $m \to n$ transition
along the jth classical trajectory and Δ_j is the phase shift along the trajectory.
Before one can use equation (7.23) one needs to define $P_{mn}^{(j)}$ or, equivalently,
define what one means by 'the $n \to m$ transition' in a classical context.
Miller[106, 193] and Marcus[105, 194] have recently suggested that n should be
taken as the appropriate classical action variable. For a harmonic oscillator
this would be an initial internal state with an action variable $\bar{n} = (n+\frac{1}{2})h$.
For the transition $n \to m$ one considers classical trajectories of initial action
variable $\bar{n}$ and angle variable (phase) ϕ. The classical equations of motion
relate $\bar{m}$ to ϕ. Since the initial phase is random $P(\phi)d\phi = (2\pi)^{-1}d\phi = P_{mn}$
$d\bar{m}$ or $P_{mn} = (2\pi)^{-1} \mid \partial\phi/\partial\bar{m} \mid_{\bar{n}}$. The use of this method has been demonstrated
by Miller[89, 106, 193]. When averages over the final (or the initial) states are
required, the oscillatory terms in $\mid S_{mn} \mid^2$ (due to interference between
different trajectories) tend to average out, thereby leading to a pure-classical
result[89, 90].

When detailed, state-to-state probabilities are required, the classical
limit as expressed in equation (7.22) is not always sufficient. This is partly
indicated by the fact that the Jacobian $(\partial\phi/\partial\bar{m})$ may diverge, and by the fact
that some $\bar{n} \to \bar{m}$ transitions may be 'classically forbidden' in that there are
no classical trajectories connecting these precise initial and final action
variables[105, 193, 195]. Levine and Johnson have suggested[196] that instead of
considering only the classically allowed $\bar{n} \to \bar{m}$ transitions, one use a corre-

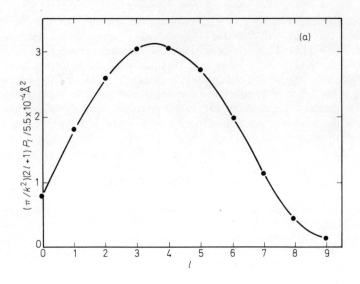

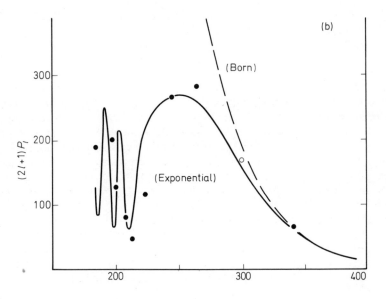

Figure 7.2 Opacity functions for rotational excitation $j = 0 \rightarrow 2 v.$ the orbital angular momentum l. (a) $He + H_2$ collision[116] under nearly adiabatic conditions. The results represent a converged, close-coupling quantal computation. (b) $Ar +$ a heavy rigid rotor under nearly sudden conditions[172]. The dots are close-coupling results. The Born and exponential Born results are based on the (analytic[149]) Born A matrix for excitation of a rigid rotor

spondence principle limit. They were only able to implement this[107, 186] using classical perturbation theory for the action[196–198], but have shown the importance of trajectories where the action change along the trajectories was not exactly $\bar{m} - \bar{n}$. Miller[195] has discussed an analytic continuation method for this regime. Marcus has recently reported[199] that he has been able to carry out such a procedure. His method is based on the representation

$$S_{nm} = \langle m, E'^- \mid n, E^+ \rangle \cdot$$

$$= \int dt d\phi \langle m, E'^- \mid \phi, t \rangle \langle \phi, t \mid n, E^+ \rangle$$

$$= \delta(E - E') \int d\phi \langle m, E^- \mid \phi \rangle \langle \phi \mid n, E^+ \rangle \tag{7.24}$$

7.2.5.4   Equivalent potentials

The reduction in the dimension of the close-coupling equation (7.10) can be effected by the introduction of an equivalent potential[9, 11, 144–146]. This formal re-writing of the problem is equivalent to the original equations but does offer a route to novel approximation schemes and to new insights. If the close coupling to *all other* channels is accounted for by the equivalent potential, one requires only the solution of a single, one-dimensional Schrödinger equation (with an unknown equivalent potential). This procedure is useful in the study of elastic[131, 144, 200–202] or total non-elastic[144–146, 203, 205] scattering. If a particular $n \to m$ transition is of interest, one can include all other channels (neither n nor m) in equivalent potentials and solve a (modified) 2×2 close-coupling equation[145].

In principle, equivalent potentials are non-local (i.e. integral operators), energy dependent and complex. (In view of the latter property they are sometimes called 'optical' potentials.) Approximate (local) equivalent potentials have been employed in decoupling approximations[38, 131, 147, 180, 181, 203], optical model analysis[200–202], and the elimination of rotational states in vibrational excitation[145].

7.2.5.5   Variational principles

Variational principles[9, 11] have not been extensively employed as a computational tool in molecular collision theory, since, in general, they do not provide a bound on the transition probabilities. They do provide a bound on the phase shift[38, 131–133] or on the reaction matrix[143] and have been so employed. Otherwise, their main use has been to justify[9, 131–133, 203] and derive[143, 203, 204] approximation techniques.

7.2.6   Studies of molecular collision processes

The methods discussed in Sections 7.2.3–7.2.5 have been applied to a variety of problems. The following is just a very brief progress report.

7.2.6.1 Rotational excitation

All the available methods have been applied to this problem, which can cover the whole gamut of coupling regimes. For H_2 the rotational spacings are sufficiently large to allow the use of close-coupling quantal calculations, while for the heavier diatoms one is often in the sudden regime, where the duration of the collision ($\propto E^{-\frac{1}{2}}$) is short compared to the rotational period. The adiabaticity of the collision, measured roughly by the ratio of the collision time to the rotational period, can be estimated as $\mu a^2/I$. (μ is the reduced mass, a is the (energy-dependent) effective range of the potential, and I is the rotor moment of inertia.) The strength of the coupling (determining the validity of perturbation theory) is given roughly by the reduced velocity D^{-1}, $D = B/A$, where B is the dimensionless reduced mass and A is the reduced wave-number[32].

Figure 7.2 shows the opacity function[9, 61] $P_l(0 \rightarrow 2)$ for rotational excitation of $j = 0 \rightarrow 2$ v. the orbital angular momentum l. The cross-section is[61]

$$\pi^{-1}k^2\sigma(0 \rightarrow 2) = \sum_l (2l+1)P_l(0 \rightarrow 2) \tag{7.25}$$

Two contrasting examples are shown: (a) the low-energy rotational excitation of H_2 by He [116, 206] where the coupling is sufficiently weak for (distorted wave) perturbation theory to be semi-quantitatively valid[207], and (b) the near-sudden rotational excitation of a heavy diatomic (mass of TlF) by Ar [172]. Here the close-coupling results are compared with first-order (Born) calculation (equations (7.16) and (7.17)) and with an exponential approximation (using the Born A matrix).

7.2.6.2 Vibrational excitation

Progress since the review of Rapp and Kassal[80] has been in the developments of semi-classical[104, 172, 186—188, 190, 193] and exponential-type approximations[172, 174, 186, 208], in the applications of second-order perturbation theory to near-resonant processes[178, 179] $(V - R)$ and in the use of equivalent potentials[146, 209, 210] and close-coupling computations[211, 354].

The use of the exponential approximation and near-classical methods has clearly established the coupling regimes for collinear vibrational excitation[106, 107, 186]. If g is the (classical) action (in units of $\hbar$) imparted to the oscillator during the collision, then weak coupling corresponds to $g \ll 1$ and strong coupling to $g > 1$. The change in the vibrational quantum number from n to m is most efficient when the resonance condition $|n-m| \simeq g$ is satisfied. In the near-classical limit one can use the approximation formulae[89, 90, 186]

$$\Delta E = (g^2/2)\hbar\omega \tag{7.26}$$

and[186, 212]

$$P_{mn} = |J_{|n-m|}(2g)|^2 \tag{7.27}$$

where ΔE is the average energy transfer to the oscillator, P_{mn} is the $m \rightarrow n$ transition probability and J_n is the Bessel function of order n. These results can be used as a rough approximation for off-centre collisions[186–188, 190, 212] by evaluating g as a function of the impact parameter b.

7.2.6.3 *Pressure broadening and collision-induced absorption*

In the impact theory, pressure broadening is treated as the perturbation of the molecular (or atomic) transition dipole moment due to collisions[213–215].

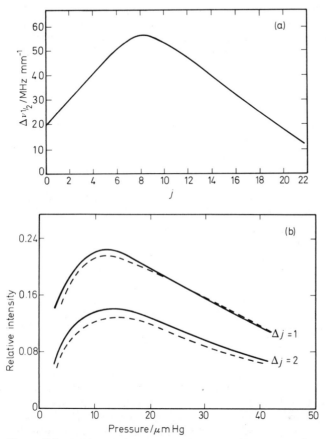

Figure 7.3 Self-broadening (a) and modulated double resonance line intensity (b) in HCN[173]. (a) The half-width of the (self-broadened) line is shown as a function of j for $\Delta j = \pm 1$. (b) The relative intensity of the transitions $\Delta j = 1$ and 2 is compared with (a least-square fit (dashed line) to) experiments, as a function of pressure

The change in any operator μ due to collisions can be written as[9]

$$\Delta\mu = S^\dagger \mu S - \mu \qquad (7.28)$$

where S is the scattering matrix. The methods of collision theory are thus directly applicable, and lead to useful predictions and correlations[2, 173, 216]. Figure 7.3 shows a comparison between experiments and results of second-order perturbation theory[173]. Alternatively one can adopt an adiabatic point of view. During the collision, the electronic states of the atom (or molecule) are modified due to the colliding species, and one can obtain the adiabatic molecular states of the combined (AB) system[217, 218]. This point of view is particularly useful in interpreting the satellite lines[219] and in collision-induced absorption[220, 221]. Here there is no transition dipole of the individual (A or B) species. During the collision, a transient dipole $\mu(R)$ is present, and can be calculated from the adiabatic electronic state of AB. Using the classical-path theory one obtains[220] $\mu(t \mid b)$. The time variation of this transient dipole is then related to the light emission, and hence (via Kirchhoff's law) to the absorption[221]. Orientational relaxation due to collisions can also be observed for nuclear spins[213, 222–224] and in depolarising collisions[225].

7.2.6.4 Quasi-bound states

The study of shape resonances, due to a hump in the effective radial potential, has progressed in two ways: (i) detailed studies were carried out of the WKB approximation, both in general, and for various potential forms[226–232]; (ii) computational studies of the energies and widths for a variety of systems of light reduced mass (of the AH type) have been discussed[233–238]. In particular, criteria for the location of these resonances through the delay time and other observables in the H + H collision were presented[238]. As yet, there is no clear-cut experimental demonstration of such resonances in the velocity dependence of the cross-section[239].

For internal excitation resonances[240–243], a complete partial-wave analysis has been presented[244] including a discussion of approximate calculations of the widths[245, 246]. Alternative formulations are based on the integral equations for the resonance parameters[247] and on the stabilisation method[248]. Such resonances appear in classical trajectory calculations as 'snarled' trajectories[249, 250].

Due to the longer interaction time when resonances are present ('compound' collisions), there is a higher probability to 'randomise' any observable that is not conserved. This includes in particular the orientation of the relative separation[1] and spin orientation[225, 251–253]. These would randomise when lifetimes exceed the rotational period. For longer-living species, the internal energy will also tend to equilibrate[1, 254]. Such quasi-bound pairs can also act as intermediates in a third-body-assisted recombination[157, 255, 256] and may give rise to new absorption lines[252, 257, 258] (sharper than ordinary collisions-induced absorption), and deviations from ideal-gas behaviour[233].

7.2.6.5 Electronic energy transfer and ionisation

There is a large family of electronically non-adiabatic processes involving the collisional quenching of an electronically excited state A*. It is customary to divide these processes according to the outcome of the collision, i.e. according to the fate of the excitation energy of A*, primarily whether or not ions are formed as products. A secondary division is made on the basis of the internal energy balance. If the initial internal excitation is equal (or nearly so) to the final internal energy, the process is a resonant (or a near-resonant) process. For such a process there is no (or only a small) change in the translational energy.

Excitation transfer processes include

$$A^* + A \rightarrow A + A^* \qquad \text{Resonant transfer} \qquad (7.29)$$

$$A^* + B \rightarrow A + B^* \qquad \text{May be near resonant} \qquad (7.30)$$

$$A^* + X \rightarrow A + X \qquad \text{Quenching} \qquad (7.31)$$

Ionisation processes include

$$A^* + B \rightarrow A + B^+ + e \qquad \text{Penning ionisation} \qquad (7.32)$$

$$A^* + B \rightarrow A^+ + B + e \qquad \text{Collisional ionisation} \qquad (7.33)$$

$$A^* + B \rightarrow AB^+ + e \qquad \text{Associative ionisation} \qquad (7.34)$$

It is convenient to distinguish between those cases where the excitation energy of A^* exceeds the dissociation energy of $(AB)^+$, so that processes (7.32) and (7.33) are possible at all energies, and those where it does not. When A and/or B are not atoms, many additional final outcomes are also possible[259]. The process of charge transfer

$$A^+ + B \rightarrow A + B^+ \qquad (7.35)$$

also comes under this heading. The measurements of the threshold behaviour of (endoergic) transfer processes is an increasingly important source of thermochemical data[260]. A better theoretical understanding of the threshold region (including the role of any 'activation barrier') is required for a full utilisation of this method.

Several recent reviews[16, 17, 261–268] have discussed the main theoretical approaches in this field. Progress in the theoretical computation of potentials between excited states[269, 270] has also led to more accurate predictions than are otherwise possible. Conversely, improved understanding of the theory has led to better inversion procedures[271–273].

In general, for the excitation transfer processes, the theory works best when the interaction is long range and (or) when the process is an exact or near-resonant case. A long-range interaction is expected when A^* is an optically allowed excited state of A. For a long-range potential of the type $C_s R^{-s}$, the quenching or transfer cross-section is expected to be of the form[261] $\sigma = A(C_s/\hbar v)^{2/(s-1)}$ where v is the relative velocity. For a dipole–dipole interaction mechanism[274] $s = 3$ and σ is proportional to C_s. While this result is an immediate conclusion[17] from the use of perturbation theory with a unitary cut-off (equation (7.19), Section 7.2.5.1), it also follows from more refined, classical-path-type exponential approximations[261–268]. The situation is far less satisfactory for short-range interactions, and, in general, for processes which are not near-resonant. Here, apart from the general conclusion that for a large internal energy imbalance, the cross-section is exponentially small[262, 268, 275, 276], usually of the adiabatic type, $\sigma \propto \exp(-v_m/v)$, one requires a numerical solution. For model potentials, however, Nikitin[16, 262, 268] and others[261, 273, 277–280] have obtained, usually in the classical-path method, analytical results for the transition probabilities for most of the coupling regimes.

The opacity analysis of the problem implies that[9] the total inelastic cross-section must be of the form

$$\sigma = (\pi/k^2) \sum_l (2l+1) \left[1 - \exp(-4\eta_l)\right] \tag{7.36}$$

where η_l is the imaginary part of the phase shift (due to the complex equivalent potential, Section 7.2.5.4). The estimation of η_l and the partitioning of this total inelasticity (the opacity function) into the different final channels have received much recent attention[276, 277, 281—285], particularly for the ionisation processes. Here, one notes that the state that can be identified as $(AB)^*$ (dissociating into $A^* + B$) is in the electronic continuum of $(AB)^+$. The problem is thus treated in a diabatic[38, 285—288] basis in that one is not using a basis that diagonalises the electronic Hamiltonian but working with diabatic states that correlate smoothly with the final states. An alternative point of view[289] is to use the adiabatic electronic eigenstates and work the problem as a breakdown of the Born–Oppenheimer approximation. It should be clear that these alternative points of view correspond simply to different choices of the basis set in the close-coupling matrix equations (cf. Section 7.2.5.1). This has been emphasised by Nikitin[262, 268], who has presented a general discussion of the two alternatives for realistic models of interaction potentials. Depending on the coupling regime, the adiabatic or the diabatic basis may be better as a start of an approximation procedure. It should also be clear that similar remarks (and methods) apply not only to electronically non-adiabatic processes but also to other, say vibrationally non-adiabatic, collisions[131, 180, 181, 209, 210, 241].

7.3 DYNAMICS OF REACTIVE COLLISIONS

7.3.1 Potential energy surfaces and trajectories

There is a growing interest in the computation of *ab initio* potential energy surfaces for reactive collisions[4, 290]. While many such calculations are of the (restricted) Hartree–Fock type, provided only that the dissociation is correctly described, one hopes that the qualitative features of the surface will be correct. However, the quantitative verification of this hope is not yet complete. Presently available surfaces include: H_3^+ [291, 292], H_3 [293, 294], H_4 [295, 296], HeH_2^+ [297], H_2Li [298, 299], H_2O [300], H_2F [301], $HFLi$ [302—304], H_2Cl [305], LiF_2 [306]. Private communications indicate that work on a large number of other systems will be generally available in the near future.

Qualitative features of such surfaces can also be gleaned using a variety of semi-empirical approaches[307]. The traditional method is an LEPS surface, with adjustable parameters, which are varied to improve the features of the surface[308, 309]. Of course, such methods are also applicable to model surfaces[311, 312]. Other procedures include the use of pseudo-potentials[313], perturbation theory[314], approximate MO theory[315], and the method of diatomics in molecules[316]. Ultimately, on the basis of the observed dynamics, one can try to predict the main features of the surface[317].

The qualitative and (averaged) quantitative dynamics on a surface are most easily explored by classical trajectory computations[88, 318, 319]. When only very averaged information is required (say, total non-reactive or reactive cross-sections), the calculations can be performed very efficiently. Such calculations can be performed not only for comparison with experiment but also to provide a classical 'standard' for more approximate theories[310,320,321]. On the other hand, the absence of accurate potential energy surfaces for most systems of experimental interest implies that such calculations cannot predict experimental results. Their main utility is in correlating existing results[317, 322] and in relating the observed dynamics to the expected main features of the surface.

Polanyi's group has published a series of studies relating the main topo-graphy of the surface to the dominant features of the dynamics[323–325]. Other recent studies include: $H+H_2$ [326], $H^+ + H_2$ [327, 328], $Cl+HI$ [308], $Cl+Br_2$ [329], $M+X_2$ [311, 320, 330, 331], $T+HR$ [332], $T+CH_4$ [333], $H+HF$ [334], $F+H_2$ [309, 335], $X+H_2$ [336], H_2+I_2 [336].

There have been several comparisons of classical and quantal calculations (for collinear collisions)[337–340]. It is almost invariably the case that for reac-tions with activation energy, the classical threshold energy for reaction is significantly higher than the quantal results. Since the thermal rate constant is rather sensitive to the threshold region, the collinear rate constants show significant differences between the classical and quantal results[339–341]. It seems premature however to draw a general conclusion. Most of the comparisons are for a collinear $H+H_2$ collision, where the small reduced mass, the absence of a centrifugal barrier, and the large vibrational spacing tend to accentuate the differences. Comparisons of three-dimensional quantal and classical calculations[342, 343] are available only for idealised (hard shoulders) potential energy surfaces, with large vibrational spacings. It is not altogether surprising that the agreement between the detailed probabilities was found to be comparatively poor, as sharp potential edges are known to lead to highly 'non-classical' features, when detailed information is required. On energy averaging, many of these features disappear, however. Final judgment should be withheld until the semi-classical theory[344] (Section 7.2.5.3) is compared with quantal results for reactive collisions. In any case, the comparison should not be made on the level of state-to-state probabilities (or cross-sections), but on the level of averaged quantities[89, 90], such as the total reactive cross-section out of a given state, the average internal energy of the products, or the isotope effect; or on the level of such diagnostic features as the influences on the reaction rate of the different ways of parti-tioning a given total energy, or the coupling[1] between product angular distributions and energy distribution.

7.3.1.1   Kinematical models

Here the potential energy surface is idealised to be, usually, of a hard-sphere type, with impulsive collisions[342, 343, 345–351]. Such models have the considerable advantage that analytical classical trajectories can be obtained using the energy and momentum conservation relations, a method introduced by Hirschfelder[346]. Moreover, with some ingenuity in constructing the surface, even compound processes can be simulated in some of these models. To my mind, the major importance of such models is in the ability to correlate the outcome of the collision with the impulses imparted during the collision. After all, a soft potential can be approximated as a series of hard shoulders. Hence averaged distributions obtained from such models provide a readily accessible way of determining the influence of the different kinematical factors (energies, masses, etc.) on the collision. Most of the detailed applications[345, 347, 350, 351] have been to reactions of hot atoms or ion–molecules where the translational energy is comparatively high.

7.3.2 Models of reactive collisions

In principle, the dynamics of the collision are determined by an (exact or approximate) solution for the motion (classical or quantal) of the colliding species. It is of some interest to formulate theories which by-pass the necessity to solve a dynamical problem. This can be achieved by the introduction of a model which specifies (i) a reaction criterion and (ii) an algorithm for computing the reaction probability in such a model. Three such models are considered here. A list of reactions which were studied with molecular beams and literature references to their theoretical analysis was recently compiled elsewhere[1].

7.3.2.1 The statistical approximation

Here one assumes that all collisions which can surmount the activation barrier for the radial motion between the reactants reach the dominant coupling regime and can react. The reaction probability is the fraction of the number of channels (states of the non-interacting collision products) that lead to the desired outcome. Recent work has indicated that while complex formation does imply statistical behaviour[60, 181, 352], it is not a necessary condition[9, 60, 61, 353]. Dominant coupling results from the strong interactions which are possible when there are no barriers. Dominant coupling is basically a non-adiabatic regime, in which a quantum number is either conserved or randomised. Statistical tests have demonstrated[353] such behaviour for results of close-coupling computations. For reactions without an activation barrier, or at high enough energies in general, the statistical model contains no adjustable parameters, apart from the need for the radial potentials for the reactants and products, which are not always available. Recent applications include the following reactions: $C^+ + D_2$ [355], $K + HCl$ [356], $H_2^+ + H_2$ [357], $K + SO_2$ [358], $H^- + O_2$ [359], and $He + N_2^+$ [360, 361]. Applications of the theory to the computation of differential cross-sections have also been considered[362]. It is to be expected that future developments would include refined criteria for the dominant coupling regime and extension of the theory to cover such processes as collision-induced dissociation[363].

In terms of the scattering matrix, one can distinguish three regimes[9, 61]: (a) $\langle S \rangle = 0$, where the brackets denote averaging over a small energy interval. This is the necessary condition for most model theories and is usually satisfied. (b) $\langle |S_{ij}|^2 \rangle \simeq N^{-1}$, where N is the total number of channels. This is the dominant coupling regime, where the statistical theory is valid. And (c) $\langle S_{ji}^* S_{jl} \rangle \simeq N^{-1} \delta_{ij}$, which is the condition for the absence of interference terms in the statistical theory for angular distributions[9, 60, 362]. Since the phase of the amplitude $S_{ji}^* S_{jl}$ is $(-\delta_i + \delta_l + \delta_r(ij) - \delta_r(jl))$, where $\delta_r(ij)$ is the phase due to the inner, strong interactions, the above condition is usually satisfied when complex formation takes place so that δ_r is varying rapidly with energy. Otherwise, (c) is not expected to hold in general. Of course, as in other applications of statistical methods, when one is not concerned with state-to-state information, assumption (c) may appear to work. In other words, while (c) in itself may not be valid, the interference terms will tend to average out when different (state-to-state) angular distributions are superimposed, and hence (c) may be used when only averaged (i.e. angular distribution of reactive scattering) information is required.

7.3.2.2 Quasi-equilibrium theories

These[9, 364-373] are essentially an extension of the transition-state theory (TST) and/or of the RRKM theory of unimolecular breakdown to bimolecular collisions, at a given total energy. (TST is concerned with collisions at a given temperature; RRKM deals with unimolecular breakdown, at a given energy.) It is assumed that one can define a 'configuration of no return' such that systems which reach this configuration will react. The reaction rate is given by the rate of departure (or arrival) at the critical configuration. It is assumed that this rate can be computed by the methods of equilibrium statistical mechanics. It should be stressed that this assumption *does not* necessarily imply (or require) that there is an equilibrium between the reactants and the transition state, or that the collision is necessarily adiabatic. It is our (but not everybody's) opinion that very often collisions are not vibrationally (or rotationally) adiabatic, and that the validity of the equilibrium assumption stems from the strong coupling (rather than the weak coupling) between the internal degrees of freedom.

To formulate the theory it is convenient to introduce a micro-canonical ensemble of collision pairs, each with a total energy in the range E to $E + dE$. The rate constant in such an ensemble (of dimensions (energy $\times$ time)$^{-1}$) is written as $Y(E)/h$ where $Y(E)$ is the yield function[9, 366]

$$Y(E) = \pi^{-1} \sum_n k_n^2 \sigma_n(E) \tag{7.37}$$

n labels the channels (states of the reactants) and $\sigma_n(E)$ is the total reaction cross-section from the state n, of internal energy E_n. $k_n^2 = (2\mu/\hbar^2)(E - E_n)$. If one sums over levels $\bar{n}$ instead of over states n (each level corresponding to $g_{\bar{n}}$ degenerate states n),

$$Y(E) = \pi^{-1} \sum_{\bar{n}} k_{\bar{n}}^2 g_{\bar{n}} \sigma_{\bar{n}}(E) \tag{7.38}$$

where $\sigma_{\bar{n}}(E)$ is the degeneracy-averaged cross-section.

In the quasi-equilibrium theory[364-366]

$$Y(E) = N_{\ddagger}(E) \tag{7.39}$$

where $N_{\ddagger}(E)$ is the number of *internal* states of the 'transition state' at an energy up to E. Equation (7.39) provides not just an expression for $Y(E)$. Using (7.38) it can be inverted[364, 365, 369, 373] to yield expressions for the individual reactive cross-sections.

The theory can be compared with trajectory calculations[365, 369, 373], to examine the validity of the assumptions. In particular, Morokuma *et al.*[369, 373], have separately examined the equilibrium assumption and the assumption of a 'point of no return'. While they and Marcus[365] found good agreement on both points for the $H + H_2$ reaction, considerable deviations were reported for the $H_2 + Br$ reaction, for which the saddle point is very much in the $H + HBr$ valley. (In principle, however, there is no reason why the point of no return should be at the saddle point. In fact, for reactions without an activation barrier, one should probably select the transition-state configuration to be at the maximum of the radial potential[9]. This is well known[9] to lead to agreement between the quasi-equilibrium theory and the optical model, for simple cases.)

7.3.2.3 Optical model analysis

The optical model analysis is an attempt to make predictions which are, as much as possible, independent of the detailed dynamics in the strong interaction region where the chemical (or physical) change is taking place. This approach is useful mainly in correlating experimental information and in providing diagnostic statements. The use of the model in studying reactive collisions through the observation of non-reactive scattering has been described elsewhere[1, 374].

The starting assumption of the optical model analysis is the relation

$$d\sigma_T = 2\pi b db \tag{7.40}$$

for the total cross-section of collisions with an impact parameter in the range b to $b + db$. Strictly speaking, this is not exact since it neglects the (forward) shadow-elastic scattering[9]. This quantal shadow (or diffraction) scattering is, however, always elastic, and when[354] $\langle S \rangle = 0$, simply contributes an *additional* factor of $2\pi b db$ to $d\sigma_T$. The non-shadow factor $2\pi b db$ can be partitioned, in a manner which depends on the dynamics, between the different possible final states. A fraction, $P(b)$, might go to reactive scattering and the remainder, $1 - P(b)$, will be non-reactive. Irrespective of the magnitude of $P(b)$, the total will add up to equation (7.40) (cf. Section 7.2.2). Note also that irrespective of the magnitude of $P(b)$, a collision cannot lead exclusively to products, as the shadow-elastic contribution is ever present[9].

One can now try to determine $P(b)$ from the available experimental information[1, 374] or introduce an additional tool, namely, the restrictions imposed by the initial and final channels on the dynamics. The motion of the reactants as they approach and the motion of the products as they recede have a twofold influence, which is best seen from equation (7.5). Both the magnitude of the distortion phase δ and the probability to reach the central interaction region depend on the motion in the initial and final channels. It follows from equation (7.5) that under semi-classical conditions, the change in any observable in a collision is a sum of three contributions[9, 90, 375]:

(i), the initial change, during the approach motion; (r), the change in the reaction region during the switch-over from the initial to the final channel; and (f) the change during the final, receding motion.

Further progress requires the introduction of assumptions regarding the dynamics. For direct reactions, these can be formally summarised by the condition that S' be a real (and hence, orthogonal) matrix. This is a formal way of stating that the switch-over from the reactant to the product channel is instantaneous[90, 376, 377] and so is not accompanied by any deflections[377]. The deflections of non-reactively and reactively scattered molecules are thus the sums of their initial and final deflections[375, 377]. For reactive scattering, the final deflection is determined, of course, by the radial potential between the products.

When one is interested in the integrated cross-sections, a criterion for the range of initial impact parameters (say $0 < b < b_m$) that can lead into the reaction region and/or the range of final impact parameters that can lead to the release of the products are required. Various criteria have been discussed[9, 359, 377–379] for different physical models. The reaction cross-section can be written in terms of either the initial or the final impact parameter sums

$$\sigma_R = 2\pi \int_0^{b_m} bP(b)\mathrm{d}b = 2\pi \int_0^{b'_m} b'P(b')\mathrm{d}b' \tag{7.41}$$

in a simplistic analysis[9, 359, 380–382] the integral is performed using an integration by parts, assuming that the variation of $P(b)$ with b is small, $(\mathrm{d} \ln P/\mathrm{d}b) \ll b^{-1}$, so that, simplistically,

$$\sigma_R \simeq \pi P(b_m)b_m^2 \tag{7.42}$$

Formally, equation (7.42) is just the mean-value theorem. Unfortunately, it only determines the product $P(b_m)b_m^2$ and hence, one cannot determine b_m directly from σ_R, unless, of course, σ_R is large, when the assumption that $P(b_m) \simeq 1$ is reasonable. The simplistic result expressed in equation (7.42) has been applied to a variety of situations[1, 359, 374, 377, 380–383]. For simple situations, it can be justified using detailed arguments. It provides a zeroth-order lever on the dynamics. Another simplistic result that provides useful zeroth-order information is obtained by combining the assumptions about deflections in direct reactions with the result for $\mathrm{d}\sigma_R$ to yield[9, 318, 375, 384]

$$\frac{\mathrm{d}\sigma_R}{\mathrm{d}\omega} = (b/\sin \theta_R)P[b(\theta)]/|\mathrm{d}\Theta_R/\mathrm{d}b| \tag{7.43}$$

where $\theta_R = |\Theta_R|$ is the deflection for the reactively scattered molecules. Among the several features correctly described by equation (7.43) is the gradual shift of the maximum of $\mathrm{d}\sigma_R/\mathrm{d}\omega$ to lower angles as the energy is increased[318].

7.3.3 Chemical dynamics

With the exception of classical trajectory calculations, exact and approximate solutions for the dynamics of reactive collisions are in their infancy. However, 'the baby has rosy cheeks and is growing daily'. In principle, all the methods described in Sections 7.2.4–7.2.6 can be used. The essential difficulty, however, is that for a reactive collision $A + BC \rightarrow AB + C$ there is no unique 'relative separation' coordinate. The AB distance, which is initially unbounded, is bounded for the products and conversely for the BC distance, etc. Attempts to define a coordinate system ('natural collision coordinates') which passes smoothly from those appropriate to describe the reactants to those appropriate to describe the products have been made, most notably by Marcus[385]. The concept is very appealing, but the practical implementation requires some caution[386]. There have been various suggestions of how to improve the natural collision coordinates or how to use other sets[133, 386–392], but the problem is still open.

In principle, one can by-pass the difficulties above and also incorporate the proper boundary conditions by working with the Faddeev equations[393–396]. This has several very attractive features, one being the absence of the worry regarding the role of the continuum in a close-coupling calculation. The distinction between reactant and product states is also clear-cut. This approach has its share of difficulties. In any practical application, the method requires a novel form of the potential energy[396] and the partial-wave analysis is also not trivial. It may be, however, that this is a worthwhile price.

With the exception of Section 7.3.3.2, this is an 'open problems' type of discussion. Much has been done in this field, but much more is to be done before it can be regarded as closed.

7.3.3.1 Computational studies

These were recently reviewed thoroughly and extensively by Light[386]. Very recent compilations will be found in studies of the $H + H_2$ reaction[339–341, 390, 397–399]. Progress since Light's review has been partly in improved computational results, but mainly in the breaking from the collinear world, into the world of rotating molecules[400–402]. This development, which paralleled a similar but earlier one in the classical trajectory calculations, is still too young to stand for judgement. Reports of the enormous effort (and computing time) that went into these $H + H_2$ calculations must be taken in perspective. The early three-dimensional classical trajectory calculations were also not trivial, yet today the method is almost routine.

Approximate calculations have also gone three-dimensional, as indicated by the recent work on $H + H_2$ [403, 404], $A + BC$ [405], $H + Br$ [406], $Ar^+ + HD$ [407], and $K + Br_2$ [429].

In view of the large amount of labour (and papers[339–341, 390, 392, 397–399]) devoted to the $H + H_2$ collinear collision, one should try to draw some general conclusions. Not everyone, however, would agree with the observations below, which apply to the collinear case that serves as a testing ground in this field.

(a) If tunnelling is defined as a reaction occurring at an energy below the potential energy barrier height (without any zero-point energy of H_3), then there is no tunnelling to any significant extent. The apparent threshold energy in classical trajectory calculations is higher (by nearly 2 kcal mol^{-1}) than the apparent quantal threshold due to a higher dynamic barrier[339, 408, 409], but there is no tunnelling, as defined above.

(b) If a vibrationally adiabatic reaction is defined as one proceeding without a significant change in the vibrational quantum number, then the $H+H_2$ reaction is not adiabatic in the post-threshold energy regime. It is more nearly statistical, practically as soon as the threshold for vibrational excitation is crossed, and certainly at energies exceeding 25 kcal mol^{-1}.

(c) When a reaction coordinate is used, the centrifugal potential due to the curvature of the reaction coordinate is a dominant effect[408], both physically and numerically. An approximation that neglects the curvature is restricted to very low velocities.

(d) The oscillations in the computed transmission probability as a function of the energy are probably due to internal excitation resonances[409].

Although the resonances are seen[409] in classical calculations as snarled trajectories, the oscillations in the transmission probability, which are due[181] to interferences between the direct and compound processes, cannot be reproduced by the classical trajectory computation.

7.3.3.2   Microscopic reversibility

In principle, this is an exact method[9, 359, 370, 382, 410–414]. Instead of obtaining directly the cross-section of interest, one looks for an (experimental or theoretical) cross-section for a process related by symmetry considerations[9] to the one under study. This is standard in high-energy physics, but has only recently[359, 370, 382, 411–414] come into use in molecular collision theory, where the most useful symmetry (but not the only one[9, 133]) is time-reversal invariance. Of course, on the level of fully averaged thermal rate constants, this symmetry (detailed balance) has been used traditionally to relate the forward and backward rate constants through the equilibrium constant.

Let $\sigma(\bar{n} \quad \bar{m})$ be the cross-section for the transition between the *levels* $\bar{n}$ and $\bar{m}$ (i.e., summed over final degenerate states m and averaged over initial degenerate states n). Then microscopic reversibility implies that at a given *total* energy E

$$Y_{\bar{n}\bar{m}}(E) = \pi^{-1} k_{\bar{n}}^2 g_{\bar{n}} \sigma(\bar{n} \to \bar{m}) = \pi^{-1} k_{\bar{m}}^2 g_{\bar{m}} \sigma(\bar{m} \to \bar{n}) \qquad (7.44)$$

(The yield *function* $Y(E)$, $(Y(E) = \Sigma_{\bar{m},\bar{n}} Y_{\bar{n}\bar{m}}(E))$, is not identical to the transition rate $\bar{\omega}(E)$ defined by Kinsey[412]. $Y(E)/h$ is the state-to-state transition rate *summed* over both initial and final sets of state while $\bar{\omega}(E)$ is the state-to-state transition rate *averaged* over both initial and final sets of states. Both rates are, however, fully symmetric). There are several ways in which equation (7.44) can be used:

(a) It implies that there is an optimal amount of information content in any measurement of averaged cross-sections[412].

(b) It enables the cross-section (or yield) in one direction to be determined from the cross-section in the opposite direction[359, 370, 382, 410–414].

(c) It enables the theoretician (in the optical model or otherwise) to consider the restrictions imposed by the final channel on the reaction cross-section[359, 362, 414].

(d) It provides a consistency check on both theoretical models and experimental results.

As a trivial example[414] consider a reaction which in the forward (f) direction is exothermic and without a barrier. Let a reaction occur whenever the reactants approach to within a separation d so that the forward (exoergic) reaction cross-section is $\sigma_f = \pi d_f^2$. At a given total energy E, the forward and backward wave numbers are related by $E = (\hbar^2 k_b^2/2\mu_b) = (\hbar^2 k_f^2/2\mu_f) + E_0$, where E_0 is the reaction exothermicity. The yield function is $Y(E) \propto (E - E_0)$ (and $\bar{\omega}(E) \propto (E - E_0)[(E - E_0)E]^{-\frac{1}{2}}$) and so the reaction cross-section in the backward, endothermic direction is[414] $\sigma_b = \pi d_b^2(1 - E_0/E)$. σ_b is an 'Arrhenius-type' cross-section, leading to the rate constant $k(T) = AT^{\frac{1}{2}} \exp(-E_0/kT)$.

7.3.3.3 Non-adiabatic processes

Are we living in a 'fool's paradise' shaped like an adiabatic potential energy surface[415]? Wherever one looks[416-425], electronically non-adiabatic processes contribute, and significantly so, as soon as they are energetically allowed. (See also Section 7.2.6.5 and references therein.) That the $H + H_2$ reaction appears to proceed (electronically) adiabatically is not surprising since the surface crossing in that case[294] appears at the equilateral configuration which does not lead to reaction. Initial theoretical studies on 'surface hopping' are just beginning to appear[16, 262, 278, 328, 426-428]. Whether one likes it or not, we shall have to be prepared to deal with this phenomenon.

7.3.3.4 Energy and chemical change

This section is just a statement of the problem. That chemical reactions produce energy and that energy can be used to promote the chemical change are two of the more fundamental observations of physical chemistry. Yet only too often all that is available is a highly averaged, equilibrium type of information, in the nature of ΔH for the reaction or the activation energy of the reaction rate constant. 'How is energy liberated?' and 'How is energy used in promoting the chemical change?' are two of the most interesting current questions, both intellectually and technologically[430, 431]. Such questions can only be answered experimentally by working under non-equilibrium conditions[430, 431] and the answers are not always the expected ones. Energy is quite often released preferentially into either internal *or* translational degrees of freedom[430-432]. Conversely, by making either internal *or* translational energy available preferentially to the reactants, one can effect a spectacular increase in the reaction rate[411, 430-438]. Only partial conceptual answers are currently available from theory[382, 408, 411, 436-441] and much work should be forthcoming on this topic.

7.4 CONCLUDING REMARKS

The recent theoretical developments in the field of molecular collisions have been mostly (but not exclusively) along the headings of the table of contents at the start of this chapter. The problem of a computational solution of the Schrödinger equation for an elastic collision (or an inelastic collision involving fewer than 100 coupled states) can be regarded as essentially closed. For systems involving densely spaced levels, semi-classical methods will usually suffice to reproduce the kind of averaged information that is to be expected from experimental studies in the near future. Further fundamental work is necessary, however, in understanding the relation between the potential energy surface and the observed scattering behaviour, particularly so for adiabatic processes and other collisions which depend in a critical fashion on the short-range forces. There is still no complete theory of the coupling constants for inelastic collisions.

Is, then, the lack of accurate potential energy surfaces the current bottleneck in molecular collision theory? Only partly. The primary aim of the theory, after all, is not to mimic experiment but to provide that which is not available from experiment, namely, the sound and useful conceptual framework that can show the correlation between the different experimental results, provide the proper medium for the formulation of working hypotheses, and predict new phenomena. Many useful results have been obtained and many are still to come, independently of a detailed knowledge of the potential energy surfaces.

The theory of reactive collisions is not with us, yet. Most (but by no means all) of the necessary tools are becoming available. The forthcoming detailed potential energy surfaces should stimulate the development of the theory and provide for it the analogue of the happy hunting ground that the Lennard-Jones (and other) potentials have provided for elastic scattering. The theory is in a 'go' position. The next few years will show if it can also deliver.

Acknowledgement

I would like to acknowledge the benefit of discussions with Professor R. B. Bernstein and Dr. B. R. Johnson, and to thank Professor Bernstein for his comments on the first draft of this manuscript.

References

1. Kinsey, J. L. (1972). *MTP Int. Rev. Sci.: Reaction Kinetics, Phys. Chem., Ser. I*, Ed. by J. C. Polanyi, Vol. 9, Chap. 6 (London: Butterworths)
2. Steinfeld, J. I. (1972). *Int. Rev. Sci.: Reaction Kinetics, Phys. Chem., Ser. I*, Ed. by J. C. Polanyi, Vol. 9, Chap. 8. (London: Butterworths)
3. Certain, P. R. and Bruck, L. (1972). *Int. Rev. Sci.: Theoretical Chemistry, Phys. Chem., Ser. I*, Vol. 1, Chap. 4. Ed. by W. Byers Brown. (London:Butterworths)
4. Lester, W. A., Jr. (Ed.). (1971). *Potential Energy Surfaces in Chemistry*. (San Jose: I.B.M.)
5. (1967). *Molecular Dynamics of the Chemical Reactions of Gases*. (London: The Faraday Society)
6. (1971). *Electronic and Atomic Collisions*. (Amsterdam: North-Holland)
7. Stevens, B. (1967). *Collisional Activation in Gases*. (Oxford: Pergamon Press)
8. Hartmann, J. (Ed.). (1968). *Chemische Elementarprozesse*. (Berlin: Springer-Verlag)
9. Levine, R. D. (1969). *Quantum Mechanics of Molecular Rate Processes*. (Oxford: Clarendon Press)

10. Burnett, G. M. and North, A. M. (Eds.). (1969–1971). *Transfer and Storage of Energy by Molecules*, Volumes 1–3. (London: Wiley-Interscience)
11. Geltman, S. (1969). *Topics in Atomic Collision Theory*. (New York: Academic Press)
12. Laidler, K. J. (1969). *Theories of Chemical Reaction Rates*. (New York: McGraw-Hill)
13. Geltman, S., Mahanthappa, K. T. and Brittin, W. E. (Eds.). (1969). *Lectures in Theoretical Physics*, XI-C: *Atomic Collision Processes*. (New York: Gordon and Breach)
14. Hochstim, A. R. (Ed.) (1969). *Kinetic Processes in Gases and Plasmas*. (New York: Academic Press)
15. McDaniel, E. W., Čermak, V., Dalgarno, A., Ferguson, E. E. and Friedman, L. (1970). *Ion–Molecule Reactions*. (New York: Wiley)
16. Nikitin, E. E. (1970). *Theory of Elementary Atomic and Molecular Processes*, (in Russian). (Moscow: Publishing House "Chemistry")
17. Massey, H. S. W. (1971). *Electronic and Ionic Impact Phenomena*, Volume III. (Oxford: Clarendon Press)
18. McDowell, M. R. C. and Colleman, J. P. (1970). *Introduction to the Theory of Ion–Atom Collisions*. (Amsterdam: North-Holland)
19. Schlier, Ch. (Ed.). (1970). *Molecular Beams and Reaction Kinetics*. (New York: Academic Press)
20. Hirschfelder, J. O. and Henderson, D. (Eds.). (1971). *Chemical Dynamics*. (New York: Wiley-Interscience)
21. Bates, D. R. and Estermann, I. (Eds.). (1965–). *Advances in Atomic and Molecular Physics*. (New York: Academic Press)
22. Prigogine, I. and Rice, S. A. (Eds.). (1958–). *Advances in Chemical Physics*. (New York: Wiley-Interscience)
23. Löwdin, P.-O. (Ed.) (1968–). *Advances in Quantum Chemistry*. (New York: Academic Press)
24. Eyring, H. (Ed.) (1950–). *Annual Review of Physical Chemistry*. (Palo Alto: Annual Reviews Inc.)
25. McDaniel, E. W. and McDowell, M. R. C. (Eds.) (1969–). *Case Studies in Atomic Collision Physics*. (Amsterdam: North-Holland)
26. (1967). *Proceedings V International Conference on the Physics of Electronic and Atomic Collisions*. (Leningrad: Nauka)
27. (1968). *Ber. Bunsenges. phys. Chem.*, **72**, No. 8
28. (1969). *Proceedings VI International Conference on the Physics of Electronic and Atomic Collisions*. (Cambridge: M. I. T. Press)
29. (1969). *Israel J. Chem.*, **7**, No. 2
30. (1971). *Appl. Optics*, **10**, No. 8
31. Barnett, C. F. (Ed.) (1963–). *Bibliography of Atomic and Molecular Processes*. (Springfield, Va.: U.S. Dept of Commerce)
32. Bernstein, R. B. (1966). *Advan. Chem. Phys.*, **10**, 75
33. Pauly, H. and Toennies, J. P. (1968). *Methods of Experimental Physics*, Bederson, B. and Fite, W. L. (Eds.), Vol. 7A, Chap. 3.1. (New York: Academic Press)
34. Beck, D. (1970). *Molecular Beams and Reaction Kinetics*, Schlier, Ch. (Ed.), 15. (New York: Academic Press)
35. Starkschall, G. and Gordon, R. G. (1971). *J. Chem. Phys.*, **54**, 663
36. Das, G. and Wahl, A. C. (1971). *Phys. Rev.*, **A4**, 825
37. Kutzelnigg, W. and Gelus, M. (1970). *Chem. Phys. Lett.*, **7**, 296
38. Levine, R. D., Johnson, B. R. and Bernstein, R. B. (1969). *J. Chem. Phys.*, **50**, 1694
39. Anderson, B. and Nielsen, S. E. (1971). *Mol. Phys.*, **21**, 523
40. Bernstein, R. B. and Muckerman, J. T. (1967). *Advan. Chem. Phys.*, **12**, 389
41. Buck, U. and Pauly, H. (1968). *Z. Physik*, **208**, 390
42. Schafer, T. P., Siska, P. E. and Lee, Y. T. (1971). *Electronic and Atomic Collisions*, 546. (Amsterdam: North-Holland)
43. Newton, R. G. (1966). *Scattering Theory of Waves and Particles*, 610. (New York: McGraw-Hill)
44. O'Brien, T. J. P. and Bernstein, R. B. (1969). *J. Chem. Phys.*, **51**, 5112
45. Gerber, R. B. (1971). Private communication; Gerber, R. B. and Karplus, M. (1970). *Phys. Rev.*, **D1**, 998
46. Firsov, O. B. (1953). *Zh. Eksp. Teor. Fiz.*, **24**, 279
47. Sabatier. P. C. (1966). *J. Math. Phys.*, **7**, 1515

48. O'Brien, T. J. P. (1968). *Ph.D. thesis,* U. of Wisconsin. (WIS-TCI-321)
49. Volmer, G. (1969). *Z. Physik,* **226,** 423
50. Miller, W. H. (1971). *J. Chem. Phys.,* **54,** 4174
51. Buck, U. (1971). *J. Chem. Phys.,* **54,** 1923
52. Bernstein, R. B., Dalgarno, A., Massey, H. S. W. and Percival, I. C. (1963). *Proc. Roy. Soc.,* **A274,** 427
53. Levine, R. D. (1969). *Chem. Phys. Lett.,* **4,** 309
54. Herschbach, D. R. and Kwei, G. H. (1964). *Atomic Collision Processes,* 972. (Amsterdam: North-Holland)
55. Gislason, E. A. and Kwei, G. H. (1967). *J. Chem. Phys.,* **46,** 2338
56. Helbing, R. K. B. and Rothe, E. W. (1968). *J. Chem. Phys.,* **48,** 3945
57. Olson, R. E. and Bernstein, R. B. (1968). *J. Chem. Phys.,* **49,** 162
58. Arthurs, A. M. and Dalgarno, A. (1960). *Proc. Roy. Soc.,* **A256,** 540
59. Levine, R. D. (1968). *Chem. Phys. Lett.,* **2,** 76. Also Ref. 9, Section 3.5.2
60. Levine, R. D. and Johnson, B. R. (1969). *Chem. Phys. Lett.,* **4,** 365
61. Levine, R. D. and Bernstein, R. B. (1970). *J. Chem. Phys.,* **53,** 686
62. Massey, H. S. W. and Mohr, C. B. O. (1934). *Proc. Roy. Soc.,* **A144,** 188
63. Richman, E. and Wharton, L. (1970). *J. Chem. Phys.,* **53,** 945
64. Olson, R. E. and Bernstein, R. B. (1969). *J. Chem. Phys.,* **50,** 246
65. Cross, R. J., Jr. (1968). *J. Chem. Phys.,* **49,** 1976
66. Miller, W. H. (1969). *J. Chem. Phys.,* **50,** 3124
67. Levine, R. D. (1969). *Chem. Phys. Lett.,* **4,** 211
68. Helbing, R. K. B. (1969). *J. Chem. Phys.,* **51,** 3628; (1970). ibid., **53,** 1547
69. Bennewitz, H. G., Kramer, K. H., Paul, W. and Toennies, J. P. (1964). *Z. Physik,* **177,** 84
70. Lawley, K. P. (1967). *Mol. Phys.,* **13,** 131
71. Reuss, J. and Stolte, S. (1969). *Physica,* **42,** 111
72. Bennewitz, H. G. and Haerten, R. (1969). *Z. Physik,* **227,** 399
73. Miller, W. H. (1969). *J. Chem. Phys.,* **50,** 3868
74. Bennewitz, H. G., Gengenbach, R., Haerten, R. and Müller, G. (1969). *Chem. Phys. Lett.,* **3,** 374
75. Gordon, M. D. and Secrest, D. (1970). *J. Chem. Phys.,* **52,** 120
76. Lester, W. A., Jr. (1970). *J. Chem. Phys.,* **53,** 1511
77. Lester, W. A., Jr. (1971). *J. Chem. Phys.,* **54,** 3171
78. Kutzelnigg, W. (1970). Private communication
79. Krauss, M. and Mies, F. H. (1965). *J. Chem. Phys.,* **42,** 2703
80. Rapp, D. and Kassal, T. (1969). *Chem. Rev.,* **69,** 61
81. Kelley, J. D. and Wolfsberg, M. (1970). *J. Chem. Phys.,* **53,** 2967
82. LaBudde, R. A. and Bernstein, R. B. (1971). *J. Chem. Phys.,* in the press
83. Pauly, H. and Toennies, J. P. (1965). *Advan. Atom. Mol. Phys.,* **1,** 195
84. Buckingham, A. D. (1967). *Advan. Chem. Phys.,* **12,** 107
85. Kestner, N. R. and Margenau, H. (1969). *Theory of Intermolecular Forces.* (Oxford: Pergamon Press)
86. Van de Ree, J. and Okel, J. G. R. (1971). *J. Chem. Phys.,* **54,** 589
87. Tang, K. T. and Karplus, M. (1968). *J. Chem. Phys.,* **49,** 1676
88. Bunker, D. L. (1970). *Molecular Beams and Reaction Kinetics,* Schlier, Ch. (Ed.), 355. (New York: Academic Press)
89. Miller, W. H. (1971). *J. Chem. Phys.,* **54,** 5386
90. Levine, R. D. (1972). *J. Chem. Phys.,* in the press
91. Pattengill, M. D., Curtiss, C. F. and Bernstein, R. B. (1971). *J. Chem. Phys.,* in the press
92. Garrido, L. M. (1960). *Proc. Phys. Soc.,* **76,** 33
93. Miles, J. R. N. and Dahler, J. S. (1970). *J. Chem. Phys.,* **52,** 616; ibid., **52,** 1292
94. Marcus, R. A. (1970). *J. Chem. Phys.,* **52,** 4803
95. Zeleznik, F. J. and Svehla, R. A. (1970). *J. Chem. Phys.,* **53,** 632
96. Eu, B. C. (1971). *J. Chem. Phys.,* **54,** 559
97. Cohen, A. O. and Marcus, R. A. (1968). *J. Chem. Phys.,* **49,** 4509; (1970). ibid., **52,** 3140
98. Altermeyer, M. and Marcus, R. A. (1970). *J. Chem. Phys.,* **52,** 393
99. Bjerre, A. (1971). *Physica,* **52,** 377
100. Van de Ree, J. (1971). *J. Chem. Phys.,* **54,** 3249
101. Feldman, D., Sutton, D. and Steinfeld, J. I. (1971). *Electronic and Atomic Collisions,* 638. (Amsterdam: North-Holland)

102. Mahan, B. H. (1970). *J. Chem. Phys.*, **52**, 5221
103. Holdy, K. E., Klotz, L. C. and Wilson, K. R. (1970). *J. Chem. Phys.*, **52**, 4588
104. Heidrich, F. E., Wilson, K. R. and Rapp, D. (1971). *J. Chem. Phys.*, **54**, 3885
105. Marcus, R. A. (1971). *J. Chem. Phys.*, **54**, 3965
106. Miller, W. H. (1971). *Accounts Chem. Res.*, **4**, 161
107. Levine, R. D. (1971). *Mol. Phys.*, **22**, 497
108. Burke, P. G. and Smith, K. (1962). *Rev. Mod. Phys.*, **34**, 458
109. Marriott, R. (1964). *Proc. Phys. Soc.*, **83**, 159; ibid., **84**, 877
110. Barnes, L. L., Lane, N. F. and Lin, C. C. (1965). *Phys. Rev.*, **137**, A388
111. Smith, K., Henry, R. J. W. and Burke, P. G. (1966). *Phys. Rev.*, **147**, 21
112. Johnson, B. R. and Secrest, D. (1966). *J. Math. Phys.*, **7**, 2187
113. Allison, A. C. and Dalgarno, A. (1967). *Proc. Phys. Soc.*, **90**, 609
114. Lester, W. A., Jr. and Bernstein, R. B. (1967). *Chem. Phys. Lett.*, **1**, 207; ibid., **1**, 347
115. Diestler, D. J. and McKoy, V. (1968). *J. Chem. Phys.*, **48**, 2941; ibid., **48**, 2951
116. Johnson, B. R. and Secrest, D. (1968). *J. Chem. Phys.*, **48**, 4682
117. Lester, W. A., Jr., and Bernstein, R. B. (1968). *J. Chem. Phys.*, **48**, 4896
118. Lester, W. A., Jr. (1968). *J. Comput. Phys.*, **3**, 322
119. Chan, S. K., Light, J. C. and Lin, J. L. (1968). *J. Chem. Phys.*, **49**, 86
120. Riley, M. E. and Kuppermann, A. (1968). *Chem. Phys. Lett.*, **1**, 537
121. Gordon, R. G. (1969). *J. Chem. Phys.*, **51**, 14
122. Cheung, A. S. and Wilson, D. J. (1969). *J. Chem. Phys.*, **51**, 3448; ibid., **51**, 4733
123. Sams, W. N. and Kouri, D. J. (1969). *J. Chem. Phys.*, **51**, 4809; ibid., **51**, 4815
124. Sams, W. N. and Kouri, D. J. (1970). *J. Chem. Phys.*, **52**, 4144
125. Gordon, R. G. (1970). *J. Chem. Phys.*, **52**, 6211
126. Sams, W. N. and Kouri, D. J. (1970). *J. Chem. Phys.*, **53**, 496
127. Wilson, D. J. (1970). *J. Chem. Phys.*, **53**, 2075
128. Allison, A. C. (1970). *J. Comput. Phys.*, **6**, 378
129. Hayes, E. F. and Kouri, D. J. (1971). *J. Chem. Phys.*, **54**, 878
130. Locker, D. J. (1971). *J. Chem. Phys.*, **54**, 1799
131. Levine, R. D. (1968). *J. Chem. Phys.*, **49**, 51
132. Coulson, C. A. and Gerber, R. B. (1968). *Mol. Phys.*, **14**, 117
133. Miller, W. H. (1969). *J. Chem. Phys.*, **50**, 407
134. Calogero, F. (1967). *Variable Phase Approach to Potential Scattering*, Chapter 19. (New York: Academic Press)
135. Kouri, D. J. and Curtiss, C. F. (1965). *J. Chem. Phys.*, **43**, 1919
136. Johnson, B. R. (1969). Private communication
137. Bellman, R., Kalaba, R. and Wing, G. M. (1960). *J. Math. Phys.*, **1**, 280
138. Goel, N. S. and Buff, F. P. (1970). *J. Math. Phys.*, **11**, 508
139. Stokes, G. G. (1862). *Proc. Roy. Soc.*, **11**, 545
140. Zemach, Ch. (1964). *Nuovo Cimento*, **33**, 939
141. Weare, J. H. and Thiele, E. (1968). *Phys. Rev.*, **167**, 11
142. Mortensen, E. M. and Gucwa, L. D. (1969). *J. Chem. Phys.*, **51**, 5695
143. Gerber, R. B. and Karplus, M. (1969). *J. Chem. Phys.*, **51**, 2726
144. Micha, D. A. (1969). *J. Chem. Phys.*, **50**, 722
145. Micha, D. A. and Rotenberg, M. (1970). *Chem. Phys. Lett.*, **6**, 79
146. Rotenberg, M. (1971). *Phys. Rev.*, **A4**, 220
147. Roberts, R. E. (1971). *J. Chem. Phys.*, **55**, 100
148. Gray, C. G. and Van Kranendouk, J. (1966). *Can. J. Phys.*, **44**, 2411
149. Cross, R. J., Jr. and Gordon, R. G. (1966). *J. Chem. Phys.*, **45**, 3571
150. Mahan, B. H. (1967). *J. Chem. Phys.*, **46**, 98
151. Reuss, J. (1967). *Physica*, **34**, 413
152. Stephenson, J. C., Wood, R. E. and Moore, C. B. (1968). *J. Chem. Phys.*, **48**, 4790
153. Thompson, S. L. (1968). *J. Chem. Phys.*, **49**, 3400
154. Shin, H. K. (1968). *J. Chem. Phys.*, **49**, 3964
155. Miller, W. H. (1968). *J. Chem. Phys.*, **49**, 2373
156. Sharma, R. D. (1969). *Phys. Rev.*, **177**, 102
157. Curtiss, C. F. and Bernstein, R. B. (1969). *J. Chem. Phys.*, **50**, 1168
158. Yardley, J. T. (1969). *J. Chem. Phys.*, **50**, 2464
159. Fenstermaker, R. W., Curtiss, C. F. and Bernstein, R. B. (1969). *J. Chem. Phys.*, **51**, 2439
160. Shin, H. K. (1969). *J. Phys. Chem.*, **73**, 4321

161. Child, M. S. (1970). *Proc. Roy. Soc.*, **A315**, 259
162. Takayanagi, K. (1965). *Advan. Atom. Mol. Phys.*, **1**, 149
163. Callaway, J. and Bauer, E. (1965). *Phys. Rev.*, **140**, A1072
164. Pechukas, P. and Light, J. C. (1966). *J. Chem. Phys.*, **44**, 3897
165. Cross, R. J., Jr. (1967). *J. Chem. Phys.*, **47**, 3724; (1968). ibid., **48**, 4838
166. Fenstermaker, R. W. and Bernstein, R. B. (1967). *J. Chem. Phys.*, **47**, 4417
167. Levine, R. D. (1969). *J. Phys. B.*, **2**, 839
168. Thiele, E. and Weare, J. H. (1968). *J. Chem. Phys.*, **48**, 2324
169. von Seggern, M. and Toennies, J. P. (1970). *Chem. Phys. Lett.*, **5**, 613
170. Tsien, T. P. and Pack, R. T. (1970). *Chem. Phys. Lett.*, **6**, 54; ibid., **8**, 579
171. Roberts, R. E. (1970). *J. Chem. Phys.*, **53**, 1937
172. Levine, R. D. and Balint-Kurti, G. G. (1970). *Chem. Phys. Lett.*, **6**, 101; ibid., **7**, 107
173. Rabitz, H. A. and Gordon, R. G. (1970). *J. Chem. Phys.*, **53**, 1815; ibid., **53**, 1831
174. Roberts, R. E. (1971). *J. Chem. Phys.*, **54**, 1224
175. Rabitz, H. A. (1971). *J. Chem. Phys.*, **55**, 407
176. Cross, R. J., Jr. (1971). *J. Chem. Phys.*, **55**, 510
177. Gordon, R. G. (1971). *Cambridge Conference on Molecular Energy Transfer*
178. Sharma, R. D. (1970). *Phys. Rev.*, **A2**, 173
179. Sharma, R. D. and Kern, C. W. (1971). *J. Chem. Phys.*, **55**, 1171
180. Levine, R. D. (1969). *Israel J. Chem.*, **7**, 237
181. Levine, R. D., Shapiro, M. and Johnson, B. R. (1970). *J. Chem. Phys.*, **52**, 1755
182. Bernstein, R. B. and Kramer, K. H. (1966). *J. Chem. Phys.*, **44**, 4473
183. Cross, R. J., Jr. (1968). *J. Chem. Phys.*, **49**, 1753
184. Curtiss, C. F. (1968). *J. Chem. Phys.*, **49**, 1952; (1970). ibid., **52**, 4832
185. Anderson, P. W. (1949). *Phys. Rev.*, **76**, 647
186. Levine, R. D. and Johnson, B. R. (1971). *Chem. Phys. Lett.*, **8**, 501
187. Shin, H. K. (1971). *J. Phys. Chem.*, **75**, 923
188. Morse, R. I. and LaBrecque, R. J. (1971). *J. Chem. Phys.*, **55**, 1522
189. Eu, B. C. (1970). *J. Chem. Phys.*, **52**, 1882; ibid., **52**, 3903
190. Wartell, M. A. and Cross, R. J., Jr. (1970). *Chem. Phys. Lett.*, **5**, 477
191. Cross, R. J., Jr. (1969). *J. Chem. Phys.*, **51**, 5163
192. Pechukas, P. (1969). *Phys. Rev.*, **181**, 166; ibid., **181**, 174
193. Miller, W. H. (1970). *J. Chem. Phys.*, **53**, 1949; ibid., **53**, 3578
194. Marcus, R. A. (1970). *Chem. Phys. Lett.*, **7**, 525
195. Miller, W. H. (1970). *Chem. Phys. Lett.*, **7**, 431
196. Levine, R. D. and Johnson, B. R. (1970). *Chem. Phys. Lett.*, **7**, 404
197. Beĭgman, I. L., Vaĭnshtein, L. A. and Sobel'man, I. I. (1969). *Zh. Eksp. Teor. Fiz.*, **57**, 1703 [*Sov. Phys.*, **30**, 920]
198. Percival, I. C. and Richards, D. (1970). *J. Phys. B.*, **3**, 315; ibid., **3**, 1035
199. Marcus, R. A. (1971). *Cambridge Conference on Molecular Energy Transfer*
200. Bernstein, R. B. and Levine, R. D. (1968). *J. Chem. Phys.*, **49**, 3872
201. Eu, B. C. (1970). *J. Chem. Phys.*, **52**, 3021
202. Harris, R. M. and Wilson, J. F. (1971). *J. Chem. Phys.*, **54**, 2088
203. Levine, R. D. (1969). *J. Chem. Phys.*, **50**, 1
204. Gerber, R. B. (1969). *Proc. Roy. Soc.*, **A309**, 221
205. Roberts, R. E. and Ross, J. (1970). *J. Chem. Phys.*, **52**, 1464
206. Eastes, W. and Secrest, D. (1971). *Chem. Phys. Lett.*, **9**, 508
207. Tang, K. T. (1969). *Phys. Rev.*, **187**, 122. Hayes, E. F., Wells, C. A. and Kouri, D. J. (1971). *Phys. Rev.*, **A4**, 1017
208. Clarke, J. H., Weare, J. H. and Thiele, E. (1971). *J. Chem. Phys.*, **55**, 3201
209. Nyeland, C. and Hunding, A. (1970). *Chem. Phys. Lett.*, **5**, 143
210. Thiele, E. and Katz, R. (1971). *J. Chem. Phys.*, **55**, 3195
211. Kolker, H. J. (1971). *Chem. Phys. Lett.*, **10**, 498
212. Levine, R. D. (1971). *Chem. Phys. Lett.*, **11**, 109
213. Gordon, R. G. (1968). *Advan. Mag. Resonance*, **3**, 1
214. Ben-Reuven, A. (1969). *Advan. Atom. Mol. Phys.*, **5**, 201
215. Cooper, J. (1969). *Lectures in Theoretical Physics*, **XIC**, 241
216. Gordon, R. G., Klemperer, W. and Steinfeld, J. I. (1968). *Ann. Rev. Phys. Chem.*, **19**, 215
217. Lewis, E. L., McNamara, L. F. and Michels, H. H. (1971). *Phys. Rev.*, **A3**, 1939
218. Royer, A. (1971). *Phys. Rev.*, **A4**, 499

219. Takeo, M. (1970). *Phys. Rev.*, **A1**, 1143
220. McQuarrie, D. A. and Bernstein, R. B. (1968). *J. Chem. Phys.*, **49**, 1958
221. Birnbaum, G. (1967). *Advan. Chem. Phys.*, **12**, 487
222. Bloom, M. and Oppenheim, I. (1967). *Advan. Chem. Phys.*, **12**, 549
223. Kinsey, J. L., Riehl, J. W. and Waugh, J. S. (1968). *J. Chem. Phys.*, **49**, 5269
224. Parks, E. K., Kinsey, J. L. and Riehl, J. W. (1970). *J. Chem. Phys.*, **52**, 5970
225. Bouchiat, C. C., Bouchiat, M. A. and Pottier, L. C. L. (1969). *Phys. Rev.*, **181**, 144
226. Miller, W. H. (1968). *J. Chem. Phys.*, **48**, 1651
227. Connor, J. N. L. (1968). *Mol. Phys.*, **15**, 621; (1969). ibid., **16**, 525
228. Berry, M. V. (1969). *Sci. Prog.*, **57**, 43
229. Dickinson, A. S. and Bernstein, R. B. (1970). *Mol. Phys.*, **18**, 305
230. Dickinson, A. S. (1970). *Mol. Phys.*, **18**, 444
231. Baylis, W. E. (1970). *Phys. Rev.*, **A1**, 990
232. Mahan, G. D. (1970). *J. Chem. Phys.*, **52**, 258
233. Buckingham, R. A. and Gal, E. (1968). *Advan. Atom. Mol. Phys.*, **4**, 37
234. Gersh, M. E. and Bernstein, R. B. (1969). *Chem. Phys. Lett.*, **4**, 221
235. Allison, A. C. (1969). *Chem. Phys. Lett.*, **3**, 371
236. Jackson, J. L. and Wyatt, R. E. (1971). *J. Chem. Phys.*, **54**, 5271
237. Johnson, B. R., Balint-Kurti, G. G. and Levine, R. D. (1970). *Chem. Phys. Lett.*, **7**, 268
238. LeRoy, R. J. and Bernstein, R. B. (1971). *J. Chem. Phys.*, **54**, 5114
239. Stwalley, W. C. (1969). *VI International Conference on the Physics of Electronic and Atomic Collisions*, 51. (Cambridge: M.I.T. Press)
240. Micha, D. A. (1967). *Phys. Rev.*, **162**, 88
241. Levine, R. D. (1970). *Accounts Chem. Res.*, **3**, 273
242. Erlewein, W., von Seggern, M. and Toennies, J. P. (1968). *Z. Physik*, **211**, 35; (1969). ibid., **218**, 341
243. Burke, P. G., Scrutton, D., Tait, J. H. and Taylor, A. J. (1969). *J. Phys. B.*, **2**, 1155
244. Muckerman, J. T. and Bernstein, R. B. (1970). *J. Chem. Phys.*, **52**, 606
245. Muckerman, J. T. (1969). *J. Chem. Phys.*, **50**, 627
246. Bandrauk, A. D. and Child, M. S. (1970). *Mol. Phys.*, **19**, 95
247. Sams, W. N. and Kouri, D. J. (1970). *J. Chem. Phys.*, **52**, 2556; (1971). ibid., **55**, 1248
248. Hazi, A. U. and Taylor, H. S. (1970). *Phys. Rev.*, **A1**, 1109. Fels, M. F. and Hazi, A. U. (1971). *Phys. Rev.*, **A4**, 662
249. Csizmadia, I. G., Polanyi, J. C., Roach, A. C. and Wong, W. H. (1969). *Can. J. Chem.*, **47**, 4097
250. Bunker, D. L. and Pattengill, M. (1968). *J. Chem. Phys.*, **48**, 773
251. Bersohn, R. (1969). *Comments Atom. Mol. Phys.*, **1**, 84
252. Gersten, J. and Foley, H. M. (1968). *J. Chem. Phys.*, **49**, 5254
253. Mitchell, J. K. and Fortson, E. N. (1968). *Phys. Rev. Lett.*, **21**, 1621
254. Ryndrandt, J. D. and Rabinovitch, B. S. (1971). *J. Chem. Phys.*, **54**, 2275
255. Roberts, R. E. and Bernstein, R. B. (1970). *Chem. Phys. Lett.*, **6**, 282
256. Roberts, R. E. (1971). *J. Chem. Phys.*, **54**, 1422
257. Mahan, G. D. and Lapp, M. (1969). *Phys. Rev.*, **179**, 19
258. McKellar, A. R. W. and Welsh, H. L. (1971). *J. Chem. Phys.*, **55**, 595
259. Cundall, R. B. (1969). *Transfer and Storage of Energy by Molecules*, Burnett, G. M. and North, A. M. (Eds.), Volume 1, 1. (London: Wiley-Interscience)
260. Chupka, W. A., Berkowitz, J. and Gutman, D. (1971). *J. Chem. Phys.*, **55**, 2724
261. Watanabe, T. (1968). *Advances in Chemistry: Radiation Chemistry II*, 176. (Washington: American Chemical Society)
262. Nikitin, E. E. (1968). *Chemische Elementarprozesse*, Hartman, J. (Ed.), 43. (Berlin: Springer-Verlag)
263. Callaway, J. (1969). *Lectures in Theoretical Physics*, **XIC**, 119
264. Smith, F. T. (1969). *Lectures in Theoretical Physics*, **XIC**, 95
265. Bauer, E. (1969). *Kinetic Processes in Gases and Plasmas*, Hochstim, R. A. (Ed.), 381. (New York: Academic Press)
266. Berry, R. S. (1970). *Molecular Beams and Reaction Kinetics*, Schlier, Ch. (Ed.), 193, 229. (New York: Academic Press)
267. Steinfeld, J. I. (1970). *Accounts Chem. Res.*, **3**, 313
268. Nikitin, E. E. (1970). *Advan. Quant. Chem.*, **5**, 135
269. Miller, W. H. and Schaefer, H. F. (1970). *J. Chem. Phys.*, **53**, 1421

270. Slocomb, C. A., Miller, W. H. and Schaefer, H. F. (1971). *J. Chem. Phys.*, **55**, 926
271. Olson, R. E. (1969). *Phys. Rev.*, **187**, 153
272. Olson, R. E. (1970). *Phys. Rev.*, **A2**, 121
273. Olson, R. E. and Smith, F. T. (1971). *Phys. Rev.*, **A3**, 1607
274. Gordon, R. G. and Chiu, Y. N. (1971). *J. Chem. Phys.*, **55**, 1469
275. Matsuzawa, M. (1967). *J. Phys. Soc. Jap.*, **23**, 1383
276. Nakamura, H. (1968). *J. Phys. Soc. Jap.*, **24**, 1353
277. Nakamura, H. (1971). *J. Phys. Soc. Jap.*, **31**, 574
278. Child, M. S. (1969). *Mol. Phys.*, **16**, 313
279. Child, M. S. (1971). *Mol. Phys.*, **20**, 171
280. Thorson, W. R., Delos, J. B. and Boorstein, S. A. (1971). *Phys. Rev.*, **A4**, 1052
281. Kodaira, M. and Watanabe, T. (1969). *J. Phys. Soc. Jap.*, **27**, 130
282. Nakamura, H. (1969). *J. Phys. Soc. Jap.*, **26**, 1473
283. Miller, W. H. (1970). *J. Chem. Phys.*, **52**, 3563
284. Micha, D. A., Tang, S. Y. and Muschlitz, E. E. (1971). *Chem. Phys. Lett.*, **8**, 587
285. Cohen, J. S. and Lane, N. F. (1971). *Chem. Phys. Lett.*, **10**, 623
286. Anderson, B. and Nielsen, S. E. (1971). *Mol. Phys.*, **21**, 523
287. Smith, F. T. (1969). *Phys. Rev.*, **179**, 111
288. O'Malley, F. T. (1967). *Phys. Rev.*, **162**, 98
289. Berry, R. S. and Nielsen, S. E. (1970). *Phys. Rev.*, **A1**, 383; ibid., **A1**, 395
290. Krauss, M. (1970). *Ann. Rev. Phys. Chem.*, **21**, 39
291. Conroy, H. (1969). *J. Chem. Phys.*, **51**, 3979
292. Csizmadia, I. G., Kari, R. E., Polanyi, J. C., Roach, A. C. and Robb, M. A. (1970). *J. Chem. Phys.*, **52**, 6205
293. Shavitt, I., Stevens, R. M., Minn, F. L. and Karplus, M. (1968). *J. Chem. Phys.*, **48**, 2700
294. Porter, R. N., Stevens, R. M. and Karplus, M. (1968). *J. Chem. Phys.*, **49**, 5163
295. Wilson, C. W. and Goddard, W. A., III. (1969). *J. Chem. Phys.*, **51**, 716
296. Rubinstein, M. and Shavitt, I. (1969). *J. Chem. Phys.*, **51**, 2014
297. Edmiston, C., Doolittle, J., Murphy, K., Tang, K. C. and Wilson, W. (1970). *J. Chem. Phys.*, **52**, 3419
298. Krauss, M. (1968). *J. Res. Nat. Bur. Stand.*, **72A**, 553
299. Wahl, A. C. and Das, G. (1971). *Potential Energy Surfaces in Chemistry*, Lester, W. A., Jr. (Ed.), 83. (San Jose: I.B.M.)
300. Miller, K. J., Mielczarek, S. R. and Krauss, M. (1969). *J. Chem. Phys.*, **51**, 26
301. Bender, C. F. and Schaefer, H. F., III. (1971). UCRL-73273
302. Lester, W. A., Jr., and Krauss, M. (1970). *J. Chem. Phys.*, **52**, 4775
303. Lester, W. A., Jr. (1970). *J. Chem. Phys.*, **53**, 1611
304. Balint-Kurti, G. G. and Karplus, M. (1969). *J. Chem. Phys.*, **50**, 478
305. Rothenberg, S. and Schaefer, H. F., III. (1971). *Chem. Phys. Lett.*, **10**, 565
306. Balint-Kurti, G. G. and Karplus, M. (1971). *Chem. Phys. Lett.*, **11**, 203
307. Karplus, M. (1970). *Molecular Beams and Chemical Kinetics*, Schlier, Ch. (Ed.), 320. (New York: Academic Press)
308. Anlauf, K. G., Polanyi, J. C., Wong, W. H. and Woodall, K. B. (1968). *J. Chem. Phys.*, **49**, 5189
309. Muckerman, J. T. (1971). *J. Chem. Phys.*, **54**, 1164
310. Morokuma, K. and Karplus, M. (1971). *J. Chem. Phys.*, **55**, 63
311. Godfrey, M. and Karplus, M. (1968). *J. Chem. Phys.*, **49**, 3602
312. Bunker, D. L. and Parr, C. A. (1970). *J. Chem. Phys.*, **52**, 1970
313. Roach, A. C. and Child, M. S. (1968). *Mol. Phys.*, **14**, 1
314. Nyeland, C. and Ross, J. (1971). *J. Chem. Phys.*, **54**, 1665
315. Claydon, C. R., Segal, G. A. and Taylor, H. S. (1971). *J. Chem. Phys.*, **54**, 3799
316. Preston, R. K. and Tully, J. C. (1971). *J. Chem. Phys.*, **54**, 4297
317. Herschbach, D. R. (1971). *Potential Energy Surfaces in Chemistry*, Lester, W. A., Jr. (Ed.), 44. (San Jose: I.B.M.)
318. Karplus, M. (1970). *Molecular Beams and Reaction Dynamics*, Schlier, Ch. (Ed.), 372. (New York: Academic Press)
319. Kuntz, P. J. (1971). *Electronic and Atomic Collisions, Invited Papers VII ICPEAC.* (Amsterdam: North-Holland)
320. Kuntz, P. J., Mok, M. H. and Polanyi, J. C. (1969). *J. Chem. Phys.*, **50**, 4623
321. Wu, S.-F. and Marcus, R. A. (1970). *J. Chem. Phys.*, **53**, 4026

322. Mahan, B. H. (1971). *J. Chem. Phys.*, **55**, 1436
323. Polanyi, J. C. and Wong, W. H. (1969). *J. Chem. Phys.*, **51**, 1439
324. Mok, M. H. and Polanyi, J. C. (1969). *J. Chem. Phys.*, **51**, 1451
325. Mok, M. H. and Polanyi, J. C. (1970). *J. Chem. Phys.*, **53**, 4588
326. Brumer, P. and Karplus, M. (1971). *J. Chem. Phys.*, **54**, 4955
327. Czismadia, I. G., Polanyi, J. C., Roach, A. C. and Wong, W. H. (1969). *Can. J. Chem.*, **47**, 4097
328. Krenos, J., Preston, R., Wolfgang, R. and Tully, J. C. (1971). *Chem. Phys. Lett.*, **10**, 17
329. Blais, N. C. and Cross, J. B. (1970). *J. Chem. Phys.*, **52**, 3580
330. Blais, N. C. (1968). *J. Chem. Phys.*, **49**, 9
331. Kuntz, P. J., Nemeth, E. M. and Polanyi, J. C. (1969). *J. Chem. Phys.*, **50**, 4607
332. Kuntz, P. J., Nemeth, E. M., Polanyi, J. C. and Wong, W. H. (1970). *J. Chem. Phys.*, **52**, 4654
333. Bunker, D. L. and Pattengill, M. D. (1970). *J. Chem. Phys.*, **53**, 3041
334. Anderson, J. B. (1970). *J. Chem. Phys.*, **52**, 3849
335. Jaffe, R. L. and Anderson, J. B. (1971). *J. Chem. Phys.*, **54**, 2224
336. Raff, L. M., Sims, L. B., Thompson, D. L. and Porter, R. N. (1970). *J. Chem. Phys.*, **53**, 1606; (1970). *J. Amer. Chem. Soc.*, **92**, 3208
337. Russell, D. and Light, J. C. (1969). *J. Chem. Phys.*, **51**, 1720
338. Mortensen, E. M. (1968). *J. Chem. Phys.*, **49**, 3526
339. Wu, S.-F. and Levine, R. D. (1972). *Mol. Phys.* In the press
340. Truhlar, D. G. and Kuppermann, A. (1972). *J. Chem. Phys.* In the press
341. Truhlar, D. G. and Kuppermann, A. (1971). *Chem. Phys. Lett.*, **9**, 269
342. Kleinman, B. and Tang, K. T. (1969). *J. Chem. Phys.*, **51**, 4587
343. Baer, M. (1971). *J. Chem. Phys.*, **54**, 3670
344. Rankin, C. C. and Miller, W. H. (1971). *J. Chem. Phys.*, **55**, 3150
345. Suplinskas, R. J. (1968). *J. Chem. Phys.*, **49**, 5046
346. Hirschfelder, J. O. (1969). *Int. J. Quant. Chem.*, **IIIS**, 17
347. Baer, M. and Amiel, S. (1969). *Israel J. Chem.*, **7**, 341
348. Marron, M. T. (1970). *J. Chem. Phys.*, **52**, 4060
349. Kuntz, P. J. (1970). *Trans. Faraday Soc.*, **66**, 2980
350. Chang, D. T. and Light, J. C. (1970). *J. Chem. Phys.*, **52**, 5687
351. George, T. F. and Suplinskas, R. J. (1971). *J. Chem. Phys.*, **54**, 1037; ibid., **54**, 1046
352. Miller, W. H. (1970). *J. Chem. Phys.*, **52**, 543
353. Lester, W. A., Jr., and Bernstein, R. B. (1970). *J. Chem. Phys.*, **53**, 11
354. Fremerey, H. and Toennies, J. P. (1971). *Electronic and Atomic Collisions*, 249. (Amsterdam: North-Holland)
355. Truhlar, D. G. (1969). *J. Chem. Phys.*, **51**, 4617
356. Truhlar, D. G. (1971). *J. Chem. Phys.*, **54**, 2635
357. Wolf, F. A. and Haller, J. L. (1970). *J. Chem. Phys.*, **52**, 5910
358. Ham, D. O. and Kinsey, J. L. (1970). *J. Chem. Phys.*, **53**, 285
359. Levine, R. D., Wolf, F. A. and Maus, J. A. (1971). *Chem. Phys. Lett.*, **10**, 2
360. Fullerton, D. C. and Moran, T. F. (1971). *J. Chem. Phys.*, **54**, 5221
361. Fullerton, D. C. and Moran, T. F. (1971). *Chem. Phys. Lett.*, **10**, 626
362. White, R. A. and Light, J. C. (1971). *J. Chem. Phys.*, **55**, 379
363. Moran, T. F. and Fullerton, D. C. (1971). *J. Chem. Phys.*, **54**, 5231
364. Marcus, R. A. (1966). *J. Chem. Phys.*, **45**, 2139; ibid., **45**, 2630
365. Marcus, R. A. (1967). *J. Chem. Phys.*, **46**, 959
366. Coulson, C. A. and Levine, R. D. (1967). *J. Chem. Phys.*, **47**, 1235
367. Eu, B. C. and Ross, J. (1966). *J. Chem. Phys.*, **44**, 2467
368. Marcus, R. A. (1968). *J. Chem. Phys.*, **49**, 2617
369. Morokuma, K., Eu, B. C. and Karplus, M. (1969). *J. Chem. Phys.*, **51**, 5193
370. Marcus, R. A. (1970). *J. Chem. Phys.*, **53**, 604
371. Truhlar, D. G. (1970). *J. Chem. Phys.*, **53**, 2041
372. Tweedale, A. and Laidler, K. J. (1970). *J. Chem. Phys.*, **53**, 2045
373. Koeppl, G. W. and Karplus, M. (1971). *J. Chem. Phys.*, **55**, 4667
374. Ross, J. and Greene, E. F. (1970). *Molecular Beams and Chemical Kinetics*, Schlier, Ch. (Ed.), 86. (New York: Academic Press)
375. Levine, R. D. (1970). *Israel J. Chem.*, **8**, 13
376. Smith, F. T. (1960). *Phys. Rev.*, **118**, 349

377. Herschbach, D. R. (1966). *Adv. Chem. Phys.,* **10,** 319
378. Grice, R. (1970). *Mol. Phys.,* **19,** 501
379. Grice, R. and Hardin, D. R. (1971). *Mol. Phys.,* **21,** 805
380. Johnston, H. S. (1966). *Gas Phase Reaction Rate Theory.* (New York: Ronald Press)
381. Bell, K. L., Dalgarno, A. and Kingston, A. E. (1968). *J. Phys. B.,* **1,** 18
382. Levine, R. D. and Bernstein, R. B. (1971). *Chem. Phys. Lett.,* **11,** 552
383. Beuhler, R. J., Jr., and Bernstein, R. B. (1969). *J. Chem. Phys.,* **51,** 5305
384. Miller, W. H. (1968). *J. Chem. Phys.,* **49,** 2373
385. Marcus, R. A. (1968). *J. Chem. Phys.,* **49,** 2610
386. Light, J. C. (1971). *Advan. Chem. Phys.,* **19,** 1
387. Connor, J. N. L. and Marcus, R. A. (1970). *J. Chem. Phys.,* **53,** 3188
388. Shipsey, E. J. (1969). *J. Chem. Phys.,* **50,** 2685
389. Kouri, D. J. (1969). *J. Chem. Phys.,* **51,** 5204
390. Diestler, D. J. (1971). *J. Chem. Phys.,* **54,** 4547
391. Snider, N. S. (1969). *J. Chem. Phys.,* **51,** 4075
392. Crawford, O. H. (1971). *J. Chem. Phys.,* **55,** 2563
393. Faddeev, L. D. (1965). *Mathematical Aspects of the Three-Body Problem in Quantum Scattering Theory.* (Jerusalem: Israel Science Press)
394. Ross, J. (1970). *Molecular Beams and Chemical Kinetics,* Schlier, Ch. (Ed.), 392. (New York: Academic Press)
395. Micha, D. A. (1970). *Bull. Amer. Phys. Soc.,* **15,** 437
396. Lovelace, C. (1964). *Phys. Rev.,* **135,** B1225
397. McCullough, E. A. and Wyatt, R. E. (1971). *J. Chem. Phys.,* **54,** 3578
398. Diestler, D. J. and Karplus, M. (1971). *J. Chem. Phys.,* **55,** 5832
399. Johnson, B. R. (1972). *Chem. Phys. Lett.,* **13,** 172
400. Saxon, R. P. and Light, J. C. (1971). *J. Chem. Phys.,* **55,** 455
401. Wolkeen, G. and Karplus, M. (1971). *Electronic and Atomic Collisions,* 302. (Amsterdam: North-Holland)
402. Baer, M. and Kouri, D. J. (1971). *Chem. Phys. Lett.,* **11,** 238
403. Karplus, M. (1970). *Molecular Beams and Reaction Kinetics,* Schlier, Ch. (Ed.), 407. (New York: Academic Press)
404. McGuire, P. and Mueller, C. R. (1971). *Phys. Rev.,* **A3,** 1358. McGuire, P. and Micha, D. A. (1972). To be published
405. Connor, J. N. L. and Child, M. S. (1970). *Mol. Phys.,* **18,** 653
406. Pirkle, J. C. and McGee, H. A., Jr. (1968). *J. Chem. Phys.,* **49,** 3532; ibid., **49,** 4504
407. Gelb, A. and Suplinskas, R. J. (1970). *J. Chem. Phys.,* **53,** 2249
408. Hofacker, G. L. and Levine, R. D. (1971). *Chem. Phys. Lett.,* **9,** 617
409. Levine, R. D. and Wu, S. F. (1971). *Chem. Phys. Lett.,* **11,** 557
410. Ross, J., Light, J. C. and Shuler, K. E. (1969). *Kinetic Processes in Gases and Plasmas,* Hochstim, A. R. (Ed.), 281. (New York: Academic Press)
411. Anlauf, K. G., Maylotte, D. H., Polanyi, J. C. and Bernstein, R. B. (1969). *J. Chem. Phys.,* **51,** 5716
412. Kinsey, J. L. (1971). *J. Chem. Phys.,* **54,** 1206
413. Klots, C. E. (1971). *J. Phys. Chem.,* **75,** 1526
414. Levine, R. D. and Bernstein, R. B. (1972). *J. Chem. Phys.,* in the press
415. Planck, M. (1971). Private communication
416. Helbing, R. K. B. and Rothe, E. W. (1969). *J. Chem. Phys.,* **51,** 1607
417. Baede, A. P. M., Moutinho, A. M. C., DeVries, A. E. and Los, J. (1969). *Chem. Phys. Lett.,* **3,** 530
418. Krenos, J. and Wolfgang, R. (1970). *J. Chem. Phys.,* **52,** 5961
419. Lacmann, K. and Herschbach, D. R. (1970). *Chem. Phys. Lett.,* **6,** 106
420. Fluendy, M. A. D., Horne, D. S., Lawley, K. P. and Morris, A. W. (1970). *Mol. Phys.,* **19,** 659
421. Holiday, M. G., Muckerman, J. T. and Friedman, L. (1971). *J. Chem. Phys.,* **54,** 1058
422. Ewing, J. J., Milstein, R. and Berry, R. S. (1971). *J. Chem. Phys.,* **54,** 1752
423. Struve, W. S., Kitagawa, T. and Herschbach, D. R. (1971). *J. Chem. Phys.,* **54,** 2759
424. Maier, W. B., II, and Murad, E. (1971). *J. Chem. Phys.,* **55,** 2307
425. Moutinho, A. M. C., Aten, J. A. and Los, J. (1971). *Physica,* **53,** 471
426. Nikitin, E. E. (1968). *Chem. Phys. Lett.,* **1,** 266
427. McMillan, W. L. (1971). *Phys. Rev.,* **A4,** 69

428. Tully, J. C. and Preston, R. K. (1971). *J. Chem. Phys.*, **55**, 562
429. Eu, B. C., Huntington, J. H. and Ross, J. (1971). *Can. J. Chem.*, **49**, 966
430. Polanyi, J. C. (1971). *App. Optics*, **10**, 1717
431. Bernstein, R. B. (1971). *Israel J. Chem.*, in the press
432. Anlauf, K. G., Charters, P. E., Horne, D. S., MacDonald, R. G., Maylotte, D. H., Polanyi, J. C., Skrlac, W. J., Tardy, D. C. and Woodall, K. B. (1970). *J. Chem. Phys.*, **53**, 4091
433. Schmeltekopf, A. L., Ferguson, E. E. and Fehsenfeld, F. C. (1968). *J. Chem. Phys.*, **48**, 2966
434. Chupka, W. A. and Russell, M. E. (1968). *J. Chem. Phys.*, **49**, 5426
435. Odiorne, T. J., Brooks, P. R. and Kasper, J. V. V. (1971). *J. Chem. Phys.*, **55**, 1980
436. O'Malley, T. F. (1970). *J. Chem. Phys.*, **52**, 3269
437. Mahan, B. (1970). *Accounts Chem. Res.*, **3**, 393
438. Wolfgang, R. (1970). *Accounts Chem. Res.*, **3**, 48; (1969). ibid., **2**, 248
439. Polanyi, J. C. (1971). *Potential Energy Surfaces in Chemistry*, Lester, W. A., Jr. (Ed.), 10. (San Jose: I.B.M.)
440. Levine, R. D. (1971). *Chem. Phys. Lett.*, **10**, 510
441. George, T. F. and Ross, J. (1971). *J. Chem. Phys.*, **55**, 3851

8
Transport Theory and its Application to Liquids

A. R. ALLNATT and L. A. ROWLEY
University of Western Ontario, London, Canada

8.1 INTRODUCTION

This review covers an approximately 5 year period. Section 8.2 is about developments in very general aspects of transport theory which have stemmed from the well-known exact master equations[1-5]. Sections 8.3 and 8.4 are specialised to classical monatomic liquids. It is hoped that the material in Section 8.4 will provide a link between the kinetic equation methods of Sections 8.2 and 8.3 and the time-correlation formalism which is reviewed in detail elsewhere[107, 108, 145]

8.2 SOME GENERAL ASPECTS OF DISTRIBUTION FUNCTION THEORY

8.2.1 Introduction

Much of transport theory is about the derivation of kinetic equations valid on a time-scale which is long in comparison with some elementary mechanical time such as the duration of a collision. A kinetic equation is a time-irreversible equation for a reduced distribution function for a few degrees of freedom. The work of Severne[6] on non-uniform classical systems with pair-wise forces, and earlier work by the Brussels school on homogeneous systems[7, 8], illustrate a very general approach. The generalisation of this scheme to other situations and further developments of it are described below. These results also provide a framework for some of the approximate theories of liquids in Section 8.2. It is therefore appropriate to begin with an outline of this procedure.

First, the distribution function $f(t)$ is split into two parts: the vacuum part, f_0, is essentially a product of functions depending on one degree of freedom only, whereas the correlation part, f_c, does not have this property. It is shown that $\partial_t f_0(t)$ is related to $f_0(\tau)$, $\tau \leqslant t$, and to $f_c(0)$ in what is called a master equation. Another equation relates f_c to its initial value and past values of the vacuum. The equations are exact. Next, the asymptotic solutions, $\tilde{f}_0, \tilde{f}_c$ are found for long times. $\tilde{f}_0$ obeys a closed differential equation, whereas $\tilde{f}_c$ is related to an integral of an operator acting on $\tilde{f}_0$. This step can be justified for certain classes of system and initial conditions if the vacuum state is adequately chosen. Finally, kinetic equations can be derived from $\tilde{f}_0(t)$ by appropriate integration over excess degrees of freedom and manipulations.

There were difficulties with the original formulations:

(i) The Laplace transformation of the time variable proved a natural tool for taking the asymptotic time limit for homogeneous systems[8] but was not fully used for inhomogeneous systems[6].

(ii) Infinite-order perturbation theory was used.

(iii) The extension to inhomogeneous systems of coupled oscillators, electromagnetic fields etc. was not apparent.

(iv) One might wonder whether the study of the hierarchy of Fourier components of f always employed is really easier than direct study of the hierarchy of reduced distribution functions.

Some progress on these points is reviewed below.

The main notation used is the following. The distribution function for a classical system of N degrees of freedom is $f(p^N, q^N, t) \equiv f(p_1, \ldots p_N, q_1, \ldots q_N, t)$ where p_i, q_i are canonical variables. Reduced distribution functions are defined by

$$f_{s,r}(p_1,\ldots p_s,q_1,\ldots q_r,t) = \frac{N!}{(N-r)!}\int (dp)^{N-s}(dq)^{N-r}f \tag{8.1}$$

$$f_s(1,\ldots,s) \equiv f_{s,s}(p_1,\ldots q_s) \tag{8.2}$$

$$\phi_s(p^s,t) = \int (dp)^{N-s}(dq)^N f \tag{8.3}$$

$$n_s(q^s,t) = \frac{N!}{(N-s)!}\int (dp)^N(dq)^{N-s}f \tag{8.4}$$

The Liouville equation and its formal solution are written:

$$i\partial_t f(t) = Lf(t) \tag{8.5}$$

$$f(t) = e^{-itL}f(0) \equiv U(t)f(0) \tag{8.6}$$

The formal results in Section 8.2 generally hold in quantum statistics with appropriate interpretations[5, 9] of f and L.

Much use is made of projection operators. For example we can divide f uniquely by operators P and Q:

$$f(t) = Pf(t) + Qf(t) \tag{8.7}$$

$$P+Q = 1, \qquad P^2 = P, \qquad Q^2 = Q \tag{8.8}$$

The accepted usage is to call such operators projection operators. Many of their applications stem from adroit use of the identity

$$e^{(A+B)t} = e^{Bt} + \int_0^t d\tau\, e^{B\tau} A e^{(A+B)(t-\tau)} \tag{8.9}$$

for arbitrary operators A and B. For example from equations (8.6) and (8.9) we can obtain

$$f(t) = U_{LQ}(t)f(0) - \int_0^t d\tau\, U_{LQ}(\tau)iLPf(t-\tau) \tag{8.10}$$

or by operating on this with iLQ and putting $Q = 1-P$ on the left

$$-\partial_t f(t) = iLPf(t) + iLU_{QL}(t)Qf(0) - \int_0^t d\tau\, iLU_{QL}(\tau)QiLPf(t-\tau) \tag{8.11}$$

$$U_A(t) \equiv e^{-iAt}, \quad A = LQ, QL \tag{8.12}$$

8.2.2   Exact and asymptotic master equations

8.2.2.1   *Exact equations and some realisations of projection operators*

If operators can be found which separate suitable vacuum and correlation components

$$f_0(t) = Pf(t), \quad f_c(t) = Qf(t) \tag{8.13}$$

then application of P to equation (8.11) and Q to (8.10) gives results with the structure discussed above; namely the master equation for f_0

$$\partial_t f_0(t) + PiLP f_0(t) = \int_0^t d\tau \psi(\tau) f_0(t-\tau) + \mathscr{D}(t) f_c(0) \tag{8.14}$$

$$\psi(t) = PiLU_{QL}(t)QiLP \tag{8.15}$$

$$\mathscr{D}(t) = -PiLQU_{QL}(t)Q \tag{8.16}$$

and an integral relation for the correlations

$$f_c(t) = f_c'(t) + f_c''(t) \tag{8.17}$$

$$f_c''(t) = \int_0^t d\tau \mathscr{C}(\tau) f_0(t-\tau) \tag{8.18}$$

$$f_c'(t) = U_{QL}(t) f_c(0) \tag{8.19}$$

$$\mathscr{C}(t) = -U_{QL}(t)QiLP \tag{8.20}$$

Equation (8.14), often loosely[10] called non-markovian because of the time convolution, is identical[5] to the earliest master equations[1–4] and is easily generalised when L is time-dependent[11, 12]. We might equally use P on equation (8.10) and Q on equation (8.11), a kind of symmetry emphasised by Balescu[13].

An alternative form[13] makes more direct contact with the resolvent formalism and perturbation theory. We can re-write equation (8.6)

$$f(t) = (2\pi i)^{-1} \int_c dz \, e^{-izt} R(z) f(0); \quad t > 0 \tag{8.21}$$

where $R(z) \equiv (z-L)^{-1}$ is the resolvent operator, and c is a line antiparallel to the real axis in the upper half-plane and a large semi-circle in the lower half-plane[8]. Suppose L can be split into an unperturbed part L_0 and a perturbed part L_1. The identity

$$R(z) = R_0(z) + R_0(z)L_1 R(z) \tag{8.22}$$

where $R_0 \equiv (z-L_0)^{-1}$ is easily checked. Define E by

$$E(z) = L_1(1 - QR_0(z)L_1)^{-1} \tag{8.23}$$

and eliminate L_1 from equations (8.22) and (8.23). The result is

$$R(z) = R_0(z)E(z)PR(z) + R_0 E(z)QR_0(z) + R_0(z) \tag{8.24}$$

Substitution of R in equation (8.21) gives equations for f and $\partial_t f$ which replace equations (8.10) and (8.11) in Balescu's formalism[13]. Operation by P and Q gives the previous results with

$$\psi(z) = P(E(z) - L_1)P, \quad \mathscr{D}(z) = PE(z)R_0(z)Q,$$

and

$$\mathscr{C}(z) = QR_0(z)E(z)P \tag{8.25}$$

provided that $QR_0 = R_0 Q$ in equation (8.24). The meaning of this condition is discussed below. These results were originally obtained using perturbation expansions[13] and later non-perturbatively[14]. The functions $-i\psi(iz)$, $\mathscr{D}(iz)$, and $\mathscr{C}(iz)$ are defined to equal the conventional Laplace transforms of $\psi(t)$, $\mathscr{D}(t)$, and $\mathscr{C}(t)$, e.g. $-i\psi(iz) \equiv \int_0^\infty dt \, e^{-zt}\psi(t)$.

The total distribution $(f = f_0 + f_c)$ can be found by solving equations (8.14) and (8.17) by Laplace transformation, equation (8.21). The result is equivalent to the basic identity[14]

$$R = (\mathscr{C} + 1)\mathscr{H}(P + \mathscr{D}) + \mathscr{P} \tag{8.26}$$

$$\mathscr{P} \equiv Q(z - QLQ)^{-1}Q\mathscr{H} \equiv (z - PLP - \psi)^{-1} \tag{8.27}$$

Equations (8.13)–(8.27), except (8.25), are identities whatever realisation of P and Q is chosen.

We review next examples of realisations which make f_0 and f_c true vacuum and correlation. Generally, L_0 is a sum of operators each depending on a single degree of freedom, and so R_0 describes the evolution of uncoupled degrees of freedom (free particles, uncoupled oscillators). For classical systems the p variables (momenta, actions) are invariants of that motion. A Fourier expansion of f can be made using orthonormal basis functions which are eigenfunctions of L_0. In the Dirac notation[8]:

$$f(t) \equiv |f(t)\rangle = \sum_k |k\rangle\langle k|f(t)\rangle \equiv \sum_k |k\rangle\rho_k(p,t) \tag{8.28}$$

with $k \equiv (k_1, ..., k_N)$. Convenient notations are

$$\rho_{0,k_1,0...k_s,0}(p,t) \equiv \rho_{k_1...k_s}(1,...,s|...) \tag{8.29}$$

$$\rho_{k_1...k_s}(1,...,s) \equiv \int (dp)^{N-s}\rho_{k_1...k_s}(1,...,s|...) \tag{8.30}$$

The $\rho_{k_1...k_s}$ terms are rather simply related[7] to s-particle cluster functions of the reduced distribution functions. $\rho_0(p,t)$ refers to the state with all Fourier variables zero and is just $\phi_N(p,t)/V^{N/2}$.

For classical gases the condition that for a large system the particles are not correlated at large separations leads[9] to

$$\phi_s(p^s) = \prod_{i=1}^{s} \phi_1(p_i) \tag{8.31}$$

$$\rho_{k_1...k_s}(1,..,s|...) = \rho_{k_1...k_s}(1,...,s) \prod_{i=s+1}^{N} \phi_1(p_i) \tag{8.32}$$

$$f_{s,r} = f_r \prod_{i=r+1}^{s} \phi_1(p_i); \quad s > r \tag{8.33}$$

For an anharmonic solid (angle-action) these 'molecular chaos' results do not hold since correlations in angle space do not have a 'finite range' properly.

For a homogeneous gas, translational invariance requires $\rho_k = 0$ unless $\Sigma_i k_i = 0$[9]. States with $\rho_k \neq 0$ relate to spatial molecular correlations and one chooses ρ_0 as the vacuum[7]. The linear operator

$$P = |0\rangle\langle 0| \tag{8.34}$$

leads directly to the Prigogine–Résibois equations[3, 15]. The operators ψ, $\mathscr{D}$, $\mathscr{C}$, $\mathscr{P}$ are 'irreducible'; they contain $U_{QL}Q$ and hence have no intermediate vacuum states. It is supposed that in consequence they are analytical functions of z in the whole complex plane except for a finite discontinuity along the real axis, and that values on the real axis and in the lower half-plane can be obtained by analytic continuation of the operators defined in the upper half-plane. These continuations are taken to be regular except for poles at a finite distance from the real axis. We have therefore

$$\psi, \mathcal{D}, \mathcal{C}, \mathcal{P} \rightarrow 0 \text{ for } t \rightarrow \infty \tag{8.35}$$

When equation (8.35) is valid $\psi, \mathcal{D}, \mathcal{C}, \mathcal{P}$ are called the collision, destruction, creation and irreducible operators and the well-known asymptotic equations, e.g. equation (8.42), then result. Their existence reflects a time-scale separation, $t_r > t_c$; the relaxation time is longer than the duration of the collision process. For the validity of equation (8.35) there are restrictions[8, 16] on the initial condition (finite range of correlations in configuration space) and intermolecular potential (short range). Finally, $\rho_0 \equiv \phi_N$ factorises, equation (8.31), as is essential to obtain closed kinetic equations.

No general recipe for construction of a vacuum state can be given but the objective is to regain the structure of the above results (factorisation of f_0, and the possibility of an asymptotic time limit, equation (8.35)). Other homogeneous systems have been considered. For a gravitational plasma there is a logarithmic singularity in $\psi(z)$ for $z \rightarrow 0$ so equation (8.35) fails[17]. For light–matter interactions equation (8.35) is plausible but not established in detail[18]. For the anharmonic solid the approach to equilibrium is well established[19] but lack of equation (8.31), or a substitute, halts the complete analysis.

For an inhomogeneous particle system f_1 reflects macroscopic inhomogeneities. There is a hydrodynamic length scale L_h (lower bound of $f_1/\nabla f_1$) and a molecular scale L_m (range of interactions). It can be shown[9] that if $L_h \gg L_m$ then ρ_{k_i} is non-zero for $|k_i| = 0(L_h^{-1})$ but zero for $|k_i| \gg L_h^{-1}$ and further

$$\rho_{k_1 \ldots k_s} = \prod_{j=1}^{s} \rho_{k_j}(j); \quad |k_j| \sim L_h^{-1}, \ j = 1, \ldots s \tag{8.36}$$

One requires that f_c relates as before to microscopic correlations varying over L_m and f_0 to macroscopic inhomogeneities varying slowly over L_h. Severne[6] did not use projection operators but assumed equations (8.32) and (8.33) at $t = 0$. He derived an equation for ρ_{k_i} and from this an equation for f_1 employing the decomposition

$$\rho_{k_1 \ldots k_s} = \prod_{j=1}^{s} \rho_{k_j} + \Gamma_{k_1 \ldots k_s}^{(s)} \tag{8.37}$$

The fully factorised term is the vacuum and the irreducible operators have no factorised intermediate states. The non-linearity (with respect to singlet coefficients) of the decomposition means the vacuum cannot be defined by a linear projection operator alone. However, Baus[15] proposed

$$P = \sum_{k_0} |k_0\rangle \langle k_0| \tag{8.38}$$

where k_0 denotes states with all wave vectors in the hydrodynamic range, $|k_j| \leqslant L_h^{-1}$, for which equation (8.36) holds. Equations (8.36) and (8.37) lead to Severne's equation but for a truncated f_1

$$f_1(i; L_h^{-1}) = \int d\mathbf{k}_i e^{i\mathbf{k}_i \cdot \mathbf{q}_i} \theta(L_h^{-1} - |k_i|)\rho_{k_i}(i) \tag{8.39}$$

where θ is the Heaviside step function. This can be used for strong gradients with the limit $L_h \to 0$ taken after the calculation[15].

Balescu[13] emphasised that the intuitive idea of a vacuum state implies that correlations cannot be created out of the vacuum or vice versa by the action of L_0, hence

$$L_0 P - P L_0 = L_0 Q - Q L_0 = 0 \tag{8.40}$$

He further proposed the non-linear realisation

$$P f_N(t) = \phi_N(t) \prod_{j=1}^{N} (f_1(j,t)/\phi_1(j,t)) \tag{8.41}$$

If molecular chaos (equation 8.31) holds, the right hand is just $\prod f_1(j)$ which is Severne's choice. The more general result might be applied however to the angle-action case, though no detailed results are available. The correct manipulation of non-linear realisations requires some care[20].

Thus the present situation is not entirely satisfactory. The Baus realisation depends on widely separated length scales and use of the Fourier representation: that of Balescu is non-linear and difficult to apply. Since this article was completed Severne[164] has considered the possibilities of working on the reduced distribution functions (see also Section 8.3).

8.2.2.2 Asymptotic equations

For systems with a well-chosen vacuum state such that the irreducible operators satisfy equation (8.35), the asymptotic master equation valid for long times is from equation (8.14)

$$\partial_t \tilde{f}_0(t) = F\tilde{f}_0(t) + \int_0^\infty d\tau \psi(\tau)\tilde{f}_0(t-\tau) \tag{8.42}$$

$$F = -PiLP \tag{8.43}$$

where the tilde denotes a distribution satisfying an asymptotic equation. This can be written in markovian form[8]

$$\partial_t \tilde{f}_0(t) = \theta \tilde{f}_0(t); \quad \tilde{f}_0(t) = e^{\theta t}\tilde{f}_0(0) \tag{8.44}$$

where the time-independent operator θ satisfies[13, 21]

$$\theta = F + \int_0^\infty d\tau \psi(\tau) e^{-\tau\theta} \tag{8.45}$$

It is straightforward to show[6, 21]

$$\theta = \Omega\theta_0 \equiv \Omega(F + \int_0^\infty d\tau\psi(\tau)) \equiv \Omega(F - i\psi^{(0)}) \tag{8.46}$$

where θ_0 is the 'dominant solution' corresponding to the stationary approximation for $\tilde{J}_0$ in the collision term, and

$$\Omega = 1 + \psi^{(1)} + (\psi^{(1)})^2 + \tfrac{1}{2}\psi^{(2)}(\psi^{(0)} + iF) \tag{8.47}$$

$$\psi^{(n)} \equiv (D^n\psi(z))_{z\,=\,+i0}; \quad D \equiv \partial/\partial z \tag{8.48}$$

Ω represents corrections arising from the finite duration of collisions. A recurrence formula[6, 21] and alternative forms[21, 22] can be derived for Ω. For homogeneous systems ($F = 0$) θ is called the $\Omega\psi$ operator. The asymptotic correlation equation is similarly

$$\tilde{J}_c(t) = \int_0^\infty d\tau \mathscr{C}(\tau)\tilde{J}_0(t-\tau) \tag{8.49}$$

The so-called 'post-initial' condition, $\tilde{J}_0(0)$, $\tilde{J}_c(0)$ can be fixed from the exact expressions for $Pf(t)$, $Qf(t)$ from equations (8.21) and (8.26). The asymptotic limit is taken, as in the derivation of equation (8.42)[3], by assuming the already noted properties of the irreducible operators and only poles at the origin are retained. For homogeneous systems the results are[23, 24]

$$\tilde{\rho}_0(t) = e^{\theta t}\tilde{\rho}_0(0) = e^{\theta t}A[\rho_0(0) + \sum_k D_k\rho_k(0)] \tag{8.50}$$

$$\tilde{\rho}_k(t) = C_k\tilde{\rho}_0(t) \tag{8.51}$$

$$A = A(1 + \sum_k D_k C_k)A \tag{8.52}$$

$$A = \sum_{n=0}^{\infty} (n!)^{-1}(D^n\psi_{00}^n(z))_{z=0} \tag{8.53}$$

$$AD_k = \sum_{n=0}^{\infty} (n!)^{-1}(D^n[\psi_{00}^n(z)\mathscr{D}_{0k}(z)])_{z=0} \tag{8.54}$$

$$C_kA = \sum_{n=0}^{\infty} (n!)^{-1}(D^n[\mathscr{C}_{k0}(z)\psi_{00}^n(z)])_{z=0} \tag{8.55}$$

where

$$\mathscr{D}_{0k} = <0|\mathscr{D}|k> \text{ etc.}$$

Equivalent expressions for A, C_k, D_k are often convenient[21, 25]. The importance of these very formal results will be seen in Sections 8.2.3 and 8.2.4. A rigorous, less formal, treatment of the asymptotic limit has been given[26].

8.2.2.3   Relation to invariants of motion

The characterisation of dissipative systems (ones which approach thermodynamic equilibrium) has been discussed by Prigogine[7]. For homogeneous systems, it follows from equation (8.42) that one requires $\psi(z \to i0) \neq 0$ in order that the momenta do not remain invariants of motion[7]. Extensions of these ideas[27-29] are relevant to Section 8.2.3. If Φ is an invariant of motion it may be shown[28, 29]

(a) $\lim (\psi(z) + PLP)P\Phi = 0$

(b) $\lim_{z \to i0} \mathscr{C}(z) P\Phi = Q\Phi$

$$\leftrightarrow L\Phi = 0 \tag{8.56}$$

For example (a) is simply a consequence[28] of equation (8.26) and the identity $PR(z)\Phi = P\Phi/z$. If $P\Phi$ is a solution of (a) then we can construct an invariant of motion using (b).

The distinction between dissipative systems and non-dissipative systems has been discussed for homogeneous systems ($PLP = 0$) on the basis of equation (8.56)[27]. If $\psi(z \to i0) \to 0$ then only (b) remains; the invariants of L_0 can be extended to L. This occurs if the Hamiltonian has a purely discrete non-degenerate spectrum. If $\psi(z \to i0) \neq 0$ then (a) can be satisfied for only very limited classes of quantities, e.g. $\langle 0 | F(H) \rangle$, where H is the Hamiltonian, and in general invariants of L_0 are destroyed in passing to L. Diagonalisation of H is not possible for such a system[27]. The distinction between non-dissipative ($\psi = 0$) and dissipative systems ($\psi \neq 0$) is incorporated in the formalism of independent sub-dynamics.

8.2.3   Independent sub-dynamics

In the formulation of Sections 8.2.1 and 8.2.2 the *exact* equations for vacuum and correlations are coupled. It has now been shown that vacuum and correlation parts can be redefined so that their evolutions are always independent[25, 27]. The redefined vacuum part satisfies an equation like (8.44) and describes the approach to thermodynamic equilibrium (for a dissipative system), and the modified correlations satisfy an independent master equation.

The asymptotic equations (8.50) and (8.51) can be summarised[23] in the matrix form:

$$\tilde{\rho}(t) = \begin{pmatrix} \tilde{\rho}_0(t) \\ \tilde{\rho}_c(t) \end{pmatrix} = \begin{pmatrix} e^{\theta t}A & e^{\theta t}AD \\ Ce^{\theta t}A & Ce^{\theta t}AD \end{pmatrix} \begin{pmatrix} \rho_0(0) \\ \rho_c(0) \end{pmatrix}$$

$$\equiv \Sigma(t)\rho(0) \tag{8.57}$$

where ρ_c denotes the whole set of Fourier components ρ_k, $k \neq 0$. By use of equation (8.52) one may verify the semi-group property

$$\Sigma(t)\,\Sigma(t') = \Sigma(t+t'); \quad t,t' > 0 \tag{8.58}$$

Consequently the operator Π defined by

$$\Pi = \lim_{t \to +0} \Sigma(t) \tag{8.59}$$

is a projection operator[23, 30]. Let us define $\hat{\Pi} \equiv 1 - \Pi$ and use the notation

$$\bar{X} = \Pi X; \quad \hat{X} = \hat{\Pi} X \tag{8.60}$$

for the components of any quantity X. If we consider the components of

$$\bar{\rho}(t) = \Pi\rho(t) \tag{8.61}$$

written in full, we obtain at once

$$\bar{\rho}_k(t) = C_k\bar{\rho}_0(t) \tag{8.62a}$$

and by considering $\Pi\hat{\rho}(t) = 0$ we obtain

$$\hat{\rho}_0(t) = -\sum_k D_k\hat{\rho}_k(t) \tag{8.62b}$$

because A is a regular operator. In other words, in the sub-space of Π only vacuum elements of ρ are independent parameters, and in the orthogonal sub-space of $\hat{\Pi}$ only the correlation components are independent parameters. By decomposing ρ_0 and ρ_c by equation (8.60) and using equation (8.62) they are found related by a regular transformation to the independent or 'privileged' components $\bar{\rho}_0$ and $\hat{\rho}_c$:

$$\begin{pmatrix} \rho_0 \\ \rho_c \end{pmatrix} = \Lambda \begin{pmatrix} \bar{\rho}_0 \\ \hat{\rho}_c \end{pmatrix} \text{ with } \Lambda = \begin{pmatrix} 1 & -D \\ C & 1 \end{pmatrix} \tag{8.63}$$

To find the equations of motion of $\bar{\rho}_0$, $\hat{\rho}_c$, one substitutes the division in equation (8.63) into

$$i\partial_t \rho_k(t) = \sum_{k'} \langle k \,|\, L \,|\, k' \rangle \rho_{k'}(t) \tag{8.64}$$

for the cases $k = 0$ and $k \neq 0$ and can obtain[25]

$$i\partial_t \bar{\rho}_0(t) = \sum_k \langle 0 \,|\, L \,|\, k \rangle C_k \bar{\rho}_0(t) = i\theta \bar{\rho}_0(t) \tag{8.65}$$

$$i\partial_t \hat{\rho}_k(t) = \sum_{k'} [\langle k \,|\, L \,|\, k' \rangle - \langle k \,|\, L \,|\, 0 \rangle D_{k'}] \hat{\rho}_{k'}(t)$$
$$\equiv \sum_{k'} \langle k \,|\, \mathscr{L} \,|\, k' \rangle \hat{\rho}_{k'}(t) \tag{8.66}$$

which are the exact independent equations for re-defined vacuum and correlations which are related to the original vacuum and correlations by the inverse of equation (8.63). Equation (8.65) generalises the Boltzmann equation (see also Section 8.2.4). Explicitly non-markovian effects appear only in equation (8.66). The following relations can also be found[25]

$$\Sigma(t)\,\hat{\Sigma}(t') = \hat{\Sigma}(t)\,\Sigma(t') = 0 \tag{8.67}$$

$$\Sigma(t) = \Pi U(t) = U(t)\Pi; \quad \hat{\Sigma}(t) = \hat{\Pi} U(t) = U(t)\hat{\Pi} \tag{8.68}$$

$$\hat{\Sigma}(t) \equiv U(t) - \Sigma(t) \tag{8.69}$$

Examination shows at once that $\Sigma(t)$ and $\hat{\Sigma}(t)$ generate the independent dynamics of $\bar{\rho}$ and $\hat{\rho}$ respectively, and further that $\hat{\rho}(t) \equiv \bar{\rho}(t)$.

For a non-dissipative system ($\psi = 0$) $\Sigma(t) = \Pi$ and hence ρ are independent of time. For a dissipative system each invarient Φ of L satisfies equation (8.56) and it can be shown that

$$\Pi\Phi = \Phi; \quad \hat{\Pi}\Phi = 0$$

Thus the invariants of motion, including the equilibrium distribution function and quantum states of infinite life-time, lie within Π. The exact projection, $\Pi\rho$, of ρ tends to thermodynamic equilibrium without any coarse-graining procedure and contains the information relevant for thermodynamics. $\hat{\Pi}\rho$ contains information on phase relations, and for long times contributions of $\hat{\Pi}\rho$ to ordinary observable vanish.

The equation for $\bar{\rho}_0$ contains the $\theta(\Omega\psi)$ operator which as we shall see has been extensively studied. It will be interesting to see how practically useful these results are for strongly coupled systems, $\bar{\rho}_0$ for ordinary transport and $\hat{\rho}_k$ for non-thermodynamic correlations, e.g. spin echo, turbulence. The

Π sub-space contains all the information to describe stationary transport phenomena[31]. The original results can be extended to general inhomogeneous systems[32].

8.2.4 Transformation theories

In the weak coupling limit equation (8.42) for $\tilde{\rho}_0$ reduces to the Peierls–Boltzmann equation, Pauli equation etc. dependent on the system. The collision operator is then Hermitian and involves energy-conserving processes, the relaxation times are real and there is a monotonic approach to equilibrium characterised by an 'H-theorem' for

$$H_B = \int dp \tilde{\rho}_0(p) \log \tilde{\rho}_0(p) \tag{8.71}$$

The entropy at equilibrium, $-kH_B$, is that for the unperturbed system.

In higher approximations the collision operator θ contains products of non-commuting Hermitian operators, e.g. $\psi^{(1)}\psi^{(0)}$. The number and variety of terms in typical problems are great and they do not analyse in terms of energy-conserving processes[33, 34]. The Boltzmann or entropy transformation theory seeks a transformation to redefined degrees of freedom (quasi-particles) obeying equations with the structure for weakly coupled systems[18, 23, 24, 34–42].

The development to 1967 was reviewed by Prigogine[34, 35, 40]. The asymptotic solution can be written

$$\tilde{\rho}_0(t) = \chi e^{-i\phi t} \chi^+ [\rho_0(0) + \sum_k D_k \rho_k(0)] \tag{8.72}$$

$$\chi \equiv \chi' \chi'' \tag{8.73}$$

where χ' is a Hermitian operator and χ'' a unitary operator. Comparison with equation (8.50) gives

$$A = \chi' \chi'; \; -i\phi = \chi^{-1}\theta\chi \tag{8.74}$$

The 'dressed' distribution function defined by

$$\tilde{\rho}_d(t) = \chi^{-1} \tilde{\rho}_0(t) \tag{8.75}$$

satisfies an equation containing only the *Hermitian*[23, 24, 34] operator $-i\phi$ (containing only energy-conserving processes):

$$\partial_t \tilde{\rho}_d(t) = -i\phi \tilde{\rho}_d(t) \tag{8.76}$$

The problem is to provide an extra condition to fix χ''.

Consider next the ensemble average of an observable B for a classical homogeneous system (extension to quantum theory is immediate):

$$\langle B \rangle = \sum_k \int dp B_k(p) \rho_k(p), \; B_k \equiv \langle B | k \rangle \tag{8.77}$$

In the asymptotic region one can eliminate the correlations using equation (8.51) and get

$$\langle B \rangle = \int dp \, \tilde{B}_d \tilde{\rho}_d \tag{8.78}$$

$$\tilde{B}_d = \chi^+ (B_0 + \sum_k C_k^+ B_k) \tag{8.79}$$

We could attempt to choose χ'' so that the entropy is given by a Boltzmann expression

$$S_B = -k \int (dp)^N \tilde{\rho}_d \log \tilde{\rho}_d \tag{8.80}$$

If such a relation exists the equilibrium distribution function will be proportional to $\exp(-\tilde{H}_d/kT)$ and it must be an eigenfunction of the collision operator with zero eigenvalue:

$$-i\phi \exp(-\tilde{H}_d/kT) = 0 \tag{8.81}$$

Up to order λ^6 in the kinetic operator, where λ is the strength of the intermolecular potential, a unique χ'' has been found from this equation in typical cases. A compact general scheme for calculation of χ has been developed[36] covering anharmonic solids, quantum gases etc. and the known results suggest[38] that the general result is the solution of

$$\partial_\lambda \chi^{-1}(\lambda) = \chi^{-1}(\lambda) \left[A(\lambda)\sum_k D_k(\lambda)\partial_\lambda C_k(\lambda)\right] \tag{8.82}$$

with $\chi^{-1}(0) = 1$.

The detailed calculations for anharmonic solids[37, 41] were already reviewed[34, 40]. Applications to unstable particle theory[37, 40] and to light–matter interaction[18] lie outside the scope of this article. Much more relevant is the demonstration[39] that S_B for a non-uniform particle system is consistent with the linear non-equilibrium thermodynamics. The entropy is now

$$S_B = -k \int (dp)^N (dq)^N \tilde{\rho}_d \log \tilde{\rho}_d = \int dr s(\mathbf{r},t) \tag{8.83}$$

where s is the entropy density, because in the non-uniform case $\tilde{\rho}_d$ is a functional of $\Pi_i^N f_1(p_i,q_i)$. Normal solutions of the kinetic equation can be found by the Chapman–Enskog procedure of expansion in powers of gradients

$$\tilde{\rho}_d = \tilde{\rho}_d^{(0)} + \tilde{\rho}_d^{(1)} + \cdots \tag{8.84}$$

with $\tilde{\rho}_d^{(0)}$ the local equilibrium function. The uniqueness of the expansion can be ensured in the usual way, because of the properties of ϕ, by the subsidiary conditions

$$\int (dp)^N (dq)^N \tilde{B}_d \tilde{\rho}_d^{(i)} = 0; \ i>0 \tag{8.85}$$

where $\tilde{B}_d$ can be the operator for number, momentum or quasi-particle energy density. A calculation of $\partial_t s(\mathbf{r},t)$ to second order in gradients from equations (8.83)–(8.85) yields precisely the Gibbs balance equation postulated in thermodynamics and hence a general H-theorem. In the third order an extra term appears roughly corresponding to propagation of mechanical waves. It is speculated that the general H-theorem may be true to all orders in gradients.

These results are obviously very important contributions to understanding the nature of non-equilibrium entropy and can be contrasted with the entropy-centred theories discussed in Section 8.4.3. In a series of papers still too incomplete for extended review, Prigogine et al. introduce a new transformation scheme, the 'physical particle representation'[43–49, 159]. The entropy again assumes the simple form like equation (8.80) but this is now a consequence of the theory in which the dressed Hamiltonian and the unperturbed Hamiltonian are diagonal·on the same basis.

The physical particle representation of the distribution function $^{(p)}\rho(t)$ is defined[43] by

$$^{(p)}\rho_0(t) = \alpha A^{-1}\bar{\rho}_{0(t)} \equiv \chi^{-1}\bar{\rho}_0(t) \tag{8.86}$$

$$^{(p)}\rho_c(t) = \beta B^{-1}\bar{\rho}_c(t); \ B^{-1} = 1+CD \tag{8.87}$$

The transformation from the original representation to the new one is then

$$^{(p)}\rho(t) = \Lambda_p^{-1}\rho(t); \ \Lambda_p^{-1} = \begin{pmatrix} \alpha & \alpha D \\ -\beta C & \beta \end{pmatrix} \tag{8.88}$$

α and β are to be chosen so that Λ_p obeys the star unitary condition[43]

$$\Lambda_p^{-1} = \Lambda_p'^{\dagger} \equiv \Lambda_p^* \tag{8.89}$$

where Λ_p' is Λ_p with L replaced by $-L$ and $\Lambda_p^{\dagger}$ the adjoint of Λ_p. This condition ensures that the averages of an observable O written as $Tr(O\rho)$ and $Tr(^{(p)}O^{(p)}\rho)$ are equal, where

$$^{(p)}O = \Lambda_p'^{-1}O \tag{8.90}$$

Equations (8.88) and (8.89) imply

$$\alpha\alpha^{\dagger} = A; \ \beta\beta^{\dagger} = B \tag{8.91}$$

The transformation can be split into a part diagonal in Π and a part diagonal in $\hat{\Pi}$:

$$\gamma \equiv \Lambda_p^{-1}\Pi = \begin{pmatrix} \alpha & \alpha D \\ O & O \end{pmatrix}; \ \hat{\gamma} \equiv \Lambda_p^{-1}\hat{\Pi} = \begin{pmatrix} 0 & 0 \\ -\beta C & \beta \end{pmatrix} \tag{8.92}$$

In the non-dissipative case one can establish an exact equation for γ and hence for α and $\alpha^{\dagger}$ consistent with equation (8.91)[43]. It turns out to be equivalent to Mandel's guess, equation (8.82), for χ^{-1}! The latter is then a unitary transformation which diagonalises H and is equivalent to Wigner–Brillouin perturbation theory[44-46]. However, equation (8.82) is apparently now established in the dissipative case, and $^{(p)}\rho_0$ satisfies the kinetic equation (8.76)[43]. No detailed applications are available except for equilibrium gases[47] and plasmas[48], which now become dynamical problems.

8.2.5 Further master equation identities

Fulinski and Kramarczyk[50] derived an exact 'markovian' equation and further identities of the same type can be found. An example is[51]

$$\partial_t Pf(t) = \dot{K}_2(t)K_2^{-1}(t)\left[(\alpha - P)f(0) + Pf(t)\right] \tag{8.93}$$

$$K_2(t) = \alpha + P[U(t)-1]; \ \alpha \text{ real}, \ |\alpha| > 2 \tag{8.94}$$

The operator K_2^{-1} exists and is expandable, whereas for the original identity[50] with $\alpha = 1$ the existence is doubtful[51]. The identities hold for classical and quantum systems and, with a suitable definition of U, even when H is time dependent. A gain–loss form with real transition probabilities can be written and the weak coupling Pauli equation derived in the long time limit if higher order terms are neglected. One can actually put $\alpha = 1$ in the final kinetic equation without difficulties. The method for taking the long time limit is rather different from that for the conventional master equation; not enough seems known to decide the best field of application[50, 52]. Different forms can be derived directly from the non-markovian Zwanzig identity[53, 54].

Finally, mention should be made of identities for f_1 derived by Fujita by simple diagram methods for classical[55] and quantum[56, 57] systems which appear structurally similar to the Severne[6] classical result. However no comparison is made with Severne or with earlier diagram studies of homogeneous systems, e.g. Reference 58. One-body Green's functions can be treated[59] by a different diagram expansion[60, 61] which it is claimed will avoid the notorious and well-established[62] divergences of density expansions in gases. There are no results for higher order functions.

8.3 APPROXIMATE KINETIC EQUATIONS

8.3.1 Three theories for liquids related to Severne's equations

The first stage in the classical route to the linear transport coefficients is to derive approximate kinetic equations for f_1 and f_2. Three highly developed theories for monatomic liquids are those of Rice and Allnatt (RA), Prigogine, Nicolis and Misguich (PNM), and Allen and Cole (AC). Current interest centres on clarifying and inter-relating their assumptions and techniques. The Severne[6] equations are a convenient framework. For long times the terms containing initial correlations are zero and one obtains from the master equation[63]

$$d_t f_1(1,t) = B(\Psi, f_1) \tag{8.95}$$

$$d_t f_1)1,t) \equiv \{\partial_t + v_1 \cdot \partial_{x_1} + m^{-1} F(x_1, t) \cdot \partial_{v_1}\} f_1(x_1, v_1, t)$$
$$B(\Psi, f_1) \equiv \int (dv dx)^{N-1} \Psi(\nabla; i0 + i\partial_t) \prod_{i=1}^{N} f_1(i, t) \tag{8.96}$$

where F is a Vlasov mean field term, Ψ is a collision operator evaluated in the limit $z \to i0 + i\partial_t$, and $\nabla \equiv (\partial_{x_1}, \dots \partial_{x_N})$. Similarly from the equation for correlations one finds

$$h_2(1,2,t) \equiv f_2(1,2,t) - f_1(1,t) f_1(2,t) = D(C_2, f_1) \tag{8.97}$$

$$D(C_2, f_1) \equiv \int (dv dx)^{N-2} C_2(\nabla; i0 + i\partial_t) \prod_{i=1}^{N} f_1(i, t) \tag{8.98}$$

where C_2 is a creation operator. If we expand to first order in the macroscopic gradients we have

$$f_1(i, t) = f_1^{(0)}(i, t) + \varepsilon f_1^{(1)}(i, t) \equiv f_1^{(0)}[1 + \varepsilon \phi] \tag{8.99}$$

$$\Psi = \Psi^{(0)} + \mu \Psi^{(1)}, \qquad C_2 = C_2^{(0)} + \mu C_2^{(1)} \tag{8.100}$$

where ε and μ are expansion parameters. $f_1^{(0)}$ denotes the local equilibrium distribution function, $\Psi^{(0)}$ and $C_2^{(0)}$ are operators for a homogeneous system, and $\Psi^{(1)}$ and $C_2^{(1)}$ are linear in ∇ and ∂_t. The first-order equations are then

$$d_t f_1^{(0)} = \mu B(\Psi^{(1)}, f_1^{(0)}) + \varepsilon \mathscr{A}(\phi) \tag{8.101}$$

$$\mathscr{A}(\phi) \equiv B(\Psi^{(0)}, f_1^{(0)}) \sum_{i=1}^{N} \phi(i) \tag{8.102}$$

$$h_2 = D(C_2^{(0)}, f_1^{(0)}) + \{\varepsilon D(C_2^{(0)}, f_1^{(0)}) \sum_{i=1}^{N} \phi(i) + \mu D(C_2^{(1)}, f_1^{(0)})\}$$

$$\equiv h_2^{(0)} + h_2^{(1)} \tag{8.103}$$

Finally, comparison with the exact BBGKY equation

$$d_t f_1(\mathbf{x}_1, \mathbf{v}_1, t) = -\frac{i 8\pi^3}{m} \int d\mathbf{v}_2 d\mathbf{k} V_k \, \mathbf{k} \cdot \partial_{v_1} h_2(\mathbf{v}_1, \mathbf{v}_2, \mathbf{k}, \mathbf{x}, t) \tag{8.104}$$

where V_k is a fourier component of the pair potential V, shows at once that Ψ and C_2 are simply related.

In the PNM theory[63-66] it is assumed, as in some older theories[67, 68], that deviations from local equilibrium can be neglected, i.e. $\varepsilon = 0$ and $f_1 = f_1^{(0)}$. This approximation excludes study of matter transport (including thermal diffusion) and kinetic contributions to other transport coefficients, but computer experiments[97] suggest it is fairly good for calculating h_2 at liquid densities. Now $h_2^{(1)}$ comprises contributions $h_{2, p}^{(1)}$ and $h_{2, v}^{(1)}$ which have matrix elements in which one free propagator or one interaction operator (L_1) respectively is modified from its homogeneous expression by the occurrence of a gradient operator (∂_{x_j}). The second PNM assumption[63] is that only the gradient operators at the latest time in the evolution operator are retained. Finally it is assumed that since the pair correlation function relaxes very quickly to local equilibrium this value can be used in the resultant expression. PNM[64] point out that their second assumption is in effect very similar to the linear trajectory approximation[69] in which L is replaced by L_0 in calculating an autocorrelation function. For numerical calculations[63, 65, 66] they used the RA model potential (hard sphere plus a soft continuous potential of Lennard-Jones type). The hard core part contributes a part identical with the dominant part of the Enskog result (which arises in that theory from $f_1^{(0)}$) and correction terms involving the complete radial distribution function. The contributions of the soft and hard parts are of comparable importance. We shall not attempt to summarise agreement with experiment except to say that it is fairly good for thermal conductivity and bulk viscosity along the co-existence curve for the rare gases but poor for shear viscosity.

The first assumption[72] of the AC[70-73] theory is that $\mu = 0$ and only terms in ε are retained, the converse of PNM. It is difficult to find evidence for this assumption. The second assumption is that $\Psi^{(0)}$ is dominated by the chain diagrams which lead to the Boltzmann binary collision operator[7] and the ring diagrams which dominate in a low density plasma[9]. Parallel assumptions are made for $C_2^{(0)}$. It is further argued that it is appropriate to use the total potential in the Boltzmann and plasma operators rather than use the RA model potential to make simplifications. The results for thermal conductivity are of the right order of magnitude[71] but poor for shear viscosity at high density[72]. However Davis and Dowling have estimated all the terms through fourth order in interaction and conclude that at liquid densities the.neglected diagrams are comparable to those retained[74]. Further, in the equilibrium limit the doublet distribution is simply a chain sum and drastically under-estimates the height of the first peak of the radial distribution function. Recently, Allen[72] proposed (i) modifying PNM by using equations (8.101)–(8.103) with the PNM approximations for $\Psi^{(1)}$, $C_2^{(1)}$ and AC approximations for $\Psi^{(0)}$, $C_2^{(0)}$ and also (ii) using the same equations but with $\Psi^{(1)}$, $C_2^{(1)}$ calculated using the AC assumption.

Unlike the PNM and AC theories the RA theory[75] was developed[76], reviewed[77, 78], given refined derivations[79, 81], and extended to mixtures[81, 106] mostly prior to the period under review. We shall therefore consider only very recent developments[74, 82-84]. We may first recall that four assumptions were used by RA:

(i) The model potential already referred to.

(ii) The motion of a pair of molecules is described by a dynamical event comprising a binary hard core encounter followed by quasi-Brownian motion in the fluctuating force field of the other molecules.

(iii) There is no direct correlation between long- and short-ranged components of the intermolecular force just before and during a binary encounter.

(iv) The dynamical event in (ii) is independent of other such events. This last assumption was introduced by the Kirkwood time-smoothing hypothesis and was later removed[80].

Davis[74] has now proposed a generalisation of the RA theory (see also Reference 83) which is briefly as follows. The RA model pair potential, $V = V_S + V_H$ where S and H are the soft and hard sphere components, can be substituted into Ψ at the latest interaction in each term and the two kinds of contribution summed separately and called ψ_S and ψ_H. The operators in the singlet equation (8.101) then factorise:

$$B(\Psi^{(1)}, f_1^{(0)}) \equiv B^{(1,0)} = B_S^{(1,0)} + B_H^{(1,0)}$$

$$\mathscr{A} = \mathscr{A}_S + \mathscr{A}_H \tag{8.105}$$

The proposal is (i) to approximate $B_S^{(1,0)}$ by the PNM method already mentioned, (ii) to approximate $\mathscr{A}_S$ by the RA Fokker–Planck operator, (iii) to approximate the remaining terms by the Enskog operator $J(E)$ used by RA (suitably linearised in gradients). The result is thus

$$d_t f_1^{(0)} = B_S^{(1,0)}(PNM) + \mathscr{A}_S(RA) + J(E) \tag{8.106}$$

in other words the RA singlet equation is corrected by using the PNM approximation for spatial delocalisation and non-markovian effects. It is shown that approximation (ii) corresponds to retaining only contributions from the inhomogeneity which occur at the latest time in the collision operator, the same as the formal statement of the second PNM approximation. The improved singlet equation leads to more nearly exact hydrodynamic equations than RA, and to kinetic contributions to transport coefficients which in the weak coupling limit (and neglecting hard sphere terms) are very near the exact weak coupling results.

A deeper break with the RA scheme is in calculation of the potential contributions to transport coefficients. By comparing the exact equation (8.104) with the new approximate one and equating coefficients of the soft part of V an expression results of the form

$$h_2 = h_2(PNM) + \Delta h_2(\phi) \tag{8.107}$$

Δh_2 contains the RA corrections to the PNM approximation arising from the deviation of f_1 from $f_1^{(0)}$. For shear viscosity the corrections are 3–20% of the PNM contributions. Agreement with experiment is not better than PNM or RA. Of course this scheme by-passes the solution of the Smoluchowski equation necessary in the RA scheme. A really detailed statement of the relation between new and old schemes remains to be made. Unlike PNM diffusion and thermal diffusion could be treated.

Another improvement of the RA theory, which is more important for plasmas than liquids, is to solve the equation for f_1 without further approximating the derived $\mathscr{A}_s$ by a Fokker–Planck operator with a momentum-independent friction constant[82].

The RA theory was an attempt to improve on the Kirkwood Brownian motion theory[76] by treating the hard and soft potential contributions separately. The Kirkwood theory gave an expression for the friction constant ζ (kT divided by the self-diffusion coefficient) which is rigorous only for a heavy particle. For the RA model potential the expression is of the form

$$\zeta = \zeta_H + \zeta_S + \zeta_{SH} \tag{8.108}$$

Helfand[69] reduced the hard sphere part, ζ_H, to the Enskog result, and evaluated the soft part, ζ_s, by the linear trajectory method. The RA theory gives ζ as the sum of these two results. A linear trajectory approximation to Kirkwood's ζ_{SH}[85] suggests this cross-term neglected by RA is comparable to ζ_H for liquids[86]. It is possible to extend the RA theory to more realistic potentials than the hard sphere plus soft potential so far used but this has not been carried through in detail.

8.3.2 Further kinetic equations

We may first consider whether the master equation identity, equation (8.14), can yield other useful results by other choices of P than those so far considered. For example Muriel and Dresden[86, 88] use

$$P = V^{-(N-r)} \int (dx)^{N-r} \tag{8.109}$$

Integration over $(p)^{N-r}$ and some reduction lead to a hierarchy of equations in which $f_r(p^r, x^r)$ is coupled to $f_{N,r}(p^N, x^r)$. They then derive a general kinetic equation using the assumptions (i) weak coupling in the form

$$e^{-i(1-P)Lt}(1-P) \cong e^{-i(1-P)L_0 t}(1-P) \tag{8.110}$$

and (ii) that an asymptotic limit for long times can be taken just as in passing from equation (8.14) to equation (8.42), i.e. a markovian approximation for the collision term and the assumption that the destruction term is zero. They explicitly consider only 'homogeneous or approximately homogeneous initial states' and assert that homogeneity is sufficient, but not necessary, to make the destruction term vanish. Hawker and Schieve[89] have analysed this procedure in some detail for f_1. The projection operator, equation (8.109), then singles out one particle in a very unsymmetric way. It corresponds in Fourier representation to

$$P = \sum_{k_1} |k_1, 0 \ldots 0\rangle \langle 0 \ldots 0, k_1| \tag{8.111}$$

and a vacuum state ρ_{k_1}, whereas in Severne's procedure[6] (Section 8.2) the vacuum states in the irreducible operators are symmetric in all particle indices. This implies that a unique or test particle is being considered, e.g. a Brownian particle. Finally, if P is used in the manner described then the destruction term contains contributions additional to those in the Severne formalism and carrying the effects of macroscopic inhomogeneity. These long-range (on a molecular level) effects prohibit $Qf(0)$ from vanishing for large particle separations and hence the destruction term does not vanish for $t \gg t_c$. The equation is therefore not suitable for inhomogeneous systems.

Nevertheless equation (8.14) can sometimes be plausibly employed with other choices of projection operator. The prototype problem is that of Brownian motion of a heavy particle in a fluid of light particles. (For examples outside the field of liquids see Reference 90.) It is well known that a firm basis for such studies was established earlier[91, 92], including derivation from Severne's equations[93]. Mazo[94-96] has demonstrated that the result can be regained and generalised to include interactions between the Brownian (B) particles. To derive an equation for the distribution function for n B particles in a system of $(N+n)$ particles one takes

$$P = f_N^\dagger \int (dx)^N (dp)^N \tag{8.112}$$

where $f_N^\dagger$ is the local equilibrium distribution function for the set N in the field of the set n fixed position, i.e.

$$f_N^\dagger = A \exp\{-\sum_{i=1}^N \beta(x_i)[(p_i - um_i)^2/2m_i + V_i(R_i) - \mu_i(R_i)]\} \tag{8.113}$$

where $V_i(R_i)$ is the potential energy of particle i in the field of the $(N+n-1)$ other particles, u the local hydrodynamic velocity, and A a normalisation factor.

The plausible assumption is made that

$$f_{N+n}(0)/f_n(0) = f_N^\dagger \tag{8.114}$$

i.e. set N is initially at local equilibrium in the field of set n. This makes the destruction term zero. We can write L as the sum $(L_N + L_n)$ where L_n is the part operating on coordinates of set n. If we call $\lambda^2 = m/M$, where m is the largest of the masses of the light particles and M the mass of a B particle, then L_n is on average $O(\lambda)$ smaller than L_N because $\langle P^2 \rangle / \langle P^2 \rangle = O(\lambda^2)$. It is assumed that one can retain only the leading terms of lowest order in powers of L_n and in macroscopic gradients. The identity equation (8.14) then yields for $n = 1,2$ the generalised Fokker–Planck results derived by Mazo. It is of course an assumption that the neglected terms are 'well behaved'. A discussion of this point has been given for the corresponding isothermal generalised Langevin equations from which the Fokker–Planck equations can be derived[161–163].

Although the Severne equations are very convenient, it will be clear that they have not so far suggested any radically new approximation scheme for calculating linear transport coefficients. It might still be that it is easier to introduce physically meaningful approximations by starting from the BBGKY hierarchy. The method of Raveche and Mayer[98], which is closely related[99] to an earlier formalism[100], uses this starting point. Self-diffusion in a stationary state maintained by reservoirs is discussed in detail. One may define unknown perturbation functions Ψ_i by

$$f_1(1) = f_1^{(e)}(1)\Psi_1(1); \qquad f_2(1, 2) = f_1(1)f_1(2)g_2^{(e)}(1, 2)\Psi_2(1, 2);$$
$$f_3(1, 2, 3) = f_1(1)f_1(2)f_1(3)g_3^{(e)}(1, 2, 3)\Psi_2(1, 2)\Psi_2(2, 3)\Psi_2(3, 1)\Psi_3(1, 2, 3)$$

$$(8.115)$$

where e denotes equilibrium and $g_2^{(e)}, g_3^{(e)}$ are the equilibrium correlation functions. The principle of the calculation is to expand the Ψ_i in powers of λ, the measure of the gradients, and to consider successively the terms of various order in λ of the hierarchy taking full account of the restrictions imposed on the possible forms of the Ψ_i by the form of the assumed experiment. Closure of the hierarchy was made by taking all three particle terms in Ψ_3 as zero, a dynamical analogue of the Kirkwood superposition approximation. A linearisation in momentum of the Ψ_i was also made. The final integro-differential equation for the momentum-independent part of Ψ_2 is of no greater complexity than the equations for $g_2^{(e)}$ currently being solved by computer. Unfortunately it has since been shown[99] to have no solutions of physical interest and it appears essential to retain three-particle terms in Ψ_3, thereby greatly increasing the work.

Finally, limitations of space allow us only to mention further developments of the free-volume models[101] and the Enskog theory for mixtures[102], and the continued evaluation of the Bearman–Kirkwood[103–105], Rice–Allnatt 'small step'[77, 105], and Enskog (for complete references see Reference 102) theories of mixtures.

8.4 GENERALISED HYDRODYNAMICS AND THE TIME-CORRELATION FUNCTION FORMALISM

8.4.1 Introduction

The alternative route to the classical linear transport coefficients considered in Section 8.3 is through the time-correlation function formalism[107]. In recent work, some emphasis has been put on obtaining more general results applicable to non-local, non-instantaneous, and non-linear responses of flows to the perturbations producing them. Some results in this field of 'generalised hydrodynamics' are considered in Sections 8.4.2–8.4.3. The rest of Section 8.4 is devoted to the actual evaluation of time-correlation functions in the linear response region. There is now a wide range of experimental techniques such as neutron scattering, infrared absorption, magnetic resonance etc. which probe different ranges of the response spectrum and are conveniently studied within the time-correlation formalism (for a review see References 108 and 145). Here only the classical hydrodynamic transport coefficients are considered. Detailed theories of critical-point phenomena have been excluded by lack of space although some of the formal methods in Sections 8.4.2 and 8.4.3 are designed to be applicable in this region (see also Reference 132).

8.4.2 Linearised hydrodynamic modes

The linearised hydrodynamic equations[109] (the continuity, Navier–Stokes, and temperature equations) can be solved by Fourier–Laplace transformation of space and time variables to find the five conserved quantities $n_k(w)$, $g_k(w)$, $T_k(w)$ (the transforms of the local particle density, momentum, and temperature). The problem reduces to diagonalisation of a 5×5 matrix whose eigenvalues are found to be, in the limit of small k with $\mathbf{k}$ along the x axis,

$$\lambda_y = \lambda_z = -k^2\eta/n; \quad \lambda_T = -k^2\kappa/nC_p; \quad \lambda_\pm = \pm ick - k^2\Gamma \quad (8.116)$$

$$\Gamma \equiv [4\eta/3 + \phi + (C_v^{-1} - C_p^{-1})\kappa]/2n \quad (8.117)$$

where n, c, C_v, and C_p are the equilibrium density, the velocity of sound, and the specific heats at constant volume and constant pressure respectively. η and ϕ are the shear and bulk viscosities and κ is the thermal conductivity. As is well known, these macroscopic modes enter in current theories of light scattering[110], the van Hove correlation function[111], critical point singularities of transport coefficients[112, 113], and the non-analytic density expansion of transport coefficients[114, 115]. A powerful way of improving these analyses may be by microscopic construction of the hydrodynamic modes from the singlet distribution function as described by Résibois[116, 117].

The starting point of the Résibois analysis is the asymptotic master equation which can be written in the form (cf. Section 8.2.2.2):

$$\partial_t \tilde{\rho}_k(v, t) = \theta_k \tilde{\rho}_k(v, t) = \Omega_k(-kv_x + \psi_k^{(0)}(v))\tilde{\rho}_k(v, t) \quad (8.118)$$

The collision operator is expanded through second order in the uniformity parameter k:

$$\theta_k = C^1 + kV^{(1)} + k^2V^{(2)} \quad (8.119)$$

where C^1 is the linear homogeneous collision operator. One needs to solve the eigenvalue problem

$$\theta_k \Psi_n^k(v) = \lambda_n^k \Psi_n^k(v) \quad (8.120)$$

$\theta_k - C^1$ being a perturbation. Now C^1 is in general not symmetric, but one can find five left and right eigenfunctions with zero eigenvalues corresponding to the five collisional invariants 1, $\mathbf{v}$, v^2. These can be used to construct the first five (degenerate) members of a set of biorthonormal eigenfunctions[119] of C^1. Degenerate perturbation calculus can then be used to obtain the eigenvalues, correct through order k^2, of θ_k. The five eigenvalues are the same as for the linearised hydrodynamic modes, equation (8.116), but with molecular expressions for η, κ, and ϕ. The latter can be shown to be identical with the usual time-correlation formulae.

Thus the singlet function suffices to calculate the complete transport coefficients without employing the functional relation between f_2 and f_1, e.g. equation (8.103) normally required in the calculation of the potential contributions, but at the price of considering second derivatives of the collision operator as well as the first derivatives which suffice in the traditional approach.

The potential usefulness of the method lies not in the derivation of the expressions for transport coefficients but in the fact that it shows that the inverse collision operator can now always be approximated by

$$\lim_{k \to 0} (\theta_k)^{-1} = \sum_{\alpha = 1}^{s} |\Psi_\alpha\rangle \lambda_\alpha^{-1} \langle \Psi_\alpha| \qquad (8.121)$$

where the λ_α, $|\Psi_\alpha\rangle$, and $\langle \Psi_\alpha|$ for $1 \leqslant \alpha \leqslant 5$ are five eigenvalues and eigenfunctions mentioned above. The transport coefficients in the λ_α can then be suitably approximated for the problem at hand. An example of such a replacement occurs in a derivation of the small wavenumber and small frequency limit of the van Hove function $G(r, t)$ previously treated by macroscopic methods. The destruction term must be accounted for in this calculation[118].

A related scheme for microscopic hydrodynamic modes has been derived[120] which retains only terms through second-order interaction but appears tractable at all k and w.

8.4.3   Schemes for generalised hydrodynamics

We outline first the method of Robertson[122-125] and Piccirelli[121] for the classical case (for quantum statistics see References 122 and 124). The aim is to get usable identities relatable to hydrodynamic ideas. These should reduce to the classical hydrodynamics in the appropriate limit, e.g. long wavelength, long time. It is also hoped that by a suitable choice of observables a description of phenomena that do not fall within the realm of classical hydrodynamics may be possible[122].

Consider position- and time-dependent observables calculable from microscopic fields:

$$\bar{F}_n(r,t) = \langle F_n(r,p^N,q^N)f(t)\rangle^c; \quad 1 \leqslant n \leqslant N \qquad (8.122)$$

where $\langle \ \rangle^c$ denotes integration over all coordinates. In ordinary hydrodynamics these are the momentum, mass, and energy densities. It is convenient to include $F_1 = 1$ and $\bar{F}_1 = 1$ among the observables. It is assumed that we may write

$$\bar{F}_n(r,t) = \langle F_n(r,p^N,q^N)\sigma(t)\rangle^c \qquad (8.123)$$

where $\sigma(t)$ is an unknown functional of the $\bar{F}_n$, $1 \leqslant n \leqslant N$. Now for any phase function $\sigma(t)$ which obeys

$$\partial_t \sigma(t) = P(t)\partial_t f(t) \tag{8.124}$$

where P (t) is a time-dependent projection operator, we may derive the identity[121]

$$f(t) = \sigma(t) + T(t,0)[f(0) - \sigma(0)] - \int_0^t ds\, T(t,s)Q(s)iL\sigma(s) \tag{8.125}$$

$$\frac{\partial T}{\partial s}(t,s) = T(t,s)Q(s)iL, \quad T(t,t) = 1 \tag{8.126}$$

where $Q(t) \equiv 1 - P(t)$. It is straightforward to show that equations (8.125) and (8.122) are consistent with equation (8.123) and that[121,122]

$$P(t).. = \sum_{n=1}^{N} \int d\mathbf{r} \frac{\delta\sigma(t)}{\delta \bar{F}_n(\mathbf{r},t)} \langle F_n(\mathbf{r},p^N,q^N)...\rangle^c \tag{8.127}$$

$$P(t)P(t').. = P(t).. \tag{8.128}$$

An equation of motion for the observable $\bar{F}_n$ can be constructed from equation (8.125):

$$\partial_t \bar{F}_n(\mathbf{r},t) = \nabla_{\mathbf{r}} \cdot \left(\int_0^t ds\, \langle \mathbf{j}_n(\mathbf{r})T(t,s)Q(s)iL\sigma(s)\rangle^c - \langle \mathbf{j}_n(\mathbf{r})\sigma(t)\rangle^c - \right.$$
$$\left. \langle \mathbf{j}_n(\mathbf{r})T(t,0)[f(0) - \sigma(0)]\rangle^c \right) \tag{8.129}$$

where we have assumed fluxes can be defined by

$$\nabla_{\mathbf{r}} \cdot \mathbf{j}_n(\mathbf{r}) = -iLF_n(\mathbf{r}) \tag{8.130}$$

To use this identity one must choose $\sigma(t)$ and an initial condition. A convenient choice for σ is

$$\sigma(t) = \sigma_L(t) \equiv \exp\left[-\sum_{n=1}^{N} \int d\mathbf{r}\lambda_n(\mathbf{r},t)F_n(\mathbf{r})\right] \tag{8.131}$$

The conjugate variables $\lambda_n(\mathbf{r},t)$ are determined as functionals of the observables $\bar{F}_n$ by equation (8.123). With the conventions for F_1 and $\bar{F}_1$ already noted, $\lambda_1(\mathbf{r},t)$ is a function of λ_n, $n \neq 1$, and $\sigma_L(t)$ is normalised and a homogeneous functional of the observables $\bar{F}_n$ of degree one[123]. If the observables $n = 2,3,4$ have been chosen as energy, momentum and mass densities then the formal structure suggests that one identifies $(\lambda_2 k)^{-1}$ as the local temperature, $-\lambda_3/\lambda_2$ as the local hydrodynamic velocity, and $-\lambda_4/\lambda_2$ as the local chemical potential. The formal similarity with equilibrium theory is completed by defining entropy as

$$S(t) \equiv -k\langle \sigma_L(t) \log \sigma_L(t)\rangle^c = -k\langle f(t) \log \sigma_L(t)\rangle^c \tag{8.132}$$

so that the Gibbs balance equation is satisfied. However, a demonstration that the conjugate variables do indeed correspond to what is measured requires an analysis of how the test probes used interact with the system[121,126]. Finally, attention is restricted to experiments prepared so that $f(0) = \sigma_L(0)$[121,122].

The initial value term in equation (8.129) is then zero and one can obtain

$$\partial_t \bar{F}_n(r,t) = -\nabla_r \cdot (\langle \mathbf{j}_n(r)\sigma_L(t)\rangle^c$$
$$+ \int_0^t ds \sum_{n'=1}^N \int dr K_{nn'}(r,r';t,s) \cdot \nabla_{r'} \lambda_{n'}(r',s)) \tag{8.133}$$

$$K_{nn'}(r,r';t,s) = \langle \Delta \mathbf{j}_n(r,t) T(t,s) Q(s) \Delta \mathbf{j}_n(r,s)\sigma_L(s)\rangle^c \tag{8.134}$$

$$\Delta \mathbf{j}_n(r,t) = \mathbf{j}_n(r) - \langle \mathbf{j}_n(r)\sigma_L(t)\rangle^c \tag{8.135}$$

The quantity $Q(s)\Delta \mathbf{j}_n(r,s)\sigma_L(s)$ is a generalisation of the subtracted fluxes appearing in ordinary time correlation formulae[107, 127] of which the $K_{nn'}$ are obvious generalisations. Piccirelli[121] displays in detail the non-local stress tensor and heat current vector and gives full references to earlier work of this character.

For the case just considered the conservation equations, the flux equations obtained from equation (8.133), and the equations of state form a closed description for a finite number of variables. But to be useful the kernels $K_{nn'}$ will have to be local in space and time and suitable expansion parameters found[121]. It has been argued that they will be invariant to the functional form of $\sigma(t)$[121]. For small deviations from equilibrium one can linearise in gradients; P(t) is then independent of time[122, 123] and earlier results are regained. No detailed applications when P is time-dependent are yet available. Extension to open systems is possible[125].

The philosophy of this method is very different from that in Section 8.2.4, where a plausible generalisation of the Boltzmann definition is used for entropy which then may not necessarily fulfil the Gibbs balance equation.

Zubarev has derived a rather similar scheme without employing projection operators[128-130]. Several methods have been used. One is to require that the information entropy $-\langle f(t) \log f(t)\rangle$ be an extremum subject to the constraint that the values of the observables $\bar{F}_n(t')$, $1 \leq n \leq N$, are given for all past times, $-\infty \leq t' \leq t$. The relation of the resulting distribution to the identities referred to above has not been given, although for small gradients the early results of Mori[131] are regained.

The most thoroughly developed aspect of generalised hydrodynamics has been the attempts to reproduce the computer data[97] on the dynamics of density fluctuations in liquid argon by using generalisations of the Navier–Stokes equation to include behaviour non-local in space and time. We refer the reader to a recent article[160] for details of this topic and return for the remainder of this Review to the classical transport coefficients.

8.4.4   Classical transport coefficients as time-correlation functions

8.4.4.1   Memory function identities and computer simulations

Recent reviews[108, 145] discuss the properties and evaluation of linear time-correlation functions in some detail. By means of operator identities similar to those employed in the derivation of the master equation one can obtain the well-known identity for the autocorrelation function of a dynamical variable $G(t)$:

$$\frac{d\Xi(t)}{dt} = - \int_0^t d\tau \phi(\tau)\Xi(t-\tau) \tag{8.136}$$

$$\Xi(t) \equiv \langle G(t)G \rangle \langle GG \rangle^{-1} \tag{8.137}$$

$$\phi(t) \equiv \langle f f(t) \rangle \langle GG \rangle^{-1} \tag{8.138}$$

$$f(t) \equiv e^{(\ P)iLt}(1-P)iLG, \quad G \equiv G(0) \tag{8.139}$$

where the realisation of P is given by[133]

$$PA = \langle GA \rangle \langle GG \rangle^{-1}G \tag{8.140}$$

where A is an arbitrary variable. The brackets represent an average over the equilibrium canonical ensemble and it has been assumed that $\langle G \rangle$ and $\langle G \dot{G} \rangle$ are zero. ϕ is called the memory function of the normalised autocorrelation function. Other forms for ϕ can be derived[134, 135]. Typically, a transport coefficient α is related to an autocorrelation function and a memory function by a relation of the form

$$\alpha = \tilde{\Xi}(0) = \tilde{\phi}(0)^{-1} \tag{8.141}$$

$$\tilde{\Xi}(0) = {}_s\lim_{\to +0} \Xi(s), \quad \tilde{\Xi}(s) \equiv \int_0^\infty dt e^{-st}\Xi(t) \tag{8.142}$$

G is a microscopic flux and $\dot{G} \equiv iLG$ is a generalised force.

Invaluable accurate results are available for the momentum autocorrelation function from molecular dynamics simulations. Recent reviews of the method are available[136, 137]. A most interesting recent result has been the demonstration[138–140] of the existence of significant many-body effects occurring at relatively long times. The momentum autocorrelation function, at a density of about one-third of the close-packing density, decays at long times as $t^{-3/2}$ for hard spheres (t^{-1} for hard discs). The velocity correlation between a molecule and its neighbours reveals a vortex flow pattern. This behaviour is well simulated by a hydrodynamic model[140] in which a small volume element of the order of the volume per molecule is given an initial velocity of the order of the root-mean-square velocity. Numerical solution of the Navier–Stokes equation gives agreement with the 'experimental' results after about 10 mean collision times. The long-time behaviour comprises a vortex ring with a flow pattern characterised by the shear viscosity and an accoustic wave part. Analytic results for this model, independent of particular intermolecular forces, can be found[141, 142] and the same result has also been obtained from kinetic theory[143]. The same effect occurs in the kinetic parts of the shear viscosity and thermal conductivity. For the potential and cross-parts the hydrodynamic model predicts no such behaviour although two-dimensional experiments do indicate effects of quite long duration[141].

An important consequence of the long-time behaviour of the velocity correlation is that it can contribute substantially to the diffusion coefficient. This is very clear in the elegant hydrodynamic model of Zwanzig and Bixon[144]. Results for the velocity autocorrelation for Lennard-Jones molecules are also available[97, 146]. The kind of back-flow pattern observed for hard spheres appears to be below the 5% correlation level if it exists[146]. The memory function has now been more completely characterised as a function of time, density and temperature[146]:

$$\phi(t) = \phi(0)e^{-at^2} + bt^4 e^{-ct}$$
$$a = a_0 + a_1 T, \quad c = c_0 + c_1 n \tag{8.143}$$

where a_0, a_1, c_0 and c_1 are known constants. b varies rapidly with T in the liquid and is also known.

8.4.4.2   Some approximations

We have seen in Section 8.3 that the linear trajectory assumption plays an important role in some kinetic equation methods. It is of considerable interest to express this approximation in a formal manner consistently applicable to the range of classical transport coefficients[147]. If we introduce λ as a measure of the strength of the intermolecular potential and write $L = L_0 + \lambda L_1$, and $G(t) = \lambda g(t)$ then the linear trajectory approximation is to retain only the lowest order in λ when a systematic expansion in λ is made for $\langle ff(t) \rangle$, equation (8.138). The result is[147]

$$\alpha_{LT}^{-1} = \int_0^\infty dt \langle \dot{G} e^{itL_0} \dot{G} \rangle \langle GG \rangle^{-1} \tag{8.144}$$

and the propagator moves the particles on linear trajectories. The geometric structure of the liquid has been retained by *not* expanding the equilibrium distribution function; the approximation is purely a dynamical one. A model calculation suggests that corrections are small but as the authors remark some more rigorous justification is required[147]. For self-diffusion, G is the momentum of one particle and the reduction to integrals of the pair correlation function and intermolecular force is well known[69]. For other transport coefficients the time integration requires further approximations such as the Rice–Kirkwood 'small step' approximation[148]. Numerical results are promised and will be of interest in evaluating the linear trajectory method. Some attempts have very recently been made to improve the approximation for self-diffusion[149, 150].

A very interesting but difficult question is to ask whether the basic decomposition of the resolvent operator into collision, creation, and destruction operators, equation (8.26) has any utility in evaluating $\Xi(t)$. One finds, by comparing the expression for $\dot{\Xi}(t)$ so obtained with the memory function equation, (8.136), the result[151]

$$\tilde{\phi}(s) = \tilde{\Phi}(s) - \tilde{\mathscr{F}}(s)[s - \tilde{\Phi}(s)]/[1 + \tilde{\mathscr{F}}(s)] \tag{8.145}$$

Φ depends on the collision operator ψ_{00} and represents dynamical memory effects in contrast to $\mathscr{F}$ which depends on the static memory of the initial correlations through the destruction operator $\mathscr{D}_{0k}$. A crude model is represented by

$$\Phi(t) = \gamma^2 e^{-\alpha t}; \quad \mathscr{F}(t) = \delta^2 t e^{-\beta t} \tag{8.146}$$

The form for $\mathcal{F}$ is approximately correct for a dilute gas with exponential pair potential. The exponential memory function for Φ is well known to be one of the simplest approximations qualitatively reproducing the computer experiments for liquids[135]. $\alpha, \beta, \gamma, \delta$ were fixed in terms of molecular parameters by equating the exact and approximate expressions for $\tilde{\phi}(0)$, and for $\phi(t)$ and its first two derivatives at zero time, plus a further approximation to circumvent the non-availability of the equilibrium four-body distribution function required. Modest agreement with the computer experiments for liquid argon at one temperature results. The model yields $\alpha \cong \beta/2$ and the effects of initial correlations disappear in about half a collision time. Furthermore it is found that $\gamma^2 \cong 4\delta^2$ so that the amplitude of the peak in the destruction function is only c. 10% of the maximum of $\Phi(t)$.

There are several other attempts[135, 150–157] to 'model' the memory function for diffusion and other transport coefficients by making a plausible guess at the form of $\phi(t)$ and fixing the parameters by requiring agreement with known exact results in various limits. These were made before the very complete characterisation noted in equation (8.144) was available. This topic and the intimately related subject of neutron-scattering correlation functions require a separate review. The technique certainly gives results not obtained by the kinetic equation approach. However for exact model calculations the memory equation may not be easy to use because the memory function has complicated analytical properties[158]. Thus the equation for the momentum autocorrelation function is non-markovian in the weak coupling limit whereas a naive expansion of the memory function in the coupling parameter leads to an incorrect markovian equation.

Presumably it will eventually be made clearer just what the approximations made in the kinetic equation theories of Section 8.3.1 mean in terms of making approximations on the exact time-correlation function expressions. The application of transport theory to liquids will then gain a coherence it at present lacks.

References

1. Nakajima, S. (1958). *Progr. Theoret. Phys. (Kyoto)*, **20**, 948
2. Zwanzig, R. (1960). *J. Chem. Phys.*, **33**, 1338
3. Prigogine, I. and Résibois, P. (1961). *Physica*, **27**, 541
4. Montroll, E. W. (1962). *Fundamental Problems in Statistical Mechanics* (compiled by Cohen, E. G. D.), 220. (Amsterdam: North-Holland Publishing Co.)
5. Zwanzig, R. (1964). *Physica*, **30**, 1109
6. Severne, G. (1965). *Physica*, **31**, 877
7. Prigogine, I. (1962). *Non-Equilibrium Statistical Mechanics*. (New York: Interscience-John Wiley)
8. Résibois, P. (1966). *Physics of Many-Particle Systems* (Ed. by E. Meeron). (New York: Gordon and Breach)
9. Balescu, R. (1963). *Statistical Mechanics of Charged Particles*. (New York: Interscience-John Wiley)
10. Oppenheim, I. and Shuler, K. E. (1965). *Phys. Rev. B*, **138**, 1007
11. Lugiato, L. A. (1969). *Physica*, **44**, 337
12. Muriel, A. and Dresden, M. (1968). *Phys. Lett. A*, **27**, 16
13. Balescu, R. (1968). *Physica*, **38**, 98
14. Pytte, A. (1968). *Phys. Fluids*, **11**, 522
15. Baus, M. (1967). *Bull. Classe Sci. Acad. Roy. Belg.*, **53**, 1291

16. De Pazzis, O. (1970). *Physica*, **40**, 229
17. Prigogine, I. and Severne, G. (1966). *Physica*, **32**, 1376
18. Henin, F. (1968). *Physica*, **39**, 599
19. Prigogine, I. and Henin, F. (1960). *J. Math. Phys.*, **1**, 349
20. Balescu, R. (1969). *Physica*, **42**, 464
21. Baus, M. (1967). *Bull. Classe Sci. Acad. Roy. Belg.*, **53**, 1332
22. George, C. (1964). *Physica*, **30**, 1513
23. George, C. (1967). *Physica*, **37**, 182
24. George, C. (1967). *Bull. Classe Sci. Acad. Roy. Belg.*, **53**, 623
25. Prigogine, I., George, C. and Henin, F. (1969). *Physica*, **45**, 418
26. Lanz, L. and Lugiato, L. (1969). *Physica*, **44**, 532
27. Prigogine, I., George, C. and Henin, F. (1970). *Proc. Nat. Acad. Sci. U.S.*, **65**, 789
28. Grecos, A. P. (1971). *Physica*, **51**, 50
29. Balescu, R., Clavin, P., Mandel, P. and George, C. (1969). *Bull. Classe Sci. Acad. Roy. Belg.*, **55**, 1055
30. Turner, J. W. (1968). *Bull. Classe Sci. Acad. Roy. Belg.*, **54**, 1526
31. Balescu, R., Brenig, L. and Wallenborn, J. (1971). *Physica*, **52**, 29
32. Balescu, R. and Wallenborn, J. (1971). *Physica*, **54**, 477
33. Henin, F., Prigogine, I., George, C. and Mayné, F. (1966). *Physica*, **32**, 1828
34. Prigogine, I. and Henin, F. (1967). *Proceedings of the International Union of Pure and Applied Physics meeting, Copenhagen, 1966*, 421. (New York: Benjamin)
35. Prigogine, I. (1969). *Advan. Chem. Phys.*, **16**, 11
36. George, C. (1968). *Physica*, **39**, 251
37. Prigogine, I., Henin, F. and George, C. (1968). *Proc. Nat. Acad. Sci. U.S.*, **59**, 7
38. Mandel, P. (1968). *Physica*, **48**, 397
39. Nicolis, G., Wallenborn, J. and Velarde, M. G. (1969). *Physica*, **43**, 263
40. Prigogine, I. (1968). *Topics in Nonlinear Physics* (Ed. by N. Zabusky), 216. (New York: Springer-Verlag New York Inc.)
41. Prigogine, I., Henin, F. and George, C. (1966). *Physica*, **32**, 1873
42. Nicolis, G. (1967). *J. Chem. Phys.*, **46**, 702
43. Prigogine, I., George, C., Henin, F., Mandel, P. and Turner, J. W. (1970). *Proc. Nat. Acad. Sci. U.S.*, **66**, 709
44. Mandel, P. (1970). *Physica*, **50**, 77
45. Turner, J. W. (1971). *Physica*, **51**, 351
46. Turner, J. W. (1970). *Bull. Classe Sci. Acad. Roy. Belg.*, **56**, 1125
47. Allen, P. and Nicolis, G. (1970). *Physica*, **50**, 206
48. Clavin, P. and Wallenborn, J. (1970). *Compt. Rend.*, **270**, 717
49. Rae, J. and Davidson, R. (1971). *J. Statistical Phys.*, **3**, 135
50. Fulinski, A. and Kramarczyk, W. J. (1968). *Physica*, **39**, 575
51. Lugiato, L. A. (1970). *Physica*, **49**, 615
52. Gesetzi, T. (1967). *Phys. Lett. A*, **25**, 12
53. Fulinski, A. (1967). *Phys. Lett. A*, **24**, 63
54. Shimizu, T. (1970). *J. Phys. Soc. Jap.*, **28**, 1088
55. Fujita, S. (1970). *Pure Appl. Chem.*, **22**, 249
56. Fujita, S. (1969). *J. Phys. Soc. Jap.*, **26**, 505
57. Lee, D., Fujita, S. and Wu, F. (1970). *Phys. Rev. A*, **2**, 854
58. Weinstock, J. (1964). *Phys. Rev.*, **136**, A879
59. Fujita, S. (1969). *J. Phys. Soc. Jap.*, **27**, 1096
60. Fujita, S. and Chen, C. C. (1969). *J. Theoret. Phys.*, **2**, 59
61. Fujita, S. and Lodder, A. (1970). *Physica*, **50**, 541 ·
62. Résibois, P. and Velarde, M. G. (1970). *Physica*, **51**, 541
63. Misguich, J. (1969). *J. Physique*, **30**, 221
64. Prigogine, I., Nicolis, G. and Misguich, J. (1965). *J. Chem. Phys.*, **43**, 4516
65. Misguich, J., Nicolis, G., Palyvos, J. A. and Davis, H. T. (1968). *J. Chem. Phys.*, **48**, 951
66. Palyvos, J. A., Davis, H. T., Misguich, J. and Nicolis, G. (1968). *J. Chem. Phys.*, **49**, 4088
67. Longuet-Higgins, H. C. and Pople, J. (1956). *J. Chem. Phys.*, **25**, 884
68. Bearman, R. J., Kirkwood, J. G. and Fixman, M. (1958). *Advan. Chem. Phys.*, **1**, 1
69. Helfand, E. (1961). *Phys. Fluids*, **4**, 681
70. Allen, P. M. and Cole, G. H. A. (1968). *Mol. Phys.*, **15**, 549, 557
71. Allen, P. M. (1970). *Mol. Phys.*, **18**, 349

72. Allen, P. M. (1971). *Physica*, **52**, 237
73. Forster, M. J. and Cole, G. H. A. (1971). *Mol. Phys.*, **20**, 417
74. Davis, H. T. (1971). *Proceedings of the Conference on Statistical Mechanics held by International Union of Pure and Applied Physics, April 1971, Chicago.* (Chicago: University of Chicago Press)
75. Rice, S. A. and Allnatt, A. R. (1961). *J. Chem. Phys.*, **34**, 2144, 2156
76. Rice, S. A. and Gray, P. (1965). *The Statistical Mechanics of Simple Liquids.* (New York: Interscience-John Wiley)
77. Rice, S. A., Boon, J. P. and Davis, H. T. (1967). *Simple Dense Fluids* (Ed. by H. L. Frisch and Z. Salsburg), 252. (New York: Academic Press)
78. Nicolis, G., Dagonnier, R., Misguich, J. and Boon, J. P. (1968). *Appl. Mech. Rev.*, **21**, 215
79. Hurt. N. and Rice, S. A. (1966). *J. Chem. Phys.*, **44**, 2155
80. Popielawski, J., Rice, S. A. and Hurt, N. (1967). *J. Chem. Phys.*, **46**, 3707
81. Wei, C. C. and Davis, H. T. (1966). *J. Chem. Phys.*, **45**, 2535
82. Baleiko, M. O. and Davis, H. T. (1970). *J. Chem. Phys.*, **52**, 2427
83. Nicolis, G. and Misguich, J. (1971). *Mol. Phys.*, (in course of publication)
84. Davis, H. T. and Baleiko, M. O. (1971). *J. Statistical Phys.*, **3**, 47
85. Davis, H. T. and Palyvos, J. (1967). *J. Chem. Phys.*, **46**, 4043
86. Collings, A. F., Watts, R. O. and Woolf, L. A. (1971). *Mol. Phys.*, **20**, 1121
87. Muriel, A. and Dresden, M. (1969). *Physica*, **43**, 424
88. Muriel, A. and Dresden, M. (1969). *Physica*, **43**, 449
89. Hawker, K. E. and Schieve, W. C. (1970). *Physica*, **48**, 217
90. Haake, F. (1971). *Phys. Rev. A*, **3**, 1723
91. Lebowitz, J. and Rubin, E. (1963). *Phys. Rev.*, **131**, 2381
92. Résibois, P. and Davis, H. T. (1963). *Physica*, **30**, 1077
93. Nicolis, G. *J. Chem. Phys.*, **43**, 1110
94. Mazo, R. M. (1969). *J. Statistical Phys.*, **1**, 89
95. Mazo, R. M. (1969). *J. Statistical Phys.*, **1**, 101
96. Mazo, R. M. (1969). *J. Statistical Phys.*, **1**, 559
97. Rahman, A. (1964). *Phys. Rev.*, **136**, A405
98. Raveche, H. J. and Mayer, J. E. (1970). *J. Chem. Phys.*, **52**, 3990
99. Jordan, P. C. and Greenberg, A. D. (1970). *J. Chem. Phys.*, **53**, 4355
100. Stillinger, F. H. and Suplinskas, R. J. (1966). *J. Chem. Phys.*, **44**, 2432
101. Turnbull, D. and Cohen, M. H. (1970). *J. Chem. Phys.*, **52**, 3038
102. Gubbins, K. E. and Tham, M. K. (1971). *J. Chem. Phys.*, **55**, 268
103. Bearman, M. Y. and Bearman, R. J. (1970). *J. Chem. Phys.*, **52**, 3189
104. Al-Chalabi, H. A. and McLaughlin, E. (1970). *Mol. Phys.*, **19**, 703
105. Loflin, T. and McLaughlin, E. (1969). *J. Phys. Chem.*, **73**, 186
106. Wei, C. C. and Davis, H. T. (1967). *J. Chem. Phys.*, **46**, 3546
107. Zwanzig, R. (1965). *Ann. Rev. Phys. Chem.*, **16**, 67
108. Mountain, R. D. (1970). *Chemical Rubber Co., Critical Reviews, Solid State Sciences*, **1**, 5
109. Landau, L. and Lifshitz, E. (1963). *Fluid Mechanics*, (London: Pergamon Press)
110. Mountain, R. D. (1966). *Rev. Mod. Phys.*, **38**, 205
111. Egelstaff, P. A. (1967). *An Introduction to the Liquid State.* (London: Academic Press)
112. Kadanoff, L. and Swift, J. (1968). *Phys. Rev.*, **166**, 89
113. Kawasaki, K. (1966). *Phys. Rev.*, **150**, 291
114. Ernst, M., Haines, L. and Dorfman, J. (1969). *Rev. Mod. Phys.*, **41**, 296
115. Pomeau, Y. (1971). *Phys. Rev. A*, **3**, 1174
116. Résibois, P. (1970). *J. Statistical Phys.*, **2**, 21
117. Résibois, P. (1970). *Bull. Classe Sci. Acad. Roy. Belg.*, **56**, 160
118. Résibois, P. (1970). *Physica*, **49**, 591
119. Morse, P. and Feshbach, H. (1953). *Methods of Theoretical Physics.* (New York: McGraw-Hill)
120. Forster, D. and Martin, P. C. (1970). *Phys. Rev. A*, **2**, 1575
121. Piccirelli, R. A. (1968). *Phys. Rev.*, **175**, 77
122. Robertson, B. (1966). *Phys. Rev.*, **144**, 151
123. Robertson, B. (1967). *Phys. Rev.*, **160**, 175
124. Robertson, B. (1970). *J. Math. Phys.*, **11**, 2482
125. Robertson, B. and Mitchell, W. C. (1971). *J. Math. Phys.*, **12**, 563
126. Garcia-Colin, L. S., Green, M. S. and Chaos, F. (1966). *Physica*, **32**, 450

127. Green, M. S. (1954). *J. Chem. Phys.,* **22,** 398
128. Zubarev, D. N. (1965). *Dokl. Akad. Nauk SSSR,* **162,** 532
129. Zubarev, D. N. and Kalashnikov, V. P. (1970). *Physica,* **46,** 550
130. Zubarev, D. N. (1970). *Theoret. Math. Phys.,* **3,** 505. (*Teor. i Mat. Fiz.,* **3,** 276)
131. Mori, H. (1958). *Phys. Rev.,* **112,** 1892
132. Clarke, B. L. and Rice, S. A. (1970). *Phys. Fluids,* **13,** 271
133. Mori, H. (1965). *Progr. Theoret. Phys. (Kyoto),* **33,** 423
134. Zwanzig, R. W. (1961). *Lectures in Theoretical Physics,* **3,** 106. (New York: Interscience)
135. Berne, B., Boon, J-P. and Rice, S. A. (1966). *J. Chem. Phys.,* **45,** 1086
136. Rahman, A. (1969). *Nuovo Cimento Riv.,* (*1*), **1,** 315
137. McDonald, I. R. and Singer, K. (1970). *Quart. Rev. (London),* **24,** 238
138. Alder, B. J. and Wainwright, T. E. (1967). *Phys. Rev. Lett.,* **18,** 988
139. Alder, B. J. and Wainwright, T. E. (1968). *J. Phys. Soc. Jap. Suppl.,* **26,** 267
140. Alder, B. J. and Wainwright, T. E. (1970). *Phys. Rev. A,* **1,** 18
141. Wainwright, T. E., Alder, B. J. and Gass, D. M. (1971). *Phys. Rev. A,* **4,** 233
142. Ernst, M. H., Hauge, E. H. and van Leeuwen, (1970). *Phys. Rev. Lett.,* **25,** 1254
143. Dorfman, J. R. and Cohen, E. G. D. (1970). *Phys. Rev. Lett.,* **25,** 1257
144. Zwanzig, R. and Bixon, M. (1970). *Phys. Rev. A,* **2,** 2005
145. Berne, B. and Harp, G. (1970). *Advan. Chem. Phys.,* **17,** 63
146. Levesque, D. and Verlet, L. (1970). *Phys. Rev. A,* **2,** 2514
147. Berne, B. J., Boon, J-P. and Rice, S. A. (1967). *J. Chem. Phys.,* **47,** 2283
148. Rice, S. A. and Kirkwood, J. G. (1959). *J. Chem. Phys.,* **31,** 901
149. Hynes, J. T. and Deutch, J. M. (1970). *J. Chem. Phys.,* **53,** 4705
150. Corngold, N. and Duderstadt, J. J. (1970). *Phys. Rev. A,* **2,** 836
151. Boon, J-P. and Rice, S. A. (1967). *J. Chem. Phys.,* **47,** 2480 (1967)
152. Singwi, K. S. and Tosi, M. P. (1967). *Phys. Rev.,* **157,** 153
153. Desai, R. C. and Yip, S. (1968). *Phys. Rev.,* **166,** 129
154. Kerr, W. C. (1968). *Phys. Rev.,* **174,** 316
155. Martin, P. C. and Yip, S. (1968). *Phys. Rev.,* **170,** 151
156. Tahir-Kheli, R. A. and Wu, D. H. (1969). *Solid State Commun.,* **7,** 1235
157. Gaskell, T. (1970). *Phys. Lett. A,* **31,** 346
158. Résibois, P., Brocas, J. and Decan, G. (1969). *J. Math. Phys.,* **10,** 964
159. Lanz, L., Lugiato, L. A. and Ramella, G. (1971). *Physica,* **54,** 94
160. Ailwadi, N. K., Rahman, A. and Zwanzig, R. (1971). *Phys. Rev. A,* **4,** 1616
161. Mazur, P. and Oppenheim, I. (1970). *Physica,* **50,** 241
162. Albers, J., Deutch, J. M. and Oppenheim, I. (1971). *J. Chem. Phys.,* **54,** 3541
163. Deutch, J. M. and Oppenheim, I. (1971). *J. Chem. Phys.,* **54,** 3547
164. Severne, G. (1971). *Transport, Theory and Statistical Physics,* **1,** 145